UNDERSTANDING AND APPLYING RESEARCH DESIGN

UNDERSTANDING AND APPLYING RESEARCH DESIGN

Martin Lee Abbott
Jennifer McKinney
Seattle Pacific University

A JOHN WILEY & SONS, INC., PUBLICATION

Cover Image: Courtesy of Dominic Williamson

Published by John Wiley & Sons, Inc., Hoboken, New Jersey.
Published simultaneously in Canada.

For general information on our other products and services or for technical support, please contact our Customer Care Department within the United States at (800) 762-2974, outside the United States at (317) 572-3993 or fax (317) 572-4002.

Wiley also publishes its books in a variety of electronic formats. Some content that appears in print may not be available in electronic formats. For more information about Wiley products, visit our web site at www.wiley.com.

Library of Congress Cataloging-in-Publication Data:

Abbott, Martin, 1949-
  Understanding and applying research design / Martin Lee Abbott, Jennifer McKinney.
    p. cm.
  Includes bibliographical references.
  ISBN 978-1-118-09648-2 (cloth)
  1. Research–Methodology.   2. Research–Statistical methods.   I. McKinney, Jennifer, 1969-   II. Title.
  Q180.55.M4A236 2013
  001.4'2–dc23

                                                                                        2012010997

Printed in the United States of America.

10 9 8 7 6 5 4 3 2 1

To
Joyce and William McKinney
Hannah Mary and Jacob Hovan

CONTENTS

PREFACE

Social scientific research is the systematic and rigorous process of exploring the world around us. Good social science requires good research design and solid analytic skills. Both authors strive to teach students the methods of research design and statistical analysis in order that students learn how to pose research questions, test research questions, and draw conclusions on the research that they have conducted, as well as to critique the research they are exposed to through media, classes, and real-life situations. We have taught research methods and statistics courses at the university level for many years. In addition, we have published articles and books on the subjects and are involved in applied research projects in which we put into practice what we develop in this book.

This book grew from the need to provide a systematic but approachable book for our students. Other research design books often use a stilted approach that masks the vibrancy of research statistics and design (or they focus simply on either statistics *or* design). In this book, we hope to avoid these issues by providing a creative format and common language that will enable students to understand the content of social research at a more meaningful level.

The layout of the book is a reflection of our approach to teaching, and it targets contemporary student learning styles. We present research design material in approachable language interspersed by content that allows students the opportunity to delve as deeply as they wish in the material. Extended study units in statistical concepts and application exercises are placed strategically throughout the book to enhance the main focus of the book, research design.

We use SPSS®[1] screen shots of menus and tables by permission from the IBM® Company. IBM, the IBM logo, ibm.com, and SPSS are trademarks of International Business Machines Corp., registered in many jurisdictions worldwide. Other product and service names might be trademarks of IBM or other companies. A current list of IBM trademarks is available on the Web at "IBM Copyright and trademark information" at www.ibm.com/legal/copytrade.shtml. We include SPSS screen shots in the following chapters and sections: Chapters 1–3, 6–11, 13, 15, and 16, Statistical Procedures Unit C, and Data Management Units A–C.

---

[1] SPSS, Inc., an IBM Company. SPSS screen reprints throughout the book are used courtesy of International Business Machines Corporation, © SPSS, Inc., an IBM Company. SPSS was acquired by IBM in October 2009.

In preparing this book, we have distilled the most meaningful content from our class-tested approaches and from our published works. We use current real-world data for our examples and discussions, in particular, the 2010 GSS[2] database, a large state (Washington) database[3] that compiles school-based data on student achievement, and publicly accessible data from the U.S. Census 2010.[4] Much of the content on statistical procedures and using SPSS is adapted from Abbott's previous work.[5] We hope readers enjoy learning about the engaging world of research premises, procedures, and designs.

MARTIN LEE ABBOTT
JENNIFER MCKINNEY

---

[2] The GSS data are used by permission. Smith, Tom W, Peter Marsden, Michael Hout, and Jibum Kim. *General social surveys, 1972–2010* [machine-readable data file] /Principal Investigator, Tom W. Smith; Co-Principal Investigator, Peter V. Marsden; Co-Principal Investigator, Michael Hout; Sponsored by National Science Foundation. NORC ed. Chicago: National Opinion Research Center [producer]; Storrs, CT: The Roper Center for Public Opinion Research, University of Connecticut [distributor], 2011. (http://www3.norc.org/GSS+Website/)

[3] The data are used courtesy of the Office of the Superintendent of Public Instruction, Olympia, Washington. The Web site address is http://www.k12.wa.us/.

[4] U.S. Census, 2010.

[5] Abbott, Martin Lee, *Understanding Educational Statistics using Microsoft Excel® and SPSS®*, Wiley, 2011. Also, Abbott, Martin Lee, *The Program Evaluation Prism*, Wiley, 2010. Both are used by permission of the publisher.

Supplementary material for this book can be found by entering ISBN 9781118096482 at booksupport.wiley.com.

ACKNOWLEDGMENTS

Several people have helped to make this book possible. We would like to thank our friends and colleagues David Diekema, Sara Koenig, Paula Mitchell, Greg Moon, Kevin Neuhouser, Lorraine Shaman, Karen Snedker, Cathy Thwing, Linda Wagner, and Cara Wall-Scheffler. We thank Dominic Williamson for his graphic design that we use in the book (and on the cover) and Roger Finke for allowing us to draw so much from the ARDA. We also thank Jacqueline Palmieri for her continuing support of our efforts to publish accessible social science matter.

Finally, we thank our students who have taught us how to think about teaching statistics and design, and who help us to remember that research methods are fun!

M.L.A.
J.M.

PART I
WHEEL OF SCIENCE:
PREMISES OF RESEARCH

PART 1
WHEEL OF SCIENCE.
PREMISES OF RESEARCH

1

"DUH" SCIENCE VERSUS
"HUH" SCIENCE

HOW DO WE KNOW WHAT WE KNOW?

When we go through the education process, we each take several categories of classes, especially if we know we're headed to college. Often one of these categories is "science" and includes classes in biology, chemistry, or physics. Because of this we come to think of science as particular substantive areas rather than as a particular *process*. The process of science allows us to follow systematic steps to better understand the world around us. Whether using amino acids, elements along the periodic chart, sound waves, or people's attitudes, following the process of science allows us to see patterns in our materials. Granted, it's often harder to think of people as "materials" than it is to think of saltwater solutions as materials. Regardless of what we are looking for, following the scientific process allows us to gauge what is going on in the world.

The process of social science differs from other sciences only in that the social sciences use people to find patterns. While most of us think of people as individuals, each individual lives in a particular social context that has a surprising amount of order to it. For example, Americans drive on the right side of the road; Britons drive on the

Understanding and Applying Research Design, First Edition. Martin Lee Abbott and Jennifer McKinney.
© 2013 John Wiley & Sons, Inc. Published 2013 by John Wiley & Sons, Inc.

left. Even though both countries are made up of individuals, they each tend to transfer their cultural order to walking on the same side of the sidewalk. Even though each individual may walk in a unique way (perhaps like Monty Python's "lumberjack walk"), each tends to gravitate toward the right or left side of a sidewalk depending on country—or cultural order—of origin.

Keeping with a roadway example, have you ever thought about the only thing keeping one vehicle from hitting another in a head-on collision is a measly 6 inches of yellow paint? Think about the 6 inches of white paint that keeps cars traveling in your direction from driving into you. If you consider a large urban area with millions of people trying to travel by car into and out of the area every day, isn't it amazing how few car accidents there are? In Seattle (even with our perpetually wet weather), there are roughly four million people trying to get into and out of the metropolitan area each weekday. But there are less than a hundred vehicular accidents in a given 24-hour period, illustrating just how effective 6 inches of paint can be in regulating the behavior of millions of people. That people and social patterns have such a high degree of order allows us to study just where these patterns originate and predict when they are going to show up.

Knowing there are social rules and boundaries in place that create a high degree of social order, the task for the social scientist is to measure people's attitudes, behaviors, and experiences to find common patterns. The question becomes, however, why should you need social science when you live in the same world or social context and experience these things for yourself? Why rely on social science to generalize to a population or group of people or things? How do you know what social science says is true? How do you know what is good information? The only way to truly know about the social patterns around us is to understand the process of science.

Say, for example, your professor distributes a class exercise asking you to evaluate some research finding. You are first asked if the finding is surprising or not, and then you are asked to write down a reason or two why you believe that finding is or is not true. Let's say that you are given the finding, "Social scientists have found that opposites attract." Is this finding surprising? How do you evaluate this statement? What evidence do you have that opposites attract? Go ahead and think of or jot down why you believe that opposites attract.

What if your professor is being a bit cagey and secretly handed out two contradictory research findings? Whereas you received "opposites attract," the other half of the class received the reverse finding that "Birds of a feather flock together." As the class comes together to discuss the research finding, an interesting thing will happen. When asked how many in the class found "this" finding to be not surprising, most of the class will raise their hands to show how unsurprised they were. That a majority of the class reports their research result is true and not surprising is interesting considering the class had two very different findings. This predicament illustrates the **hindsight bias**. In hindsight, research results seem like common sense; we take for granted that research findings must be true—after they are given.

As you thought about the finding you were given, you probably searched your experience for one case (person) where "opposites attract" was true. Generally when we hear about research findings after the fact, we think of at least one case of **confirming evidence**. This means we look to our own experience and try to find one person or situation that fits the finding given. In this case, you probably thought of at least one friend or acquaintance who was in a relationship where opposites attract. Your classmates with the contradictory finding were doing the same thing, trying to find an example of someone they knew in a relationship where birds of a feather flock together. But trying to explain research findings using our own experiences and already being biased by what the result appears to be hurts our ability to see the world as a whole. If you thought of one person who served as an example of each finding, that's two people. Can two (or even 10 people you may have thought of) represent the whole social spectrum? Even in just an American context, there are well over 310 million people to consider. Do we really want to base our understanding of which adage is more true simply by finding two examples that *confirm* the finding and *conform* to our limited experience? It's highly unlikely that diametrically opposed research findings like opposites attract or birds of a feather flock together happen exactly randomly and at the same rate in a given social context. So how do we know which is more descriptive of everyday attraction?

Social scientists recognize that, while everyone's personal experience of the world is unique, there are social patterns that transcend our own experiences. Social scientists look for both the confirming evidence and the disconfirming evidence—examples where a finding would not be true—to give us direction as to how to generalize to the whole population which would be true. Can we find instances where B does not follow A? If so, that leads us into asking new questions and testing data to give us a more comprehensive picture and stop us from making a hasty conclusion based on very little evidence.

Every day we evaluate information based on our own experience. This is usually helpful for us. What if you and your friends are trying to decide which movie to see on a weekend? Do you choose whichever blockbuster is showing? Do you go to the film that may not be in the theater next week? Do you choose a movie based on what genre you tend to prefer? Do you pick a movie based on a friend's recommendation? Do you choose the film based on the critical reviews? Do you choose a film based on your schedule—whichever is showing at the closest theater at a particular time? You probably use a combination of these methods to figure out which is the best movie to see at any given time. Have you found that even when using your own judgment (based on your preferences and friends' reviews) that the movie was a dud? In effect, social scientists are trying to tease out all of the ways we can think about a particular topic. That helps us to test topics to try to find consistent answers to research questions.

Science is needed because we do not experience the world randomly. How we experience and view the world is highly influenced by our **social location**—where we fit into the social order (our social class perspective, our gender perspective, our educational perspective, our political perspective, our religious perspective, etc.). Two people viewing the exact same event could interpret it very differently, depending on their personal context (or biases). For example, bringing a homeless encampment to a local college campus could elicit a hearty "well done" from students and faculty who want to address the issues of homelessness and poverty. At the same time, parents and local residents may protest bringing a group of homeless people to stay on campus as dangerous—for their personal safety and the safety of their property. The same event is viewed very differently by people living within the same neighborhood because they have different social locations (students, faculty, parents, and homeowners have different interests and expectations of events). Wouldn't it be helpful to have some social science research that can explain and predict what really happens when a homeless encampment is brought to campus, as well as why people from different social locations respond differently to the same event?

Common Sense versus Science

Because we view the world from particular social locations, or biases, we need science to provide a baseline; what effects does one thing have on another, regardless of your perspective? Like trying to explain why opposites attract or birds of a feather flock together, social scientists are often accused of pointing out what is only common sense or what everyone already knows to be true. Of course, the hindsight bias hurts our ability to think novelly or clearly about particular relationships or facts, and it leads many to conclude that social science is just "duh" science—senseless science that points out the obvious.

"DUH" SCIENCE

Eryn Brown (2011) from the *Los Angeles Times* writes about this seemingly pointless research, enumerating studies that seem silly at best, wasteful at worst. For example, she writes of studies confirming that nose-picking is common among teens, or that college drinking is as bad as researchers believe, or that making exercise more fun may improve the fitness of teens, or that driving ability is compromised with people who have Alzheimer's disease. "Well, duh, you might think—and you wouldn't be the first," she writes. The perception that social science simply tests the obvious is widespread, and yet there is more to "duh" science than meets the eye.

Many studies have to test the so-called obvious because until there are widely established links between behaviors or attitudes and some effect, we simply cannot be sure that real links between them exist. Even when clear and reliable links are found, it may take oodles of evidence to convince others that the links are real— often because people don't understand the nature of science or they dismiss commonsense findings as "duh" science. Look at how many studies had to be done linking smoking to various cancers and lung disease before people began to believe these results were real (Brown 2011). Because of research we now understand the link between smoking and cancer, but we didn't at first (and of course, with hindsight bias it seems silly to think there isn't a link between smoking and a variety of cancers).

An Example of "Duh" Science: The 2011 Ig Nobel Prize Winners[1]
Chemistry Prize: For determining the ideal density of airborne wasabi (pungent horseradish) to awaken sleeping people in case of a fire or other emergency, and for applying this knowledge to invent the wasabi alarm. Makoto Imai, Naoki Urushihata, Hideki Tanemura, Yukinobu Tajima, Hideaki Goto, Koichiro Mizoguchi, and Junichi Murakami of Japan.

Reference: U.S. patent application 2010/0308995 A1. Filing date: February 5, 2009.

Medicine Prize: For demonstrating that people make better decisions about some kinds of things but worse decisions about other kinds of things when they have a strong urge to urinate. Mirjam Tuk, Debra Trampe, and Luk Warlop and jointly to Matthew Lewis, Peter Snyder, and Robert Feldman, Robert Pietrzak, David Darby, and Paul Maruff.

Reference: Tuk MA, Trampe D, Warlop L. Inhibitory spillover: increased urination urgency facilitates impulse control in unrelated domains. Psychol Sci 2011; 22(5):627–633; Lewis MS, Snyder PJ, Pietrrzak RH, et al. The effect of acute increase in urge to void on cognitive function in healthy adults. Neurol Urodyn 2011;30(1):183–187.

Psychology Prize: For trying to understand why, in everyday life, people sigh. Karl Halvor Teigen.

Reference: Teigen KH. "Is a sigh 'just a sigh'? Sighs as emotional signals and responses to a difficult task." Scand J Psychol 2008;49(1):49–57.

Literature Prize: John Perry of Stanford University, USA, for his Theory of Structured Procrastination, which says, "To be a high achiever, always work on something important, using it as a way to avoid doing something that's even more important."

Reference: Perry J. How to procrastinate and still get things done. Chronicle of Higher Education, February 23, 1996. Later republished elsewhere under the title "Structured Procrastination."

Biology Prize: For discovering that a certain kind of beetle mates with a certain kind of Australian beer bottle. Darryl Gwynne and David Rentz.

Reference: Gwynne DT, Renta DCF. Beetles on the bottle: Male buprestids mistake stubbies for females (Coleoptera). J Aust Entomol Soc 1983;22(1):79–80; Gwynne DT, Renta DCF. Beetles on the bottle. Antenna: Proc (A) Royal Entomol Soc London 1984;8(3):116–117.

(Continued)

[1] The Ig Nobel Prizes honor the unusual and imaginative research conducted in science, medicine, and technology (see Improbable Research 2012 at http://improbable.com/ig/).

Physics Prize: For determining why discus throwers become dizzy and why hammer throwers don't. Philippe Perrin, Cyril Perrot, Dominique Deviterne, Bruno Ragaru, and Herman Kingma.

Reference: Perin P, Perrot C, Deviterne D, et al. Dizziness in discus throwers is related to motion sickness generated while spinning. Acta Oto-Laryngol 2000;120(3):390–395.

Mathematics Prize: Dorothy Martin (who predicted the world would end in 1954), Pat Robertson (who predicted the world would end in 1982), Elizabeth Clare Prophet (who predicted the world would end in 1990), Lee Jang Rim (who predicted the world would end in 1992), Credonia Mwerinde (who predicted the world would end in 1999), and Harold Camping (who predicted the world would end on September 6, 1994, and later predicted that the world will end on October 21, 2011), for teaching the world to be careful when making mathematical assumptions and calculations.

Peace Prize: Arturas Zuokas, the mayor of Vilnius, Lithuania, for demonstrating that the problem of illegally parked luxury cars can be solved by running them over with an armored tank.

Reference: VIDEO and OFFICIAL CITY INFO.

Public Safety Prize: John Senders for conducting a series of safety experiments in which a person drives an automobile on a major highway while a visor repeatedly flaps down over his face, blinding him.

Reference: Senders JW, et al. The attentional demand of automobile driving. Highway Research Record 1967;195:15–33. VIDEO.

"HUH SCIENCE"

While it is easy to dismiss scientific findings that seem obvious, keep in mind that our biases impact how we view what is obvious and what is not. Not only do social scientists try to find a baseline of behavior that may seem obvious (regular exercise leads to longevity), they are also able to illustrate the not so obvious. In the United States, for example, most people understand that religion has been in consistent decline since the birth of the nation when all of the Pilgrims walked to church every Sunday in the deep snow, uphill both ways. We all know this is true—common sense informs us that the United States was a devoutly religious culture and is now a highly secular culture. That religion has been in consistent decline is anachronistic—obvious. Yet participation in American religion can be measured. When looking across time, actually counting religious participation, Finke and Stark (2005) found that religious participation had only increased in America until the 1960s when the total percentage of the population

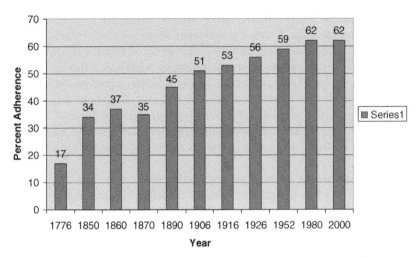

Figure 1.1. Rates of American religious adherence, 1776–2000. Source: Data from Finke and Stark (2005).

participating plateaued at approximately 62 percent of the population. Figure 1.1 illustrates the pattern in American religious adherence from 1776 to 2000.

What do these data tell us about the trend in American religious adherence? Rather than being in constant decline, American religious adherence increased until the 1960s. This is exactly the opposite of our cultural common sense that explains religion in decline. Social scientists who study American religion often argue about the nature of this relationship, but the data tell a very different story than our commonsense perceptions of religion in America.

In fact, there has been much made of another trend in American religion that seems to counter the trend illustrated in Figure 1.1. In recent years much has been made of the increasing number of people who report that they have no religious affiliation (we call these people "nones"). According to the American Religious Identification Survey (ARIS) in the 1990s, 8 percent of Americans claimed "no religious affiliation" (Kosmin and Keysar 2008). In 2001 the number of Americans claiming "no religious affiliation" almost doubled to 14 percent. Much research has been done over the course of the last decade to explain how the increasing "nones" may or may not illustrate the commonsense trend of America's religious decline (see Hout and Fischer 2002; Kosmin and Keysar 2008). By 2008, however, the nones had increased by only 1 percent, rising to 15 percent of Americans.

What do these seemingly conflicting trends tell us about American religious adherence? Consider the evidence given here. Approximately 62 percent of Americans participate in formal religion in the United States while approximately 15 percent of Americans report having no religious affiliation. When 62 percent of Americans are religious adherents, that leaves 38 percent of Americans who are not. In the past it was seen as less acceptable to claim to have no religious affiliation. People who did not actively participate with any religious group would often claim a religious affiliation (e.g., if they'd attended church with a grandparent, they may claim to be affiliated with their grandparent's religious group). Although this group of nones is growing, it doesn't really tell us that fewer people are participating in American religion. It illustrates that—of the 38 percent of American who are not religious adherents—15 percent are more comfortable reporting that they have no religious affiliation, a more accurate measure of self-reported religious affiliation. The self-report of "nones" has virtually no bearing on the aggregate data illustrating religious adherence (the religious adherence rates were collected by counting church records and census records, not self-reported data).

Different types of data, different methods, different measures, and different research questions lead to a variety of findings. Whereas hindsight bias may make us go "duh," when our cultural common sense agrees with the result, science itself gives us a window into a more complex human world, where findings may be quite different than what common sense may tell us. This is another advantage of the scientific process. Science is not about loading an argument in favor of our own opinions, but developing a baseline reflecting what the data tell us is truly happening in the world. We should not be afraid of divergent findings that show us how complex the world is and give us new avenues toward thinking about the world. Science is about looking at *all* of the evidence in a rigorous and systematic way, so that we can unravel the mysteries of human interactions by testing, measuring, and replicating studies that tell us about social patterns.

HOW DOES SOCIAL SCIENCE RESEARCH ACTUALLY WORK?

Social science research gives us tools to evaluate relationships between concepts. For example, does being religious influence generosity? Does lower socioeconomic status impact the age at first sexual experience? Does gender impact career choice? Each of us has tried to explain how one thing leads to another (e.g., "I studied really hard for that test, which is why I did so well on it"). Science gives us direction as to how to test if our assumptions are true. Following the systematic steps of the scientific method allows us to think critically about the information we consume—from the news media, within your classes, in conversations, from social media, and so on. We need to be careful, however, to learn how to critique/question research findings responsibly. Often learning "critical thinking" is interpreted as attacking or negatively assessing some piece of research/information. Critiquing research findings includes asking the appropriate questions and having the skills to assess how to answer those questions, not just tearing something apart.

One of the first questions you should ask is "How do I know what I know?" Like the taken for granted assumption of the decline of religious adherence in the United States, how do you know what you think you know? Research gives us the opportunity to step back from our cultural commitments—common sense—to test to see if relationships are true. As we noted earlier, there is broad consensus that religion in America has only been in decline. The data do not support this common perception.

Another taken for granted assumption is a link between education and income. Would you earn a college degree if you didn't think that your level of education was linked to your potential for higher earnings? Yet have you personally experienced having a higher level of education and subsequently earning higher income? Most people take for granted that there is a *relationship* between education and income, but how do you know it's true? If you only consider people who have both, are you seeing the whole picture? Look at your professor. Most professors have the highest levels of education available—PhDs. Do they make the highest levels of income? Think of those who make high incomes—sports stars, entertainers, CEOs. Do these people have the highest levels of education? Often high-profile sports stars and entertainers have less than a college degree. So why do we believe there is a link between education and income—so much so that we spend several years and thousands of dollars obtaining a college degree? Is it worth the time and money investment? Is there a link between education and income?

Luckily, social science allows us to investigate/test the relationship between education and income. Using the General Social Survey for 2010, we test the relationship between education (respondent's highest degree earned, or RS Highest Degree) and income (respondent's income category, or rincomecat) expecting that the more education a person has, the more income they will earn. Figure 1.2 shows that the respondents with the most education (people with graduate degrees) do earn higher levels of income ($25,000 or more).

Therefore, based on the evidence, there appears to be a link (or relationship) between education and income, where higher levels of education tend to be associated with higher levels of income. Not everyone who has a graduate education makes the highest level of income, but there is a clear linear trend.

		RS highest degree					
		Less than high school	High school	Junior college	Bachelor	Graduate	Total
Rincomecat 1.00 <$1000 – $14,999		65	204	24	35	17	345
		52.0%	35.1%	25.0%	14.3%	10.9%	28.7%
2.00 $15,000 – $24,999		27	118	20	22	8	195
		21.6%	20.3%	20.8%	9.0%	5.1%	16.2%
3.00 $25,000 or more		33	259	52	187	131	662
		26.4%	44.6%	54.2%	76.6%	84.0%	55.1%
Total	Count	125	581	96	244	156	1202
		100.0%	100.0%	100.0%	100.0%	100.0%	100.0%

Figure 1.2. The relationship between education (RS Highest Degree) and income (rincomecat) in the 2010 General Social Survey (GSS) data.

What Are the Basic Assumptions of Science?

The first assumption to doing social science is that order does exist in human behavior. The social sciences follow the natural sciences. After scientists were able to determine that the natural world followed predictable patterns, social scientists began to see the social world as a place that also followed predictable patterns (and just because there is a category for "natural" science, that doesn't make social science "unnatural").

The scientific revolution arguably began with the publication of Copernicus's *De revolutionibus* in 1543 (Stark 2003). Just like with "duh" science, where it takes a multitude of evidence to shift people's perceptions, it took quite a bit of time and research to shift the perception away from the earth being the center of the universe. With the clear discovery of "laws" of nature (e.g., gravity, Newton's three laws of motion, etc.) and ways to study the natural world systematically, the next logical outcome was to look at the social world.

Within a century or more of the scientific revolution, the social world was experiencing quite a bit of upheaval. Industrialization, the changing nature of economic systems (from agrarian to manufacturing capitalism), and the relationship between governments and people led the way to questioning if there were social laws that science could uncover. For example, did kings rule by divine authority? Could people govern themselves? Were the rich really superior to the poor? Was European culture superior to African culture? Was there a reason that the American colonies or the people of France revolted against systems of government that had been in place for centuries? The economic, political, and social unrest of the eighteenth century paved the way for social science.

Human interaction is highly complex, but it does have order to it. We are constantly talking to each other about how and why things go together: have you ever tried to interpret the actions of someone you like to discern their feelings for you? Have you ever interpreted these "signals" wrongly? Science allows us to measure a wide variety of phenomena so that we can more accurately reflect what is real. We need science—the rigorous and systematic study of the relationship between concepts—to guard us against our conscious and unconscious biases. Several things impact our ability to see the world in unbiased ways. Common sense and our own experiences, although valid, cannot provide us with a clear picture of a complex world.

Common Sense Is Not Enough: Errors in How We Observe

As you've read through the introduction to social science and the process of science, we hope you've begun to see the importance of science. While each of us experiences the world in unique ways, we also tend to make very un-unique (or common) errors in how we observe the world around us. We noted earlier that each of us inhabits a social location that biases the way we interpret the world. For example, Republicans and Democrats see the same piece of data, but they interpret it in different ways (both agree that the economy is in jeopardy; to fix it Republicans advocate lowering taxes to the wealthy, whereas Democrats advocate raising taxes for the wealthy). Conservative Protestants and liberal Protestants disagree on what is more important about being

Christian (conservatives feel what you believe is most important; liberals believe what you do is most important). White supremacists interpret having a biracial president as evidence that whites are discriminated against, whereas multicultural educators see it as progress that nonwhites are receiving more opportunities. Political perspectives, religious perspectives, and personal prejudices play a part in interpreting what is going on in the world. What do the data say about these things? In the past did raising taxes or cutting taxes help stabilize the economy? What beliefs and behaviors do liberal and conservative Protestants share that make them both Protestant? Are whites being discriminated against because nonwhites have more opportunities? These are empirical questions that can be addressed. Social location is one bias that hurts our ability to evaluate what is going on. Other common errors we make include observing the world inaccurately, overgeneralizing, and observing selectively.

Have you ever been driving when you're suddenly stricken with a scary thought: "Did I just run that stoplight?" As you check your rearview mirror and try to reconstruct pulling up to the light and seeing if it was green or red, you panic; you simply cannot remember seeing what color the light actually was. In effect, driving has become so commonplace that you were distracted in deep in thought and not fully conscious of your surroundings. The good news is that even in your semiconscious state, your brain likely took in the appropriate information and responded appropriately (going through a green light). The bad news is that you were operating heavy machinery while your brain was on autopilot. This semiconscious state impacts the information we take in from our surroundings and interactions. Have you ever asked someone a question and then didn't really listen to their answer? We consistently make **inaccurate observations** or take in inaccurate information, making casual or semiconscious observations. Yet at the same time, we tend to think that whatever we experience is a defining experience where we fully and consciously take in accurate information or observations.

The taking in of inaccurate observations can have significant consequences. In the classic film *Twelve Angry Men*, a jury is set to convict a man of murdering his father. The jury is charged with evaluating the evidence brought to bear on the case, often through eyewitness testimony. During the course of the film, the jury assesses the reliability of the eyewitness accounts only to discredit some of this testimony as being inaccurate (e.g., one eyewitness testifies that she saw the murder after awaking in the middle of the night. She heard screams, got up, went to her window, and looked across to the father's apartment while an elevated train ran between them). This semiconscious observation could have important consequences on the man being charged with murder.

Apart from making inaccurate observations, another common error we make in observation is by **overgeneralizing**, which assumes a wider understanding and knowledge based on very little evidence. Have you ever left a classroom after getting an exam back feeling that you did not do very well? Did you mention to your friends or family that "Everyone failed the test"? This is likely a bit of an embellishment. Although you may have kibitzed about not doing well with some others in the class, did you actually see their scores? Were you able to take a full accounting of everyone else's score to make sure when you told others that "everyone" had failed, you knew it to be true?

A local neighborhood blog illustrates the problem with overgeneralization, as well as another problem in how we observe. In the daily news blog a concerned citizen writes, "On Monday night a beige middle 2000s model Subaru wagon on 28th Avenue was driving at high speeds and uncontrollably hitting cars and other objects" (*Magnolia Voice* 2011). The blog goes on to explain that the reporter heard "squealing tires and multiple loud crashes. By the time we made it outside the Subaru was speeding off, fish-tailing uncontrollably and headed south down 28th." At this point there is no over-generalizing. The reporter describes what she is seeing and experiencing. Like many blog postings, however, it's not the article itself but the responses that are most interesting. The first responses to the story caught our eyes:

> "I give those Subaru wagons a wide berth. The people who drive them are terrible."

> "How true. I was thinking the same thing. They are either driving way too slow and holding up traffic or careening and speeding like drunken madmen."

Clearly these two people had preconceived notions regarding Subaru wagons and/or those who drive Subaru wagons (which includes one of this textbook's authors). But is there real evidence that Subaru wagon drivers are more reckless? This is clearly an overgeneralization, buoyed by another common error in how we observe: selective observation.

Oftentimes when overgeneralizations are made, we begin to see only the evidence that reinforces the understood overgeneralization. When we look around us, our brains only take note of the times we see what we expect to see—thereby strengthening our misdirected generalization. When an out-of-control car careens down a street, people make generalizations about the model of car or the people who drive that model of car. The overgeneralization takes on more legitimacy when others can attest to the experience by citing their selectively observed "evidence."

For example, let's say that one of this book's authors, who doesn't drive a Subaru wagon, purchased a red sports car. At the time, several people commented that red was the wrong color to buy because the police pulled over red cars more than any other color. As we noted earlier, the question to ask here is "How do they know that?" Did each of those people read the National Transportation Safety Board's annual report as to what cars were most likely to speed? Did they read *Police Beat* to find evidence that state troopers were in the habit of pulling over red cars rather than black cars? It was widely understood that red cars were the most likely to be pulled over for speeding. In driving, however, passengers only remarked on a car being pulled over when it was red. The passengers never remarked on silver cars, white cars, blue cars, and so on, that were pulled over: **selective observation.** Once a pattern has been determined, only the evidence that supports the pattern is consciously taken in, reinforcing a pattern that often misrepresents the full picture.

Social location, inaccurate observation, overgeneralization, and selective observation hinder our ability to see the world as it is, rather than what we selectively understand or experience. Our perceptions are powerful, but they do not always reflect what is real. Based on the number of crime dramas on television, we might perceive that the number of felons who plead innocence by means of insanity is 20 percent or higher. Yet research has shown that less than 10 percent of felony crimes go trial, and less than

1 percent of felons plead innocent by insanity (Bolt 1996). Our individual experiences are valid, but they can't represent the whole picture.

Our perception and processing of data is influenced and biased by numerous sources. We need to have more structured ways of observing the world, if it is to be considered scientific. *After all, wasn't it just common sense that told us the sun revolved around a flat earth?* The research process—the scientific research process—guards against many of the errors we just reviewed because it follows a systematic approach to study humans and their behavior. We simply cannot rely on our own perceptions and experiences to tell us about the whole context because they are not random and cannot fully measure the whole context.

Exercise: Should Marijuana Be Made Legal?
The title of this exercise probably got your attention. While the question of legalizing marijuana is not a scientific one (it is what we refer to as a normative statement—a statement (or in this case question) of opinion). Scientists cannot test whether or not marijuana should indeed be legalized, but we can see if there are certain categories of people who think that marijuana *should* be made legal.

The Association of Religion Data Archives (the ARDA) is an online resource that provides data to interested students, educators, journalists, and researchers. Let's go to the ARDA and look at the General Social Survey (GSS) for 2010 to see if there are categories of people who may be more likely to say that marijuana should be made legal.

Go to www.thearda.com. On the front page of the Web site are a variety of tabs at the top of the page. Click on the tab for "Data Archive." The Data Archive page lists all of the data and categories of data available on the ARDA. Under "U.S. Surveys," click on the link for the GSSs. Scroll down to the GSS for 2010. Click on the link to the 2010 GSS. Notice the new page also contains several tabs; click on the "Search" tab. On the Search page, type "marijuana" into the search box and click on "Search."

The GSS survey item, "GRASS," should be returned. The question from the GSS asks, "Do you think the use of marijuana should be made legal or not?" The category responses are

  0) Inapplicable
  1) Should
  2) Should not
  8) Don't know
  9) No answer

Under the category responses is a link for "Analyze Results." Click on this link.

(Continued)

Figure 1.3. The Analyze Results page from TheARDA.com for the variable GRASS.

The Analyze Results page in Figure 1.3 shows two summaries for the question, GRASS. There is a pie chart, giving a visual picture of how the data play out (notice that 28.3 percent of people said marijuana *should* be made legal, and 33.2 percent of people said marijuana *should not* be made legal). There is also a summary table with the same descriptive information.

Below the two summaries of data are several tables showing how sex, political ideology, age, region of the country, religion, race, and church attendance impact whether or not marijuana should be made legal. Focusing on the row for "should," go through each table to see what patterns you find:

Sex

	Male	Female
Should	32.1%	25.3%

The table shows that 32.1 percent of males replied that marijuana should be made legal, and 25.3 percent of females replied that marijuana should be made legal. Therefore, these data show that men are more likely to support the legalization of marijuana.

Look at the remaining tables and fill in the percentages for each category who say that use of marijuana *should* be made legal. Then summarize the results (e.g., looking at SEX, we found that "men are more likely to support the legalization of marijuana"):

Political ideology

	Extremely liberal or liberal	Slightly liberal	Moderate	Slightly conservative	Conservative	Extremely conservative
Should	41.2%	32.0%	29.4%	25.4%	17.7%	14.7%

What is the pattern evident in the data? Who is most likely to support making the use of marijuana legal? _____

_____

_____

Age

	18–29 yrs old	30–34 yrs old	45–59 yrs old	60–74 yrs old	75+ yrs old
Should	33.2%	28.3%	32.3%	24.6%	16.2%

Is there a pattern in the data? Who is most likely to support making the use of marijuana legal, people younger than 60 years old or people 60 years old and older? _____

_____

_____

Region

	Northeast	Midwest	South	West
	30.3%	27.0%	24.8%	34.3%

Is a pattern evident in the data? Which region is most likely to support making the use of marijuana legal? _____

_____

_____

(Continued)

Religion

	Protestant	Catholic	Jewish	None	Other
Should	24.3%	25.2%	42.2%	41.7%	30.1%

Is a pattern evident in the data? Which religious group is most likely to support making the use of marijuana legal? _____

_____

_____

Race

	White	Black	Other
Should	28.8%	28.7%	22.6%

Is a pattern evident in the data? Which racial group is most likely to support making the use of marijuana legal? _____

_____

_____

Church attendance

	<Once/ yr	Once/ yr	Several/ yr	1–3/ month	Nearly/ week	Once+/ week
Should	38.2%	32.8%	32.2%	26.4%	20.3%	18.3%

Is a pattern evident in the data? Which group is most likely to support making the use of marijuana legal? _____

_____

_____

Look at the preceding tables and summaries. Which findings were you expecting (that is, which findings were not a surprise)? _____

_____

_____

Which findings were surprising? _____

_____

_____

Go back to the top of the Analyze Results page, and click on the tab for "Search." Do a search to find another variable that you can autoanalyze (make sure the variable you choose has the option to "Analyze Results"). Once you find a variable choose the category response that you are interested in (e.g., we selected the *should* category for whether or not people thought the use of marijuana should be made legal). Click on the Analyze Results link and look at the tables to see if there are patterns in the data. Fill in the percentages for each category:

Sex

	Male	Female

If there is a pattern to the data, what is the pattern? _____
_____
_____

Political ideology

	Extremely liberal or liberal	Slightly liberal	Moderate	Slightly conservative	Conservative	Extremely conservative

If there is a pattern to the data, what is the pattern? _____
_____
_____

Age

	18–29 yrs old	30–34 yrs old	45–59 yrs old	60–74 yrs old	75+ yrs old
Should					

If there is a pattern to the data, what is the pattern? _____
_____
_____

Region

	Northeast	Midwest	South	West

(Continued)

If there is a pattern to the data, what is the pattern? _____
_____
_____
_____

Religion

	Protestant	Catholic	Jewish	None	Other

If there is a pattern to the data, what is the pattern? _____
_____
_____
_____

Race

	White	Black	Other

What is the pattern evident in the data? Who is most likely to support making the use of marijuana legal? _____
_____
_____
_____

Church attendance

	<Once/ yr	Once/ yr	Several/ yr	1-3/ month	Nearly/ week	Once+/ week

What is the pattern in the data? _____
_____
_____

Look at the preceding tables and summaries. Which findings were you expecting (that is, which findings were not a surprise)? _____
_____
_____

Which findings were surprising? _____
_____
_____

2

THEORIES AND HYPOTHESES

In our first chapter we illustrated ways in which our perceptions and processing of data are influenced and biased. We noted our need to have more structured ways of observing the world. The **scientific method** gives us a systematic way to understand what we observe. Is what we observe real or representative? What do our observations tell us about similar phenomena? The scientific method ensures as much objectivity as possible in how we think about and then observe the world. Within our discussion of the scientific method, we are going to focus on the "wheel of science." Wheels represent a whole, in our case a whole process. The four basic spokes to the wheel are theory, hypotheses, observation, and empirical generalization, as noted in Figure 2.1.

The Wheel of Science has two starting points, the deductive and inductive. **Deductive** science begins with the theory, statements that explain the patterns we observe. When we begin with theory, we know what we are looking for and are able to deduce—derive conclusions from the assumptions of the theory, so that we know what we expect to see. Theory drives the research, allowing us to formulate hypotheses that we can then test. The testing involves collecting observations: what do we see happening in the real world? Once we've collected the observations, we are then able to summarize what these observations tell us: do they match what we expected to find? The summary

Understanding and Applying Research Design, First Edition. Martin Lee Abbott and Jennifer McKinney.
© 2013 John Wiley & Sons, Inc. Published 2013 by John Wiley & Sons, Inc.

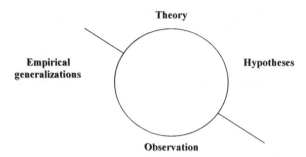

Figure 2.1. The Wheel of Science model.

is the empirical generalization. Once we've completed the wheel, we are able to take what we've observed and evaluate if our theory correctly predicted what we found.

Another way to begin the process is to start **inductively**—to induce or begin with our observations to conclude what those observations mean for the whole. When we begin inductively, we start with observations to make generalized statements about what they mean. For example, if you've ever walked into a toy store, you may have gotten the sense that there are two major categories of toys: gender categories and age categories. Toys are often organized by both categories; first they are organized by age appropriateness, and then they are organized within age by gender (usually denoted by color: pinks and pastels for girls, blue and primary colors for boys). Based on going into a few toy stores and seeing similar organizations, we may make some generalizations regarding the organization of toy stores. We might then devise a reason why toy stores are organized this way (theory) and subsequently create a hypothesis to describe how we expect to see this in a representative sample of toy stores. While both inductive and deductive approaches are valid, we begin with a deductive approach to walk through the wheel of science, beginning with theory.

WHAT ARE THEORIES?

According to Charles Ragin (1994:x), "social research involves a special dialog between ideas and evidence" resulting in generalizations about the social world that are grounded in theory and data (not just our own perceptions and/or experiences). Theory is the basic building block of social science, helping to structure the ways in which we view how and why things work together. The main task for theory is to link concepts. Look at that again. Theory links concepts. Minimally theory tells us which **concepts** should go together (e.g., education and income). Theory answers the "why" question. For example, in Chapter 1 we asked why people believe that education should be linked to income. We have general explanations as to why education should be linked to income, or how political preference impacts voting, or how marital status affects happiness. **Theory** makes assumptions about human nature and then offers us general explanations of a variety of social behaviors given certain social conditions. Once we have an explanation in place, we can predict that when one condition is present the other condition will be present also.

Several years ago the state of Montana abolished the speed limit on its highways. In reading about the speed limit change, it was easy to assume that the law would be repealed within the next year. Why? Whenever the speed limit is raised, more people die in highway accidents. When more people die as a result of the increased speed limit, states decrease the limit. People are sometimes uncomfortable when social scientists make these generalized statements (how could anyone possibly know that more people will die when the speed limit is increased?). Using the scientific method, however, it is clear that on average more people are killed in highway accidents whenever highway speed limits are increased. Stated in a scientific prediction, increasing speed limits leads to more highway deaths. This is the key to social science: theories allow us to both explain and predict that when one thing occurs, another thing will result. As predicted, within several months Montana reinstated highway speed limits because too many more people had died on Montana highways when the limit had been abolished.

Scientific theories provide us with strategies or directions to get to what we observe and have it make sense, subsequently allowing us to predict behavior (again, if there was no link between education and income, would you be in college)? Theories tell us *what to look at, when to look,* and *what to expect to see.* Theories are more than just common sense; they should be **parsimonious,** explaining the most phenomena with the least amount of theoretical assumptions. Let's walk through this process using two theories from the scientific study of religion, secularization theory, and "new paradigm" theory.

Throughout much of the twentieth century, secularization theories were the conventional wisdom when explaining the rise or fall (mostly fall) of religion in the Western world (Lechner 1991; Stark and Bainbridge 1985, 1987; Warner 1993). The assumptions of secularization theory explained that with modernization and reliance on scientific thought, the need for religion and religious organizations would eventually disappear. One famous sociologist proclaimed in the 1960s that by the twenty-first century, "religious believers are likely to be found only in small sects, huddled together to resist a worldwide secular culture". Although many secularization theorists have retreated somewhat from this position (Chaves 1993, 1994; Yamane 1997), the basic assumptions underlying the theory still hold: in a modern world where religion coexists with other meaning systems, religion loses its believability and its ability to regulate society (Berger 1967; Luckmann 1967; Tschannen 1991).

Recognize the power of this theory. Even if you've never read or heard about secularization theory, doesn't it seem familiar? Most of us have a basic understanding of a once golden era of religion in America (and the West in general)—a long past golden era. Yet in Chapter 1 we used a graph illustrating the opposite trend in the United States. Certainly, over the last few decades the percentage of American religious adherents hasn't grown; neither has it declined. In looking at the empirical data, a second theoretical tradition was born. This "new paradigm" agrees that secularization is a process that is occurring, but adds that secularization is self-limiting (Finke and Stark 2005; Stark and Bainbridge 1985, 1987; Stark and Finke 2000). What that means is that as societies change, religious organizations will also change. So while *some* religious groups and organizations may disappear, other religious organizations will take their place. In effect, this means that regardless of how modern or rational a society becomes, religious participation will not decline.

Here are two competing theories; each well thought out and each making sense if we think of our own experiences. Like trying to justify why opposites attract (or why birds of a feather flock together), by finding at least one case that makes it true (the confirming evidence), each of these theories can explain some part of what we think we see in regard to religion in America. For example, the oldest denominations in the United States have been in decline (losing market share, as it were) for over 100 years (e.g., the United Methodist Church, the Episcopal Church of America, and the United Church of Christ). The decline of these denominations seems to confirm secularization theory's prediction (these denominations tended to have the most highly educated members; perhaps as the members became more educated, or more "modern," they were less likely to believe in their church's doctrines). At the same time, the nonde-nominational movement of the 1980s and beyond is going gangbusters, seeming to confirm the new paradigm theory—that as some churches decline, others will take their place. Which of these theories is more accurate? This question has sparked an enliven-ing debate within circles that study the social sources of religious belief and behavior. Drawing on a variety of measures, social scientists who study religion research these questions, giving ample evidence of pieces of American experience that confirm and seemingly disconfirm the theories. Divergent data and the continuing testing of these theoretical perspectives help advance our knowledge of the complexities of religion in the social world.

Theory alone, however, is not enough. While theories are the beginning of deduc-tive social science, if they stand untested they are not very useful. Social scientists are interested in **empirical** facts—facts that are observable using the physical senses (sight, sound, taste, touch, smell). In science we want to *empirically test* theories to see if they indeed predict and explain social patterns and behaviors. If something cannot be observed using the five senses, then it is not within the purview of social science but perhaps of philosophy or religion. Science is about testing what can be observed—as Ragin (1994) said, a "dialog" between ideas (theory) and evidence (empirical data).

We sometimes tell our students that without knowing them at all, we can predict things about them. How can we do that? Obviously it's because we're amazing (or psychic). For example, when students use the word *spendy* to describe something that is expensive, it's an indicator that they come from a particular region of the country (the middle West Coast). Other regions of the country use the word *pricey* to describe expensive items. Another way we can generalize to our students—without knowing them directly—is about their ethnic heritage. For our students from the Pacific North-west, we can predict that, on average, they are third- or fourth-generation Nordic—they are amazed when roughly 67 percent of them can agree that they are. Social scientists are not really psychic (although some may be amazing), but as the natural world revolves in an orderly fashion, so does the social world. The Pacific Northwest is the last of the old frontier; the people who settled this region of the country were predomi-nantly Nordic and immigrated to the area in the late 1800s and early 1900s, accounting for the ethnic and generational norms of our students' heritage. The orderliness of the social world allows us, as social scientists, to learn about, describe, explain, and predict social phenomena.

WHAT ARE HYPOTHESES?

Once theory defines which concepts are linked, we move to the second step of the Wheel of Science: hypotheses. **Hypotheses** follow from theories, stating relationships between two or more concepts in such a way that they can be empirically tested. For example, theory may link the following concepts:

Gender affects occupational choice.

Social class affects voting behavior.

Religious participation affects attitudes toward abortion.

Occupation affects income.

Hypotheses state *how* these concepts are expected to work together:

Theory: Gender affects occupation.
Hypothesis: Men will be more likely to report having managerial occupations.

Theory: Social class affects political participation.
Hypothesis: Those with higher social class will be more likely to participate in politics.

Theory: Religious participation affects attitudes toward abortion.
Hypothesis: People who attend religious services more often will be less likely to support legalized abortion.

Theory: Occupation affects income.
Hypothesis: Those with white-collar occupations will earn more income.

Hypotheses allow us to test whether or not relationships between facts are supported. For example, in looking at the relationship between social class and political participation, we can predict that the relationship works in one of two directions: positive or negative. In a positive relationship, both variables vary in the same direction:

Hypothesis: The higher one's social class the more likely one is to participate in politics.

"Varying in the same direction" means that as one variable increases (as social class increases), the other variable will also increase (there is a higher likelihood of political participation). The variables are still varying in the same direction, even when positing the opposite—when social class is decreasing we expect the likelihood of political participation to decrease as well:

Hypothesis: The lower one's social class the less likely one is to participate in politics.

You will notice two things within hypotheses: hypotheses embody the relationships intended by the theory but in such a way that they can be tested, or "objectively" verified. For example, the theory linking social class to political participation is made measurable given the hypotheses stated earlier. The other thing to notice is that the hypothesis statement implies an *impact* of social class on political participation. This suggests that political behavior is increased or decreased by social class standing; the relationship is not static, it has a **directionality** to it.

Based on these hypotheses, we can obtain data that has measures of social class and political behavior to see if the hypothesis is supported by the data. Here we've **operationalized** social class with a variable from the 2010 General Social Survey (GSS) that measures "subjective class identification" or a person's conception of where they fall on the social class ladder between lower class, working class, middle class, or upper class. We've also found a variable that measures political participation, asking respondents whether or not they voted in the 2008 presidential election. Let's test the hypothesis that as the subjective class identification increases, people will be more likely to have voted.

Figure 2.2 shows how perceived social class varies with political participation (measured by voting practices). As you move along the top of the figure, you see the four categories of social class moving from lower to higher (lower, working, middle, and upper). Because we hypothesized that people of higher social class will be more likely to have voted, we then want to focus on the row of the variable for "Voted." As we move across the row, moving from 49.7 percent of the lower class reporting that they voted, to 81.8 percent of the upper class who reported voting, you can see that as

Voted08 * SUBJECTIVE CLASS IDENTIFICATION Cross-tabulation

			SUBJECTIVE CLASS IDENTIFICATION				
			LOWER CLASS	WORKING CLASS	MIDDLE CLASS	UPPER CLASS	Total
Voted 08	1.00	Voted	81	598	656	45	1380
			49.7%	67.8%	81.3%	81.8%	72.4%
	2.00	Did not Vote	82	284	151	10	527
			50.3%	32.2%	18.7%	18.2%	27.6%
Total		Count	163	882	807	55	1907
			100.0%	100.0%	100.0%	100.0%	100.0%

Figure 2.2. The relationship between subjective class identification and voting (GSS data).

you move across the categories for social class, the percentage of respondents who voted *increased*. Alternatively, you can see that the percentage of respondents who report not voting *decreased* as the social class changed from Lower Class (50.3%) to Upper Class (18.2%). These findings confirm the hypotheses statements we included earlier (a positive directionality; as social class increases, having voted also increases).

While the relationship to social class and political participation was hypothesized to be in a positive direction (meaning that as the independent variable increased, so did the dependent variable), another possibility is to have a negative direction. In a negative relationship, variables vary in opposite directions:

Hypothesis: As religious participation increases, support for abortion decreases.

Hypothesis: As religious participation decreases, support for abortion increases.

Drawing on data from the GSS using SPSS software to analyze the hypothesis statements, we find a variable that measures religious participation (frequency of religious service attendance) and a measure for whether or not respondents support abortion in cases where the woman wants an abortion for any reason.

Figure 2.3 illustrates a negative relationship captured in the hypothesis statements regarding religious participation and attitudes toward abortion. The figure shows whether respondents agree that a woman should be able to have an abortion for any reason (Abortion if woman wants for any reason) according to how often they attend religious services (How often [Respondent] attends religious services). If we focus on the top row (for Yes) of data across the attendance categories, you will see that as religious attendance increases, the percentage of respondents who agree with the statement (Yes) *decreases* from None or less than once per year (59.1%) to Weekly or more

ABORTION IF WOMAN WANTS FOR ANY REASON * Attendance Cross-Tabulation

ABORTION IF WOMAN WANTS FOR ANY REASON		Never or less than once per year	1 per year	Several times per year	Once per month	Almost weekly	Weekly or more	Total
				Attendance				
YES	Count	218	97	64	45	47	65	536
		59.1%	55.7%	47.4%	47.4%	32.9%	20.9%	43.7%
NO	Count	151	77	71	50	96	246	691
		40.9%	44.3%	52.6%	52.6%	67.1%	79.1%	56.3%
Total	Count	369	174	135	95	143	311	1227
		100.0%	100.0%	100.0%	100.0%	100.0%	100.0%	100.0%

Figure 2.3. The relationship between religious service attendance and attitude toward abortion.

(20.9%). Alternatively, you can see that the percentage of respondents who did not agree with the statement *increased* as attendance increased from None or less than once per year (40.9%) to Weekly or more (79.1%). These findings confirm the hypotheses statements we included earlier (with a negative directionality).

OPERATIONALIZING VARIABLES

Once a hypothesis tells us *how* variables go together, then we are ready to **operationalize** the variables that allow us to test the hypothesis. As we noted earlier, when testing hypotheses on social class and voting, as well as religious participation and attitudes toward abortion, we operationalized the concepts by finding variables in the GSS that measured them. Operationalizing variables means to transform them into something measureable. Hypothesis statements embody the relationships intended by the theory but do so in such a way that they can be objectively verified. By transforming variables into something measureable, we can more easily see and verify whether the relationships implied in the theory hold true empirically.

Variables are the concepts we are interested in testing. For example within a data set we find **indicators** that are observable measures of a concept. Variables should have two distinctive features. They should be exhaustive and mutually exclusive. Variables that are **exhaustive** have answer categories that allow every person or thing to be accounted for. For example, if I ask you your religious preference and the answer categories only include Protestant, Catholic, or Jewish, can everyone fit into those three categories? What categories might we be missing? If we try to enumerate all possible religious categories, our survey might run out of space. By knowing your population well, you should be able to provide the appropriate categories:

What is your religious preference?
1. Protestant
2. Catholic
3. Jewish
4. Muslim
5. None
6. Other

In a U.S. context, the addition of Muslim is helpful, as are the additions of "None" and "Other." The answer categories for religious preference are now exhaustive.

The second general rule for variables is that answer categories should also be **mutually exclusive,** where no one person or thing can fit into more than a single category. A good question or variable has answer categories that allow respondents to select only one answer. Look at the answer categories given for religious preference. Not only are these exhaustive, but they are also mutually exclusive. Given that all of the religions listed are monotheistic, people are expected to follow just one. Therefore people should be able to fit into only one of the categories, making them mutually exclusive.

Exercise: Operationalizing Concepts

When operationalizing concepts within survey data, the variables you choose should have a *description* (which tells us *what* the variable is measuring) and *response categories* (which tell us *how* the variable is measured; and of course, the response categories should be exhaustive and mutually exclusive as described in the text).

Practice operationalizing concepts by going to the Association of Religion Data Archives (www.thearda.com) and click on the "Data Archive" tab. Under the category for U.S. Surveys, click on the link for the General Social Surveys. Scroll down to the General Social Survey for 2010 and click on the link to the survey.

Let's say I'm interested in how a person's gender impacts their religiosity, health, and levels of prejudice. Go to the Search tab and do a search for gender. What do we find?

Figure 2.4 shows the first 4 of 16 results for variables with the word *gender* in them. Of these results, 14 ask for the "Gender of . . . " some other person. The other two results ask, "Do you feel in any way discriminated against on your job because of your gender?" (WKSEXISM) and "(To what extent do you agree or disagree with the following statements?) In America, people have the same chances to enter university, regardless of their gender, ethnicity or social background" (EQUALCOL). While these latter two questions address attitudes toward gender (in the workplace and entering college), they don't quite get to measuring someone's actual gender—which is what we are interested in (how does a person's gender impact religiosity, etc.).

When operationalizing concepts, we need to think outside of the box. Searches for variables are different than using Internet search engines. When using a search feature on databases (online or otherwise), the database search simply looks for words or phrases that appear in the text of the description or answer categories. So when doing a search for "gender," the respondent's actual gender does not come up. What words or phrases might result in better operationalizations?

If a search for "gender" does not result in finding out if someone is a woman or a man, then I need to do a search for "sex." Go back to the top of the page and do a search for "sex." Among the many interesting survey items that appear, the first one is just what we need:

43. SEX
Respondent's sex
1) Male
2) Female
Now we can break these down into their appropriate pieces:
Variable: _____*43. SEX*_____
Variable description: *Respondent's sex*_____
Response Categories:
 1) Male
 2) Female

Keep in mind that as you search for variables, you may need to be creative in how and what you type into the "search" box. You don't want to assume that

(Continued)

Figure 2.4. Search Results from TheARDA.com for the variable "gender" in the GSS 2010.

searching for "gender" will automatically return all of the possible operationalizations. For example, if we were interested in *attitudes toward gender*, rather
than a person's gender, WKSEXISM and EQALCOL would seem to fit the bill—
both deal with a person's attitude toward or about gender. But if we stopped
with just the two variables that have "gender" in the description, we would be
missing a host of other operationalizations of attitudes toward gender.

Other terms that would help locate additional variables operationalizing
attitudes toward gender might include *woman, man, women,* or *men.* If you
type each of these four words into the search box, you will see a plethora of
variables that measure attitudes toward gender. While there is some overlap
between them, each search yields new variables that measure attitudes toward
gender.

Now you can try to find operationalizations of the concepts for religiosity,
health, and prejudice. Based on what we've discussed for conducting search in
databases, consider what words or phrases might help you find variables that

measure each concept. Then do a search for each concept finding at least three operationalizations of each.

Religiosity
What words or phrases might be helpful in searching for operationalizations of "religiosity"? List several of these:

_____
_____
_____
_____

Using some of the words and phrases you chose, list the number, name, description, and response categories for each variable:

1. Variable: _____
   Variable description and answer categories: _____
   _____
   _____
   _____
   _____

2. Variable: _____
   Variable description and answer categories: _____
   _____
   _____
   _____
   _____

3. Variable: _____
   Variable description and answer categories: _____
   _____
   _____
   _____
   _____

Health:
What words or phrases might be helpful in searching for operationalizations of "health"? List several of these:

_____
_____
_____
_____

Using some of the words and phrases you chose, list the number, name, description, and response categories for each variable:

1. Variable: _____
   Variable description and answer categories: _____
   _____

(Continued)

_____
_____
_____

2. Variable: _____
   Variable description and answer categories: _____
_____
_____
_____
_____

3. Variable: _____
   Variable description and answer categories: _____
_____
_____
_____
_____

Prejudice:
What words or phrases might be helpful in searching for operationalizations of "prejudice"? List several of these:
_____
_____
_____
_____

Using some of the words and phrases you chose, list the number, name, description, and response categories for each variable:

1. Variable: _____
   Variable description and answer categories: _____
_____
_____
_____
_____

2. Variable: _____
   Variable description and answer categories: _____
_____
_____
_____
_____

3. Variable: _____
   Variable description and answer categories: _____
_____
_____
_____
_____

Did you encounter any problems when operationalizing religiosity, health, and/ or prejudice? What were they and would you search differently if you were going to do this again?

_____

_____

Select one variable for each concept that you operationalized. Go back to the ARDA, search for each of the three variables, and click on "Analyze Result," Using the variable for SEX, explain how (or if) sex impacts each variable (that is, do you see any patterns between males and females)?

Religiosity
Variable: _____

	Males	Females

If there is a pattern in the data, what is it? _____

_____

_____

Health
Variable: _____

	Males	Females

If there is a pattern in the data, what is it? _____

_____

_____

Prejudice
Variable: _____

	Males	Females

If there is a pattern in the data, what is it? _____

_____

_____

INDEPENDENT AND DEPENDENT VARIABLES

The testing of our two previous hypotheses was predicated on variables having a particular place or time order to them. We hypothesized that something about social class impacted political behavior, that if we knew what was happening with social class, we could anticipate what was going to happen with political behavior (that as social class increased, we could subsequently expect the percentage of those who voted to increase). When testing hypotheses, we have two kinds of variables: independent variables and dependent variables. An **independent variable** is a variable that is assumed to influence another variable; it comes first—when we know what is happening with the independent variable, we can predict what's going to happen to the other (dependent variable). Another way to think about the differences between these two variables is that the independent variable is the cause of change in the **dependent variable**, which is the effect.

Independent/Dependent and Predictor/Outcome Variables
In this book, we use the language of other researchers to refer to different variables. As we noted, independent and dependent variables are usually identified with "cause" and "effect," respectively. These designations are most particularly used when we are talking about experiments because experiments directly deal with the issue of causality (we discuss experiments more comprehensively in a later chapter). Other designs, like those using survey data, focus more on identifying patterns of relationship among variables rather than establishing causality. In these situations, some researchers may use other designations for variables, like "predictor" for the variable that comes first and is thought to be the influence on other variables, and "outcome" for the variable that is thought to be influenced by other variables. As you can see, predictor/outcome correspond to independent/dependent variables but are used differently depending on the nature of the research design.

3

OBSERVATION AND EMPIRICAL GENERALIZATION

In dealing with the building blocks of the Wheel of Science, we start with theories and hypotheses to allow us to frame research. Once theory has given us direction for research, we create hypotheses to test our theoretical propositions. When we know what concepts we're testing and how we expect them to work together, we need to consider what research design will best test our hypotheses. **Research designs** give social scientists tools that help determine what observations they are going to use to test their hypotheses.

Social researchers draw on four basic research designs from which to collect data to observe the concepts they want to test: surveys, experiments, field research, and secondary sources (which include historical documents, aggregate [or comparative] data, and content analysis). We devote individual chapters to discussing five designs: surveys, experiments, field research, and two of the secondary source designs: aggregate (or comparative) and content analysis. Each research design offers researchers unique ways to collect data about people and social things. It is also quite common for researchers to combine methods in order to **triangulate** their data.

Designs fall into two basic types, quantitative and qualitative. **Quantitative** designs rely primarily on describing or measuring phenomena in quantity—generally thought of as numerical quantity. Quantitative designs employ statistics to describe large

Understanding and Applying Research Design, First Edition. Martin Lee Abbott and Jennifer McKinney.
© 2013 John Wiley & Sons, Inc. Published 2013 by John Wiley & Sons, Inc.

populations with survey data or to use theoretical generalizations gleaned from experiments. Quantitative designs are also used with aggregate-level measures (comparative measures) to generalize statistically to larger geographic or social units (units generally composed of multitudes of individual people or things).

Qualitative designs rely on quality of description; rather than quantifying large samples of people or units, qualitative designs rely on great detail in reporting human processes. Field research is primarily a qualitative design, where researchers go into a field setting to observe people, collecting very detailed information about some smaller group, process, and/or interaction. Content analysis is another qualitative design, where cultural artifacts are examined to provide context and derive meaning from the things that people create. Both of these designs provide a wealth of detail about social patterns.

To allow you to connect to a general understanding of the full Wheel of Science, we briefly discuss the basics of what each research design entails, to give you an overview of the five designs we explain more fully in later chapters. We've organized the book to reflect the grouping of quantitative and qualitative designs, to give you a sense of moving from one type of design to another. Thus we cover surveys, aggregate data, and experimental designs first, as quantitative designs, and follow up with the qualitative designs of field research and content analysis.

QUANTITATIVE DESIGNS

Surveys

Everyone, it seems, has taken a survey. **Surveys** collect information from people using interviews or questionnaires composed of written questions. Surveys are arguably the most used tool for social scientists, market researchers, and a variety of others looking for information regarding people's attitudes, behaviors, and experiences. Yet many surveys lack the integrity of being truly scientific. Wordings of questions, question order, and question categories are important considerations for collecting survey data.

Surveys are useful for studying the world as it is—not as we *think* it is, not as we think it *should* be, but how it *actually* is. As we noted in the introduction, we do not experience or observe the world randomly. Our ideas about the world are influenced by our culture and social locations. Scientific surveys should ask unbiased, valid, and reliable questions that tell us about actual patterns of people's thinking and behaving. What is unique to surveys is their ability to ask a variety of questions to large samples of people (note: only people can take surveys). These questions can be easily replicated to the same population and/or to other populations to gauge change within populations, between populations, or over time.

In previous chapters we've had you visit the Association of Religion Data Archives (the ARDA) to search through various items in the 2010 General Social Survey (GSS). The GSS is a nationally representative survey designed to be part of a program of social research to monitor changes in American's social characteristics and attitudes (ARDA 2011; GSS 2010). Funded through the National Science Foundation and administered by the National Opinion Research Center, the GSS has been administered annually or biannually since 1972. As a general survey, the GSS asks a variety of questions on a variety of topics. For the 2010 GSS, special modules (sets of questions) were asked on the subjects of aging, the Internet, gender roles, immigration, religious identity, and sexual behavior, as well as several other topics. By going to the ARDA (www.thearda.com) you can peruse the Codebook for the 2010 GSS (www.thearda.com/Archive/Files/Codebooks/GSS10PAN_CB.asp) to get a fuller sense of the types of questions a general survey asks. You can also visit the ARDA's "Learning Center" to take a survey that allows you to compare yourself to a larger national profile. The "Compare Yourself to the Nation" survey allows you to see how you compare to others based on the results from the 2005 Baylor Religion Survey (addressing religious identity, beliefs, experiences, paranormal views, etc.).

Apart from the actual survey instrument (the questions themselves), most surveys are obtained using *samples* of *populations*. **Populations** are the set of all the people you are interested in studying with a survey. Populations can be so large as to prohibit our ability to study all of the units (people), so social scientists select only a portion or subset of people (a **sample**) to observe. This gets a bit tricky, however. Just having access to a group of people does not make that group of people a scientific sample. We'll spend a chapter talking about populations and sampling techniques, but for surveys know that generally we select a *random* subset of a population to take the survey. Randomness ensures that the sample is statistically *representative* of the population as a whole. Again, we'll discuss what that means in Chapter 7, "Learning from Populations: Censuses and Sampling."

Princeton Review's Top-Ten Party Schools 2012[1]
Each year the Princeton Review ranks the top colleges and universities in a variety of categories, including Top Party School. The rankings are derived from the results of an 80-question survey collected from 122,000 college students. Here are the 2012 top party schools according to the survey:

1. Ohio University, Athens, OH
2. University of Georgia, Athens, GA
3. University of Mississippi, Oxford, MS
4. University of Iowa, Iowa City, IA
5. University of California, Santa Barbara, Santa Barbara, CA
6. West Virginia University, Morgantown, WV
7. Penn State University, University Park, PA
8. Florida State University, Tallahassee, FL
9. University of Florida, Gainesville, FL
10. University of Texas, Austin, TX

---

[1] Joanne Viviano, "Survey Calls Ohio U Top Party School," *Seattle Times,* August 1, 2011, available at http://seattletimes.nwsource.com/html/nationworld/2015790839_apuspartyschools.html.

Aggregate Data

An **aggregate** is one unit that consists of a number of smaller units. Also known in the social sciences as "comparative" data, aggregate data are based on comparing large social units, for example, countries, cities, states, schools, clubs, or churches. Calling data "comparative" is a bit confusing when you first begin to look at research design. Isn't the entire process of science about comparing how one thing changes when something else changes? To be more clear, we use "aggregate data" to talk about this research design.

Since aggregate data are collected on large social units, rather than individual units, it's often not data that you collect yourself (while you can write your own survey and send it out, or design your own experiment to carry out, someone else has to have counted the number of visits per national park, for example). Generally aggregate data are "secondary" data—data that are collected and recorded by some official agency or organization. For example, the U.S. Census Bureau collects data on the percentage of the national population that lives in a metropolitan area, or the percentage of the population that is under 18 years old, or the ratio of male to female in states. Since the United States has a constitutional amendment prohibiting the establishment of an official religion, organizations apart from the Census Bureau collect data on religious membership.

The Religious Congregations and Membership Study (RCMS) collects data on congregations across the United States, noting the number of congregations, membership, and adherence. One benefit of aggregate data is the fact that because each aggregate unit is made up of smaller units, statistically there is less measurement error, and the relationships are more robust; it's rare to find a relationship between variables (or not find one) if a relationship doesn't exist. Relying on official statistics, however, may carry some costs. When the RCMS data are collected, how do we know if each religious group counts their members the same way? Does each group count children as members, do they even have membership (nondenominational Christian churches often do not), what counts as a religious "adherent"?

Another unique issue within aggregate data is the fact that aggregate units are composed of multiple individual units; thus the aggregates themselves represent a smaller number of cases. For example, when looking at census data for the 50 U.S. states, each state is made up of millions of individuals. When analyzing data for the states, however, there are only 50 cases (because there are 50 states). With so few cases, aggregate data sometimes have outliers. An **outlier** is an extreme case that masks the true relationship between variables (e.g., what appears to be a significant relationship only appears that way because of an outlier). Checking for outliers is important when using aggregate data, to make sure statistical generalizations accurately reflect the data.

Exercise: Using the Religious Congregations and Membership Study

Go to www.theARDA.com and click on the tab for US Congregational Membership. Under "Reports," click on the link for "counties." Select your home state and then your home county and click the "submit" button.

Home state: _____

Home county: _____

1. In looking at the religious categories of Evangelical, Mainline, Orthodox, Catholic and Other, which category has the most members? _____
   Scroll through the religious denominations/organizations listed in the report.
2. Which religious group had the most adherents in 2000? _____
3. Which religious category (e.g., Evangelical, Mainline, etc.) does the largest religious group fall under? _____

4. Which religious group had the fewest adherents in 2000? _____
5. Which religious category (e.g., Evangelical, Mainline, etc.) does the smallest religious group fall under? _____
   Click on the tab for "1990–2000 Change."
6. Which religious group had the largest percentage increase between 1990 and 2000?
   _____

7. What religious category (e.g., Evangelical, Mainline, etc.) does this religious group represent?
   _____

8. What religious group had the largest percentage decrease between 1990 and 2000?
   _____

9. What religious category (e.g., Evangelical, Mainline, etc.) does this religious group represent?
   _____

10. How do these data illustrate how religion plays out in your home county? Are there any surprises in the data? _____
    _____
    _____

Experiments

Whereas most people have taken a survey, few have taken part in a formal experiment. Students often ask why they can't just experiment on each other in a research methods course. While introductory social science classes may have students perform classic norm-breaching experiments like facing the wrong way in an elevator, not standing during a standing ovation, or walking backward all day, true experiments are a bit more complicated. **Experiments** are a unique design in that, generally, researchers have complete control over the physical surroundings or the experimental condition. In most experiments there is only one independent variable being manipulated. The independent variable is often referred to as the stimulus, where one group of people is exposed to the stimulus (the independent variable being manipulated) while another group, the control group, is not exposed to the stimulus. The groups are then compared to discover if the independent variable/stimulus actually impacts their behavior (the dependent variable). If the stimulus indeed changes individual behavior, we expect the group exposed to the stimulus (the experimental group) to respond differently than the group who was not exposed to the stimulus or independent variable (the control group). The power of the experimental design is the manipulation of the one independent variable to pinpoint changes in behavior. True experimental designs watch the manipulation of the independent variable to see if it *causes* changes in the dependent variable.

> **Elements of Experiments**
>
> 1. Independent and dependent variables
> 2. Pretest and posttest
> 3. Experimental and control groups
> 4. Random assignment of persons to an experimental or control group
> 5. Researchers manipulate (change the value) of one independent variable

As with other quantitative designs, theory and hypotheses undergird experiments— we need to know why concepts are linked in order to choose to observe people within experiments. For example, a popular experimental topic is the relationship between media exposure and cognition: how does watching television impact our ability to think and learn? Research has linked watching television with long-term attention problems in children. How might we test this experimentally? One study addressing the relationship between media exposure and learning operationalized (measured) media exposure by looking at the effects of watching the television show *SpongeBob SquarePants* on children's attention and learning. The experimental design had three *conditions*, or ways to operationalize the independent variable. The researchers randomly assigned 60 four-year-old children into three groups: a group that watched nine minutes of *SpongeBob*, a group that watched nine minutes of *Caillou* (a PBS cartoon), and a group that spent nine minutes drawing pictures. Following these activities the children were given mental-function tests. The researchers found that the children who watched *SpongeBob* did worse on measures of attention and learning than the children in either of the other groups (Lillard and Peterson 2011).

So what do the results of this experiment tell us? Many take issue with the fact that there were only 60 children in the experiment, which can be a problem although these data are robust. Others may take issue with the three experimental conditions, in that the three conditions (watching two different cartoons versus drawing) were not equivalent; that was the point of the design—vary the conditions and see what happens.

You may say, "*SpongeBob* is for older kids," which is how Nickelodeon spokesman David Bittler responded (Tanner 2011). There are many ways we can critique the study; however, the experimental design for this study was quite strong and allows us to generalize to behavior—after watching *SpongeBob*, four-year-olds are more likely to show diminished attention and/or learning (Lillard and Peterson 2011).

Experiments are considered one of the strongest research designs precisely because researchers have so much control over the experimental environment and are generally manipulating just one independent variable at a time. Where surveys allow us to take data from large groups of people to determine statistical patterns, experiments allow us to determine theoretical patterns by showing a distinctively cause-and-effect relationship. In a true experiment with one variable being manipulated, observing change in the dependent variable between the experimental group "exposed" to the independent variable and the control group who aren't "exposed" allows us to assume the difference between groups must be due to the independent variable. Depending on the type of experimental design, there are additional things to look for, including maturation effects, history effects, testing effects, and so on. We'll go into more detail on those issues when we look specifically at experimental design in Chapter 14.

The second characteristic unique to experiments is **random assignment**. When we briefly looked at surveys we talked about random sampling. Sampling and assignment are different, but if both are "random," they draw on the same mathematical theory. For surveys we are interested in randomly selecting samples from relevant populations to generalize to a whole population. Researchers doing experiments simply ask for volunteers to serve as subjects (university students are most likely to serve as subjects). Once the volunteer pool is established, researchers randomly *assign* people to the experimental or the control group. Using the mathematical concept of probability, random assignment creates equivalent groups. You may not have a perfectly divided set of groups (may not get a perfect 50 percent male/female, older/younger, etc.), but generally the characteristics of each group are representative of the whole group. As long as each subject is randomly assigned, you can attribute differences between experimental and control groups to the independent variable.

Experiments can be replicated, and if done using reliable and valid measures, they show us true causation—when the independent variable changes, we can see changes in the dependent variable. Unfortunately, the high level of control within experiments makes them limited. Experiments can be impractical and artificial. How often are children exposed to nine minutes of *SpongeBob SquarePants* and then given cognition tests? Experiments are also limited by size. It's impractical to import significant numbers of people into an experimental setting. There are ways to get around these constraints, but they inevitably involve giving up some control for a more realistic setting. In Chapter 14, when we discuss experiments in depth, we'll also evaluate settings outside of laboratories where experiments may be more realistic.

QUALITATIVE DESIGNS

As noted earlier, the previous three designs—surveys, experiments, and aggregate data—are quantitative designs, which focus on using statistics to summarize what our observations tell us. The next two designs we explore are *qualitative* designs. Field research and content analysis are unparalleled in providing rich, descriptive data on processes. These qualitative observational designs represent unique dimensions on how we collect data on or about people.

Field Research

"Field" typically refers to the setting in which research takes place. For example, people can be studied in classrooms, cities, places of worship, or in coffee shops. As such, **field research** is research conducted in "natural" settings where people are found. By contrast, many experiments are conducted in laboratories, which are often much more artificial. Field research, or observing people as they engage in activities in their natural settings, seems to be more aligned with how people envision doing social research. When majoring in a field where the focus is studying people, we often see field research as an intuitive process—aren't we already engaged in observing people? Like the other research designs, field research requires a systematic process of taking observations. While it may seem like common sense to generalize to a particular observed behavior, we noted in the book's introduction that we simply do not observe the world randomly. Thus while studying people seems more in tune with social science than dealing with statistics through surveys or aggregate data, field research requires skill to recognize both the freedom and the constraints that come with direct interaction and observation of people.

Field research provides a wealth of data about smaller groups of people, giving us rich, descriptive information about people and processes. Field research is, however, limited to smaller groups of people making it generalizable only to the group you study—in effect creating a case study. Like surveys and aggregate data, field research also cannot establish causation.

Several factors should be considered when you decide to use field research. These factors include where you will be collecting data (in a public setting or a private setting), when to go into the field as a covert versus an overt researcher, and how much observation versus participation you will employ.

When developing a research question that is appropriate for field research, where the research takes place is important. The field should be a place where the **observation** of people in their natural setting sheds light on the research question—are observations best taken in public or private settings? For collecting observations in a public setting—where anyone has access (parks, coffee shops, malls, etc.), there are fewer hurdles. Since access is open in a public setting, researchers do not need to get specific permissions to collect observations. There is ambivalence in the field, however, about doing research in public settings: who has more rights to freedom, the researcher who should be free to observe in public spaces or the people in those spaces who are unaware they are being observed for a research project? Generally, public spaces are seen as open for research observations; keep in mind, however, that there are boundaries of what and how research in public spaces can be done.

Imagine a study where research is being conducted on behavior in public restrooms. After the study begins, the researcher realizes he can obtain greater detail about the participants if he can connect them to the cars they drive. He watches as people leave the public restroom to see which cars they drive. The researcher also realizes that while he's connecting people to their cars, he can take down the cars' license plate numbers and use them to track people to their home addresses. While this study may sound farfetched, it is actually a classic field research study (*Tearoom Trade*) that we discuss in Chapter 4 on ethics. Clearly the researcher was creative in his bid to collect more and more data on men in public restrooms; however, by taking such information about them (tracking them to their homes), he violated their rights to privacy.

When choosing a site for research that is in a private space (e.g., doing research on a club, congregation, neighborhood, organization, etc.), there are more hurdles. Researchers need to determine if they are going to go into the private space as an overt (open) or covert (hidden) researcher. Since the research will take place in a private setting, the best practice is to get permission to conduct the research whenever possible. For some studies it is appropriate for the entire group to be aware of the researcher and his or her aim; for other studies it is more appropriate if permission is granted by a group leader, rather than the whole group. For example, Lofland and Stark (1965) sought to study the process of religious conversion. Finding a religious group that was new to the United States, the researchers realized that they did not know how long the process of religious conversion might take. Keeping this in mind, they sought permission from the religious group leader, asking her if they could remain unknown to the other group members. This strategy for research was entirely suitable for their research aims. Permissions may be integral to a researcher's ability to get access to a private space, requiring researchers to negotiate their access to the field site(s). We look at issues of access to the field in greater detail in Chapter 17.

As we just described, another conundrum related to entering the field is whether to be a covert or an overt researcher. **Covert observation** is when those being observed are unaware of the research. This is helpful in minimizing observer effects since researchers won't interrupt the natural setting or arouse suspicion by their observation. Going into the field covertly, however, can also hinder research because the researcher is expected to act in accord with the larger group. Covert research can also be problematic ethically. Like collecting data in public spaces, there is a question of who has more freedom—shouldn't those being observed have the right to refuse to be part of the research? Often the clearer choice is to go into the field as an **overt researcher,** where those being observed are aware they are the objects of research. Being an overt field researcher makes conducting research easier ethically, although there may be some initial issues with the *Hawthorne effect*—where subjects may change their behavior when they are aware of being observed. Again, we discuss these issues more thoroughly in Chapter 17.

One last issue pertaining to entering the field is choosing what level of participation to have with the group. Can the researcher simply observe without participating in the group, or is being a participant important to the researcher's method (e.g., being fully immersed into the group)? Participant observation yields some fascinating results while also raising some questions, For example, can a participating researcher maintain objectivity, or does the researcher's participation change the nature of the group or of an activity? What risks is the researcher prepared to accept when participating with the group? While field research allows social scientists to interact directly with people, this research design comes with a variety of decisions and constraints that researchers using other designs do not encounter.

Content Analysis

The last design we discuss is content analysis. Like aggregate data, which doesn't directly deal with humans, content analysis is an unobtrusive measure. An **unobtrusive measure** is one that is inconspicuous, meaning it does not have a direct impact on people. Instead of surveying, experimenting on, or observing people in the field, **content analysis** is the study of **cultural artifacts**—things that people have created that tell us about human life. Thus comic strips, song lyrics, Web pages, blogs, films, sermons—verbal, aural, and visual data—are the materials that are coded for meanings and patterns. As a qualitative design, content analysis provides detailed and rich descriptions of phenomena: for example, is rap music more violent than heavy metal music?

Sociologist Amy Binder (1993) undertook a content analysis to study the rhetoric surrounding harm in rap music and heavy metal music. Within five years of each other, these two music genres garnered national attention. Founded by a group of Washington, D.C., wives, the Parents Music Resource Center (PMRC) testified before the U.S. Senate in 1985 on what they described as the harmful messages in heavy metal music lyrics. The PMRC claimed that heavy metal lyrics were damaging to youth (and therefore the nation) because they focused on pornography, drug use, occultism, and suicide (among other things). Five years later in 1990, another music genre came under fire when the rap group 2 Live Crew released their album *As Nasty as They Want to Be*, which became the first album to be declared obscene by a federal court. Band members were arrested when they tried to perform from the album, and some local record store owners were arrested when they sold the album (Binder 1993).

Drawing from these high-profile cases, Binder conducted a content analysis on the rhetoric used to construct these two music genres as harmful, finding that race made a significant difference in how the lyrics were interpreted by the mainstream media. In the rhetoric that constructed harm in "white" heavy metal music, the focus was on the "anti-authority themes" within the lyrics that were touted as harmful because adolescents would become corrupted "in their attitudes about school, parents, and sex" (Binder 1993:765). The rhetoric used to frame "black" rap music, however, focused on a general outrage at the "unprecedented explicitness of rap," noting that rap music was harmful because it would "cause listeners to wreak havoc on police and women" (Binder 1993:765). Whereas the heavy metal genre was seen to be white, the impact of the lyrics was about individual adolescent attitudes. The rap genre, seen as black, was constructed as a threat to the whole society. Binder (1993:765) concludes that the mainstream media depictions of harm in heavy metal and rap music used "images of race and adolescence to tell separate stories of the dangers lurking in the cultural expressions of the two distinct social groups."

Binder's analysis addressed the rhetoric used to construct music genres as a threat. Because rhetoric does not deal directly with people, the unobtrusive nature of the analysis makes content analysis a somewhat more accessible design. Content analysis can also be a more economically feasible (i.e., cheaper) design, since it does not require a large research staff and/or any particularly specialized equipment (simply having access to the materials and a printer or photocopier is often the most access and equipment needed). Content analysis, however, has some drawbacks. For example, the design is limited to analyzing materials that have been recorded. Even when materials have been recorded, do these materials measure what you propose to analyze? Are they the appropriate materials to use for an analysis and/or are there enough of them recorded to use for analysis? If the materials are recorded, appropriate, and plentiful, how do you determine which ones to use (do you select all of the materials or do you take only a sample of the materials to analyze)? We look at these issues and others regarding content analysis in Chapter 18.

RELIABILITY AND VALIDITY

When evaluating research designs, two issues are important to assess: reliability and validity. **Reliability** is the extent to which a given measuring instrument produces the same result each time it's used. Reliability is about consistency. If an independent sample of the same population is given the same survey question, we would expect to get similar results if the survey question is reliable. So in the GSS over the course of two consecutive years, we would expect that we'd have roughly the same percentage of people who feel marijuna should be made legal. If there is a significant discrepancy between the responses for year 1 and year 2, when no significant change has happened in the larger culture, then we would question the reliability of the survey item. In looking at Figure 3.1, we can see the ebb and flow of the percentage of Americans who believe majijuana should be made legal, using GSS data from 1973 to 2010.

As expected, Figure 3.1 illustrates that attitudes change slowly over time, with declining support for legalization in the mid-1980s and then gradual increasing support through 2010. Notice the marked change between the consecutive years 2008 (where 39.8 percent of Americans supported legalizing marijuana) and 2010 (where 47.9 percent of Americans supported legalizing marijuana). Generally speaking, the slow change of opinion over time illustrates the measure to be reliable. Sometimes, however, something specific happens within the cultural context that sparks marked change. For example, due to the attacks on the World Trade Center in September 2001, we might expect to see a significant change between 2000 and 2002 in response to the GSS question, "Are the following threats to the United States greater, about the same, or less today than they were 10 years ago?—Terrorism by foreignors." Unfortunately, the question was not asked those consecutive years (a missed opportunity) in order for us to test our hypothesis.

Like reliability, validity should also be assessed. When looking at the **validity** of a measure, we're interested to know if it's accurate—in effect, does the measure measure what it's supposed to measure (or does a variable measure what we said it does)? For example, if you wanted to know the percentage of people who voted in the last election and asked the question, "Do you remember if you voted in the last election?" with the response categories being "yes" or "no," is that a valid measure of

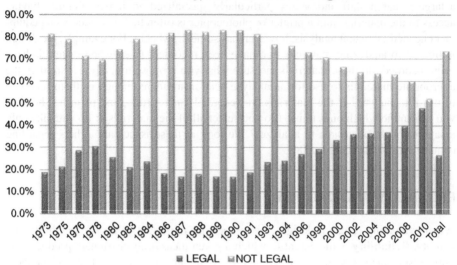

Figure 3.1. The percentage of Americans reporting marijuana should/should not be made legal.

voting behavior? Does the question as asked actually tell us if someone voted? Although the question deals with voting behavior, it doesn't actually measure whether or not a person voted. The question measures whether or not a person *remembers* voting.

It is not uncommon to see issues of validity in survey research findings. Sometimes attitude measures get confused with behavior measures. For example, the GSS sometimes asks the question, "There are different opinions as to what it takes to be a good citizen. As far as you are concerned personally, on a scale of 1 to 7, where 1 is "not important at all" and 7 is "very important," how important is it to help people in America who are worse off than yourself?" Eighty-nine percent of Americans report that "helping people in America who are worse off than yourself" is somewhat important, important, or very important. While having nearly 90 percent of Americans find it important to help others who are worse off is impressive, we cannot assume that Americans are actually helping those who are worse off. What one *thinks* of helping is an attitude, not a behavior. Unfortunately, attitudes sometimes get generalized to behaviors: "A survey showed that nearly 90 percent of Americans are helping those who are worse off."

The GSS does ask a question that better measures the *behaviors* of Americans helping those in need. In 2004 the GSS asked the previous question dealing with attitudes about helping others; a second question asked a *behavior* question: "During the past 12 months, how often have you done each of the following things? Done volunteer work for a charity." Thirty-six percent of respondents reported that they had done volunteer work for a charity more than twice in the last 12 months, with 51 percent reporting that they had done no volunteer work for a charity within the last 12 months. While doing volunteer work for a charity may not be the most comprehensive measure of helping those who are worse off, the divergent responses between an attitudinal question versus a behavioral question are noteworthy. Attitudes do not directly measure behavior and are thus invalid measures of behavior and vice versa.

EMPIRICAL GENERALIZATIONS

The last spoke in the Wheel of Science is empirical generalization. If "empirical" means observable and "generalization" means summary, the last spoke in the wheel allows us to summarize what we have observed. Moving around the wheel, starting at the top, you have a theory that links concepts (socioeconomic status impacts health). Once the concepts are linked, we create hypotheses telling us how these concepts vary together (the higher a person's socioeconomic status, the better the person's health). Based on theory and hypotheses, we choose a research design to collect observations (survey research would be an appropriate design to find indicators of socioeconomic status and health). Once the observations have been collected, we analyze them to see if there is a relationship between our concepts. The **empirical generalization** summarizes what the data (our observations) tell us—is our hypothesis supported or not supported by the results of the data?

Taking indicators (variables) from the GSS, we test the hypothesis that higher socioeconomic status leads to better health. Using cross-tabular analysis we operationalize socioeconomic status by using a variable that measures total family income and a subjective health variable (self-reported variable on how a person believes his or her health to be) and find the following results in Figure 3.2:

CONDITION OF HEALTH * incomeCat Crosstabulation

		incomeCat			
		Less than $25,000	$25,000–$74,999	$75,000 and Above	Total
CONDITION OF HEALTH	EXCELLENT	62	105	118	285
		16.5%	23.1%	37.9%	25.0%
	GOOD	152	229	151	532
		40.4%	50.4%	48.6%	46.6%
	FAIR	120	108	32	260
		31.9%	23.8%	10.3%	22.8%
	POOR	42	12	10	64
		11.2%	2.6%	3.2%	5.6%
Total		376	454	311	1141
		100.0%	100.0%	100.0%	100.0%

Figure 3.2. The relationship between health and income.

Since our hypothesis focused on people with higher socioeconomic status (the independent variable), we want to begin there. In Figure 3.2 the highest level of income is the category for those who report making $75,000 or more in total family income. We hypothesized that those making more money would report having better health. The category that reflects having the best health is that of "excellent." We want to begin interpreting what we find based on the intersection of these two categories. If our hypothesis is correct, we expect that there will be a higher percentage of people making $75,000 or more reporting having "excellent" health than people making less than $24,999 and people making $25,000 to $74,999.

When comparing the three categories of the independent variable (less than $24,999, $25,000 to $74,999, and $75,000) across the row for "excellent," we see that as income level increases, a higher percentage of people report having excellent health. While the pattern of the data indicates that our hypothesis is supported, there is more statistical information we need to definitively describe the relationship between income and health. We'll go more in depth on those additional statistics as we further explore the Wheel of Science. Suffice it to say, empirical generalizations allow us to summarize what the data tell us. For our hypothesis testing the relationship between income and health, we can clearly say that as income increases, more people report having better levels of health.

Apart from survey research, you can test the relationship between socioeconomic status and health by developing other ways to operationalize variables measuring the concepts. Near our university there is a cemetery. If we sent you into a cemetery to test the relationship between socioeconomic status and health, could you find ways to operationalize the concepts? Indicators of socioeconomic status might be found from the gravestones. Larger or more ornate gravestones may indicate higher socioeconomic status. Having a family mausoleum (a small building for a whole family) or a family grave plot may also be an indicator of higher socioeconomic status. How would we measure health? Most gravestones have the dates for birth and death engraved on them. We could calculate age as a measure of health, making the assumption that the longer a person had lived, the better that person's health. What might the results of this analysis tell us? How would you generalize what the empirical results show? Based on previous research we would expect having larger and/or more ornate headstones would be linked to having had longer lives.

CORRELATIONAL VERSUS CAUSAL RELATIONSHIPS

Much of what we describe when we summarize the results of our analyses is based on **correlation**, or naturally occurring relationships. When two factors like socioeconomic status and health vary together, we sometimes assume that one (socioeconomic status) is causing changes in the other (health). We can assume that people who have higher status will be healthier. People with higher socioeconomic status tend to have more education regarding healthy practices. They have more access to better foods, health-care providers, and can afford gym memberships. If we see a correlation between socioeconomic status and health, the correlation simply shows that the two concepts are related (that they vary together either in a positive or a negative direction). It is tempting, due to our theoretical musings (those with higher status can afford better health care, etc.) to conclude that higher status *causes* better health. That conclusion, however, is not what the correlation tells us.

Can you make the opposite argument regarding health and socioeconomic status? In what ways might better health help to increase socioeconomic status? People who are healthier are better able to complete higher levels of education, which might then impact their economic status. People who are healthier are better able to work harder and maintain a good job history, giving them access to better opportunities and higher incomes. People who are healthy live longer and accumulate more wealth. Thus we could argue that the relationship between socioeconomic status and health is the opposite, or a relationship between health and socioeconomic status. We could also argue a third alternative, that the relationship is reflexive—both variables cause changes in the other.

Correlation research allows us to predict relationships between variables; if we know the state of one variable, we can generalize what is happening with the other variable. Correlation does not tell us if changing one variable will *cause* a change in the other variable. It could be that some other factor links these concepts (e.g., family of origin social status might impact both a person's status and/or health). In his famous book *How to Lie with Statistics,* Darrell Huff (1982) discusses the problems encountered with a "post hoc fallacy." Huff cites a study that found cigarette smokers had lower grades than nonsmokers. While Huff (1982:88) describes the study as being "properly" done, he also notes that "an unwarranted assumption is being made that since smoking and low grades go together, smoking causes low grades." Like the relationship between socioeconomic status and health, couldn't the smoking/low grades relationship go the other way? Beware of the post hoc fallacy—determining causation after the fact. Just because two variables are correlated, it does not follow that one causes the other. In fact, it could be that extroverts are more likely to smoke and less likely to get good grades (Huff 1982). The point is we simply do not know that cigarette smoking *causes* lower grades (or vice versa). When making empirical generalizations we need to keep in mind that finding a correlation between variables does not equate to finding causation between them. We will look more closely at the differences between correlation and causation in later chapters, to make these distinctions more clear.

TYPES OF RESEARCH

Social scientists talk about several different types of research including "pure" research, applied research, evaluation research, and action research. While the boundaries between these types of research are somewhat blurred, the reasons why research is undertaken and the purposes for which it is used can be quite different.

Pure Research

Pure research, also known as basic research, is a designation identifying research performed solely to "advance knowledge" or to develop theoretical understanding. The object of pure research is not to respond to a specific research problem that requires an immediate answer. Examples in social science might be "Does religious competition help or hurt church membership?" or "How does poverty affect crime?" An additional example would be a study conducted by McKinney (2001; see also McKinney and Finke 2002) to determine how social network ties impacted clergy connection and involvement with denominations. While most clergy reported that their social networks did not impact their beliefs or involvement with the church, the results of the study showed exactly the opposite; the clergy most actively involved in denominational movements had close network ties (friends) also involved in the movements.

Since pure research does not directly address any societal need, most of this research occurs in universities or is funded through large foundations. It is not immediate in nature or focused on sales, so it is not likely to be funded commercially. The very nature of science is based on simply discovering more about the world. Sometimes, however, people see pure research as simply "duh" science at best or wasteful of resources at worst. Pure research allows scientists to address an unlimited array of research questions. Much of the knowledge gained by pure research serves as building blocks for future research that is more applied.

Applied Research

Applied research focuses on the actual real-world problems in a field of study. Typically, practitioners working in social science fields identify problems that need to be solved and look to researchers for assistance in ascertaining the extent of the problem and then applying solutions to it. An example might be, "Is summer learning affected by family income level?" Another example might be a study Abbott conducted several years ago in a subsidized housing project to determine the effects of community policing on drug activity. City agency leaders introduced a different model of policing than currently existed in a low-income urban housing area. Whereas previously police officers conducted typical "beat" patrols, community police officers walked among the houses, got to know residents, and relied on residents to assist police in reporting crime. In this way project leaders "experimented" by introducing community policing to see if it made an impact on incidents of drug activity. Community policing was thus being compared to the typical police patrol procedures. These newer community policing procedures were deemed successful when program leaders observed decreasing levels of crime in the low-income housing area, especially at rates compared to the city as a whole.

Evaluation Research

Some researchers and policymakers make a distinction among various types of applied research. Evaluation research is one specific type of applied research to determine whether a specific intervention was successful according to a set of criteria. This research is often found in corporate and institutional settings in which there is a concern over the impact of a specific program that has been introduced. When school districts introduce measures to lower class sizes or fund technology, for example, they are interested in what difference the program makes on some specific outcome (like student achievement) in their districts. Because of the specific nature of this type of research, it is known by many names. Prominent among these are program evaluation, effectiveness research, impact analysis, and even cost-benefit analysis.

Action Research

Action research is applied research aimed at providing insights and solutions in cascading fashion. The attempt is not so much to make a pronouncement about whether a program objective had been met (as in evaluation research), but to apply research findings in order to further refine solutions for change. When a corporation, institution, or other agency employs action research, it is for the purpose of understanding a current problem, suggesting potential solutions, and then using the findings to suggest additional avenues of change. Typically, action research involves the participation of a researcher who creates findings for a research problem, works with the sponsoring group to implement new approaches, gathers further findings to develop more specific understanding, and continues to develop change processes.

An example of action research may be found most readily in the consulting world. If hospital administrators are concerned about staffing patterns in their critical care units, for example, they might hire a researcher-consultant to evaluate their current practices. The researcher might then work with a hospital committee to refine and implement a resulting staffing strategy, which will yield further information that can lead to additional change strategies.

We have now covered the four spokes of the Wheel of Science. We've described how theories link concepts and once linked, give direction to research by helping us create hypotheses—statements of how concepts are related. Once we have hypotheses to test, we select a particular research design to test empirically whether the hypotheses are supported or not. These research designs give us structured ways of observing the world in order for us to empirically generalize what our observations tell us. While these processes take us through the Wheel of Science, there is still one area that needs to be addressed, the ethics of social research. Social research uses *people* as the basis for our observations. When researching people, there are several guidelines in place to ensure the safety of human participants. We turn, in Chapter 4, to the ethical concerns related to studying human subjects.

4

ETHICS

In the late 1960s, sociology PhD student Laud Humphreys began collecting data in "tearooms," which referred to public restrooms that served as "locales for sexual encounters without involvement" (Humphreys 1975:2). Humphreys had been studying the lives of homosexual men and found that one aspect of this subculture not well studied was the behavior that took place in these tearooms. Humphreys found that many men—heterosexual, married, single, or homosexual—often wanted "instant sex" and would seek out these public restrooms (tearooms) in order to engage in this instant, impersonal, and anonymous sex.

Humphreys describes the intricacies of his field research by stating that in order to gather data on this type of private behavior, he had to pass as a "deviant."[1] Participating in the illegal activity of public sexual encounters was considered unethical, so how could Humphreys infiltrate these public spaces to learn about the behavior associated with this subculture? Humphreys discovered a way to participate as an observer. Due to the fear encountered by men who participated in "tearoom trade," a role was created for a third party that fit the needs of a social researcher. Because acts

---

[1] In 1966 when Humphrey began collecting data on men in tearooms, homosexual acts were illegal and seen as "deviant" behavior.

Understanding and Applying Research Design, First Edition. Martin Lee Abbott and Jennifer McKinney.
© 2013 John Wiley & Sons, Inc. Published 2013 by John Wiley & Sons, Inc.

of homosexuality were criminalized at the time, men who participated in these encounters had quite a lot to fear, thus opening an opportunity for an observer, in this case a lookout. Humphreys became the lookout (or "watchqueen" as it was called in the subculture), participating in hundreds of these "impersonal" encounters within the confines of public restrooms. By choosing to play the role of the watchqueen, Humphreys gained access to tearooms without having to participate in the sexual encounters themselves.

As the watchqueen, Humphreys's role in the impersonal sex process was to keep an eye on the restroom entrance in order to determine who was coming into the restroom. If a stranger (legitimate "straight" men who were simply there to use the restroom facility), or police officer, or some other law enforcement officer was about to enter, Humphreys's job was to give a loud cough to notify participants of the possible danger. If the entering man was a "regular" (someone known to the group to be there for the sexual encounter), Humphreys would nod to the other participants, letting them know they were not in danger of being discovered.

Participating in his unique role, Humphreys gained unprecedented access to a subculture that was not well known. In the course of his data collection, Humphreys realized that he had access to more information about the men he was studying. As a regular participant, Humphreys was able to watch the other men come and go from the restroom from his vantage point as a lookout. He had noticed the wide array of men who participated in the tearooms and began to connect them to the cars that they drove. By linking each man to his car, Humphries was able to tell more about the participants (for example, were they driving an expensive car, a family car, etc.) than just through his observations in the tearooms.

Recognizing that he could match participants to their cars, Humphreys began to record several pieces of information about each participant; he recorded the license plate number on the car, a brief description of the car, and a description of the participant. Humphreys reports that in most cases he would observe the activity of participants in the tearoom, leave the tearoom, and then wait in his car until participants came out, getting into their cars to leave the area. Armed with license plate numbers and descriptions of cars, Humphreys posed as a market researcher and visited local police precincts where "friendly policemen" gave him access to the license registers where he obtained the names and addresses of tearoom participants.

After the first year of his observations, Humphreys was asked to develop a questionnaire for a social health survey for men by a social research center. With the project director's permission, Humphreys added the names and addresses of 100 tearoom participants to the survey sample in order to go to their homes and interview them about a host of topics. Since formal interviews had been part of his original research design, adding his tearoom participants to the survey sample gave him the opportunity to interview them in a nonthreatening way (since the surveys put them into the context of a random sample of men participating in a legitimate study rather than as sexual "deviants" who he had observed in anonymous sexual encounters). Subsequent to the formal health survey interviews, Humphreys took a sample of 50 of the tearoom participants and 50 of the non-tearoom participants from the health survey sample to do in-depth follow-up interviews so that he could compare his sample of tearoom participants with a "control" group of "nondeviants."

When Humphreys published his work on the tearooms, the work incited a maelstrom of controversy (see Horowitz and Rainwater 1975; Humphreys 1975, von Hoffman 1975, Warwick 1975). Had Humphreys's field research been ethical? Had Humphreys misrepresented himself by pretending to be part of the tearoom practices? Had Humphreys violated the rights of any individuals? Had he committed felonies while participating in criminal acts? These questions raised a larger question: what is permissible when collecting social scientific data?

Humphreys's *Tearoom Trade* study illustrated a conundrum within the research process: who should have more freedom, the researcher or the participant? While there are certainly benefits to doing or participating in social research, there are also costs, which include deceiving participants (who are then unaware they are being observed for the sake of social research), and subsequently invading their privacy, using research findings in harmful ways (see Warwick 1975). In the case of *Tearoom Trade: Impersonal Sex in Public Places*, Humphreys deceived his subjects in a number of ways. First he misrepresented himself as just another participant by serving as a watchqueen when he was truly participating as a researcher. Second, by connecting participants to their license plate numbers and then posing as a market researcher to obtain their names and addresses, Humphreys misrepresented himself. Finally, by adding tearoom participants' names into the sample for the men's social health survey in order to go to their homes to interview them, he also misrepresented himself. (This last misrepresentation has implications for the integrity of social research for another reason—adding his participants to the sample impacted the representativeness of the sample with consequences for the health survey.)

Much of the social scientific community lauded Humphreys for his unique and interesting contribution to the field, noting that his motives were pure and that he had done an admirable job maintaining the confidentiality of his participants while at the same time getting "needed, reliable information about a difficult and painful social problem" (Humphreys citing Horowitz and Rainwater 1975:179). Others, however, were appalled at his tactics for obtaining such detailed information about his participants.

While researchers and the public at large have a right to know about social phenomena, as individuals and groups within society, each of us also has a right to our privacy. Warwick (1975 cited by Humphreys 1975:209) summarizes the consensus on the issues Humphreys's study evokes: "Social scientists have not only a right but an obligation to study controversial and politically-sensitive subjects, including homosexuality, even if it brings down the wrath of the public and government officials. But this obligation does not carry with it the right to deceive, exploit, or manipulate people."

Warwick's concerns and Humpheys's study coincided with a renewed professional emphasis in ethics. In the early 1970s, professional associations like the American Psychological Association and the American Sociological Association were in the process of debating new codes of ethics. **Ethics** are the standards of conduct of a given profession or group. When dealing with humans in research, there are several procedures in place to protect individuals and groups participating in research. While scientific research systematically investigates questions that contribute to our general knowledge of what people think and do, no scientific goal should ever overshadow the rights an individual has in choosing to (or not to) participate in social research.

HUMAN SUBJECTS ABUSES

We have a need to be clear about what is and is not permissible when conducting research with human beings. Unfortunately, there are myriad examples of human rights abuses in the name of scientific research. Accounts of human research atrocities during wartime are sometimes stupefying in their level of depravity. The extensive human research the Nazis performed in the concentration camps during World War II serves as an example of the inhumanity of some human research. "Nazi ideology was predicated on the concept of racial supremacy" (Bogod 2004:1155) with blacks, gypsies, homosexuals, and Jews (the *untermenschen*) at the bottom rung of the racial ladder. Since these groups were considered subhuman, they were then seen as legitimate targets for medical experimentation.

The so-called medical studies performed by the Nazis subjected people to a variety of chemical injections (gasoline) and injections of live viruses (including typhus, tetanus, and streptococcus). Experiments also subjected people to submersion in ice water or leaving them strapped to stretchers naked for extended periods of time in below-freezing temperatures (part of a program to help researchers learn how to revive Luftwaffe pilots who were shot down over the North Sea). As part of these experiments children were castrated, limbs were amputated, and men and women faced mass sterilization; women were injected with "caustic substances" and men underwent irradiation (Bogod 2004; see also National Institutes of Health [NIH] 2008; Weindling 2004).

As a counterpart to Nazi experimentation, Japanese military doctors and scientists also conducted a wide range of horrific experiments on humans in a quest to develop an effective biological weapons program from the 1930s until the end of World War II (Kleinman, Nie, and Selden 2010). Like the Nazi ideology that perceived certain groups of people as not fully human, the Japanese saw the Chinese and other subjects of their experimentation as simply "experimental material." The Japanese referred to their subjects as *maruta*, meaning "logs of wood" or "lumber" (Kleinman, Nie, and Selden 2010:5). Carried out in occupied China and Southeast Asia (often through the secret Unit 731), the Japanese Imperial Army conducted human experiments including the amputation and reattachment of limbs from one person to another, the vivisection of live people, and injections of diseases including the agents that caused anthrax, cholera, the plague, and typhoid (Kleinman, Nie, and Selden 2010; see also Nie et al. 2010).

While it is easy to point fingers at other nations, one infamous human research study conducted in the United States spanned a 40-year period (1932–72). In 1929 the U.S. Public Health Service (USPHS) conducted studies in the rural South to assess the prevalence of syphilis among blacks in order to develop strategies for mass treatment of the disease. Brandt (1985) argues that social Darwinism had created a new racial ideology in America that considered blacks to be a more primitive people who could not be assimilated into complex societies. As a more "primitive" people group, blacks were seen as prone to disease, specifically venereal diseases like syphilis (Brandt 1985).

Having identified Macon County, Alabama, as having the highest rates of syphilis in the counties surveyed by the USPHS, the Tuskegee Syphilis Experiment began in 1932, seeking to track the development of untreated syphilis in 400 black men. Even when penicillin became available as a treatment for syphilis in the early 1950s, the men in the study were denied treatment (even during World War II when some of the men visited their local draft or health boards, they were denied entrance into the army and treatment for syphilis). In fact, the USPHS worked with doctors, county and state health officials, and draft boards to ensure the continued participation of the Tuskegee Syphilis Experiment subjects. As late as 1969, oversight committees (one from the federally funded Centers for Disease Control and Prevention) met to decide if the study still had relevance and should be continued—still denying treatment to the participants (see Brandt 1985; Jones 1993; NIH 2008).

Only in 1972 when details of the study came out in the popular press was the experiment halted. By then at least 28 of the men (much likely many more) had died due to complications from syphilis (Brandt 1985). The public outrage generated by the Tuskegee Study resulted in the National Research Act of 1974, as well as basic policy including the Protection of Human Research Subjects (Brandt 1985; NIH 2008). While science has the great potential to create better lives for humans, it also bears the burden of responsibility to make sure that harm does not come to humans in the name of scientific discovery.

PROTECTION OF HUMANS IN RESEARCH

From the "scientific" experiments on those in Nazi Germany and Imperial Japan to the Tuskegee experiment in the United States, research on humans has resulted in a number of truly horrific consequences. In order to guard against these wrongs, we now have federally mandated guidelines in place to protect human subjects. The U.S. Department of Health and Human Services' Office of Human Research Protections (OHRP) evaluates research, holding research institutions and individual researchers to account for basic human safety, assessing the possible risks to human subjects, the adequacy of protection from risks, and the potential benefits to research participation for subjects. These federal guidelines are instituted through human subjects committees, or institutional review boards (IRBs) located at research institutions. In a university setting, for example, IRBs are composed of panels of faculty members who review proposed research involving human subjects in order to guarantee that subjects' rights and interests are protected.

The basis for conducting human research comes from a 1979 document, the *Belmont Report: Ethical Principles and Guidelines for the Protection of Human Subjects of Research* (NIH 2008). The *Belmont Report* identifies three principles essential to ethical human research. These principles are respect for persons, beneficence, and justice (NIH 2008). **Respect for persons** rests on the principle that people should be treated as "autonomous agents" who are allowed to consider for themselves the potential harms and benefits of a situation, analyze how risks and benefits relate to his or her personal values, and then to take action based on their analysis. Social research can be intrusive; therefore research subjects have the right to evaluate whether or not they want to participate in research. Research should be voluntary, and researchers have the responsibility to provide potential subjects with adequate information so that the subjects can make an informed decision on whether or not they want to participate.[2] This requires researchers to provide informed consent.

Informed consent allows subjects to base their voluntary participation on a clear understanding of what a proposed study will do, any possible risks associated with the study, how their data will be handled (e.g., how will researchers keep their data confidential), and if there are resources available for any adverse effects (psychological or medical). Informed consent should also clearly state that participants have the right to terminate their participation at any time in the course of the research. The challenges encountered within the first principle of the *Belmont Report* include making sure that participants fully understand both the risks and the potential benefits of participating in a study, and making sure there is no coercion or any undue influence by the researcher to get a subject to participate.

Informed Consent
[List title of project here]
   Include or exclude following information as applicable.

Purpose
You are invited to take part in a research study. The purpose of this study is to _____. You have been invited to take part in this study because _____. The number of people that will be part of this research is ____.

Procedures
List all procedures, preferably in chronological order, which will be employed in the study. Point out any that are considered experimental and explain technical and medical terminology.
   State where the study will take place and the amount of time required of the participant per session and for the total duration of study.
   If applicable to your study, list:
   Information concerning taping or filming.
   Indicate which procedures are experimental

Risks and Discomforts
List the foreseeable risks or discomforts, if any, of each of the procedures to be used in the study, and any attempts that will be used to minimize the risks. Order multiple risks by magnitude and time duration.

(Continued)

---

[2] There are provisions in place to protect vulnerable populations who may not be able to make informed decisions due to their age (children) and/or condition (pregnant women, human fetuses, prisoners, people with mental disabilities, and/or people who are economically or educationally disadvantaged).

Benefits

List the **direct** benefits to participants. If there are no direct benefits (e.g. cure of the participants' disease), you must state: We do not anticipate direct benefits; however, following this statement, you may also include benefits to others, or the body of knowledge, or indirect benefits to the participant (e.g. satisfaction).

Participation and Alternatives to Participation

Your participation in this study is voluntary; you may decline to participate without penalty. If you decide to participate, you may withdraw from the study at anytime without penalty and without loss of benefits to which you are otherwise entitled. If you withdraw from the study before data collection is completed, your data will be returned to you or destroyed. Likewise, the Researcher may terminate your participation in the study at any time.

Alternatives to participation in this study include . . . . . .

If the research includes medical treatment disclose appropriate alternative procedures or courses of treatment that might be advantageous to the participant.

Emergency Medical / Psychological Treatment

(For research involving more than minimal risk add here)

The University does not offer to reimburse participants for medical claims or other compensation. If physical injury is suffered in the course of research, or for more information, please notify the investigator in charge, (list PI name and phone number). List referral number for medical or mental health counseling if there is any risk of impact from research

Confidentiality

The information in the study records will be kept confidential. Data will be stored securely and will be made available only to persons conducting the study unless you specifically give permission in writing to do otherwise. No reference will be made in oral or written reports that could link you to the study.

Your de-identified data may be used in future research, presentations or for teaching purposes by the Principal Investigator listed above.

Compensation

(if applicable add here)

For participating in this study you will receive _____. Other ways to earn the same amount of credit or compensation are _____. If you withdraw from the study prior to its completion, you will receive _____.

Subject Rights

If you have questions at any time about the study or the procedures, (or you experience adverse effects as a result of participating in this study,) you may contact the Principal Investigator, [Name], at [Office Address], and [Office Phone Number]. If you have questions about your rights as a participant, contact the University Institutional Review Board Chair at [Phone Number or Email].

> *Consent*
> **Your signature on this form indicates that you have understood to your satisfaction the information regarding participation in this research project and agree to participate in this study. In no way does this waive your legal rights nor release the investigators, sponsors, or involved institutions from their legal and professional responsibilities.**
>
> **I have read the above information and agree to participate in this study. I have received a copy of this form.**
>
> | **Participant's name (print)** | **Researcher's name (print)** |
> |---|---|
> | **Participant's signature** | **Researcher's signature** |
> | **Date** _____ | **Date** _____ |
>
> Copies to: Participant Principal Investigator

Beneficence, the second principle of the *Belmont Report*, implies that researchers have an obligation to secure the well-being of research participants. Two general rules of beneficence are to do no harm and to maximize the potential benefits of the research while minimizing the potential harm of the research. Research should never be allowed to cause any harm to participants, which means no physical, psychological, social, legal, or economic harm. Along with providing participants with informed consent, researchers are responsible to apprise participants of any possible risks—the probability that a certain harm can or will occur over the course of the research. The general rule is that no risk during the course of the research would be greater than risks encountered in daily life. The challenge in applying the second Belmont principle lies in determining when the possible benefits from the research outweigh the possible risks, and vice versa.

The third principle of the *Belmont Report* speaks to the moral requirements for research; that is, are the procedures fair, are the individuals or groups treated fairly—in effect, is the research and how the subjects are treated just? For research to fall under the principle of **justice**, researchers must make sure that the benefits and risks of the research are fairly distributed among groups, individuals, or societies. One of the concerns with justice is that sometimes deception is used in the course of a study. Although generally unethical, there are times when slight deception can be used ethically. In some social experiments, for example, in order to observe a subject's authentic responses, researchers have to redirect the gaze of the subjects.

Let's suppose that a researcher was concerned that people's behaviors were unduly influenced by the power of an authority figure. The researcher designs an experiment where a research scientist will request that a subject participate in an experiment on learning and memory. The subject will read word pairs over an intercom system to a "learner" in another room. In this other room a "learner" is strapped to a machine that doles out electric shock whenever the learner incorrectly answers the prompt from the teacher reading word pairs. In the initial phases of the experiment, the subject reading word pairs administers small amounts of electric shock to the learner in the other room. At some point, however, the subject hears the learner through the intercom system exclaim that he wants out of the experiment. The subject then turns to the authority figure (the research scientist overseeing the experiment) to ask if he should stop administering the shock to the learner and let the learner out of the experiment. The research scientist prompts the subject to continue with the experiment regardless of the response of the learner. This latter move is intended to truly test the relationship between authority and obedience. If the researcher had initially introduced the experiment as one in which subjects were being tested to see if they would follow authority, the chances that the researcher would observe true responses are quite small. Instead, the researcher redirects the gaze of the subject to focus on an experiment for the impact of memory on learning. If this experiment sounds familiar, it should; it's the famous obedience experiment conducted by Stanley Milgram, which we'll revisit in Chapter 14 on experiments.

Whenever deception is used in a study, researchers have an obligation to debrief each participant. A **debriefing** provides subjects with a full account of the study's actual goals, so that subjects are fully aware of the true goals of the study in which they participated. By providing a full description of the study, a debriefing corrects misconceptions and reduces the stress, anxiety, or concerns participants may have encountered during the research. While the three principles of respect for persons, beneficence, and justice speak directly to human subjects, there are additional ethical standards in place to regulate the professional practice of social research.

One last concern regarding ethics is how to deal with the **paradox** that can occur in research. Researchers are sometimes in a unique position to give participants access to some benefit. In one famous educational experiment, researchers sought to study the connection between teacher expectations and student performance (Rosenthal and Jacobson 1968). In creating their research design, the researchers fashioned a fictitious measure (the Harvard Test of Inflected Acquisition) to "identify" students who would be experiencing an educational growth spurt over the course of a school year. Alerting teachers to these students, researchers created a classroom environment where teacher expectations seemed to positively impact student performance. Students identified at the beginning of the school year as "spurters" did show greater intellectual development at the end of the school year (Rosenthal and Jacobson 1968).

The experimental design in this study was intended to test *if* there was a link between teacher expectations and student performance; the researchers did not know that there would be a link. This illustrates the paradox. The students who were part of the experimental group (the "spurters") benefited from their random assignment into the experimental group. If you had a child who had been assigned to the control group, would you be upset that he or she didn't benefit from higher test scores? The general rule for researchers is that study participants cannot be worse off than if they had not participated in the study. The children in the teacher expectation/student performance experiment were not worse off than if they had not been participating in the study at all. Therefore, we consider the study to be ethical.

Protecting Human Research Participants[3]: The Three Fundamental Aspects of Informed Consent

1. Voluntariness: Individuals' decisions about participation in research should not be influenced by anyone involved in conducting the research
2. Comprehension: Individuals must have the mental or decisional capacity to understand the information presented to them in order to make an informed decision about participating in research.
3. Disclosure: Federal regulations require that researchers disclose:
   a. The purpose of the study.
   b. Any reasonably foreseeable risks to the individual
   c. Potential benefits to the individual or others
   d. Alternatives to the research protocol
   e. The extent of confidentiality protections for the individual
   f. Compensation in cases of injury due to the research protocol
   g. Contact information for questions regarding the study, the participants rights, and in cases of injury
   h. The conditions of participation, including the right to refuse or withdraw without penalty

---

PROFESSIONAL ETHICAL STANDARDS

While the primary directive in doing social research is to protect those who participate, there are a variety of other ethical issues related to doing social research, including discipline-specific professional standards of competence, integrity, scientific responsibility, and social responsibility. As professional sociologists, we are subject to the American Sociological Association's (ASA) *Code of Ethics and Policies and Procedures of the ASA Committee on Professional Ethics*. The ASA code of ethics, much like other professional codes of ethics, stipulates that members maintain their competency through education, training, and experience using the appropriate resources (scientific, professional, technical, and administrative resources).

As professional social scientists, we are also charged with being honest, fair, and respectful of others in the course of both research activities as well as everyday professional activities, adhering to the highest scientific and professional standards. There is also an expectation that professional researchers have a responsibility to their communities, by making public the knowledge they learn through their research for the public good. As the ASA Code of Ethics states, "When undertaking research, [sociologists] strive to advance the science of sociology and to serve the public good" (American

---

[3] Adapted from the NIH Office of Extramural Research (see NIH 2008).

Sociological Association [ASA] 1999:5). It would be easy to pretend that it is always simple to follow these guidelines for ethical research, as if they are always clear cut. Social research, however, can be quite complex as Humphreys's *Tearoom Trade* study illustrates. As social scientists we must take to heart the professional ethical directive to strive for the advancement of science and the public good. In the case of research with human subjects there is no scientific goal, no matter how lofty, that should ever supersede the health and well-being of the people we study.

PART II
WHEEL OF SCIENCE:
PROCEDURES OF RESEARCH

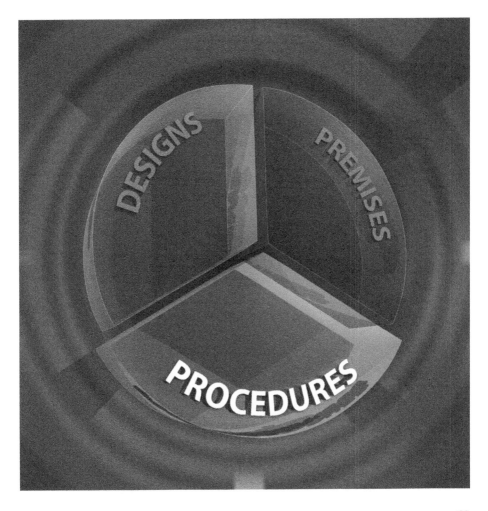

5

MEASUREMENT

Having walked through the Wheel of Science and given an overview of the process of science, we now turn to addressing some of the building blocks that help us move between theory and hypotheses to observation. Once our theories are formalized into testable hypotheses, we need to take a closer look at the variables including expanding on what variables are (versus constants), what we mean by "variance," and how to operationalize variables and to recognize levels of measurement.

VARIABLES AND CONSTANTS

The primary task of social science is to explain the connection between concepts. We've explained that theory allows us to link concepts and that this theoretical pairing is formalized into hypotheses that state relationships between the concepts—how one concept impacts another concept (e.g., we expect that higher levels of education lead to higher levels of income). In order to test our hypotheses, we need to operationalize our concepts—finding indicators that measure concepts. Let's revisit the relationship between education and income.

Understanding and Applying Research Design, First Edition. Martin Lee Abbott and Jennifer McKinney.
© 2013 John Wiley & Sons, Inc. Published 2013 by John Wiley & Sons, Inc.

OPERATIONALIZATION

Operationalizing concepts simply means that we find variables (measures) of the concepts. If we are interested in operationalizing education and income, we could turn to a national survey (like we did in Chapter 1) and look for variables that measure "education" like "What is the highest level of education you have completed?" and "What was your total family income before taxes last year?" to measure "income." These two specific measures of education and income are operationalizations: variables that measure the concepts education and/or income.

In Chapter 2 you operationalized variables for the concepts of religiosity, health, and prejudice. When operationalizing a variable two parts are necessary. Variables need to have some sort of *descriptor* telling us what the variable is measuring, as well as having certain *attributes* or answer categories. For example, when operationalizing education in our example, the question itself is the descriptor for what is being measured: "What is the highest level of education you have completed?" Since operationalization is a broader process than our focus on survey research items, another example would include our illustration from Chapter 2 of going to a graveyard to operationalize socioeconomic status and health. For this example a descriptor would be the gravestone itself (for both socioeconomic status and age/lifespan).

The second important part of a variable is the attributes or categories. The attributes or categories tell us how the variable is being measured—the descriptor tells us *what* is being measured, and the categories tell us *how* the variable is being measured. In the case of the variable we operationalized to measure education, "What is the highest level of education you have completed?" appropriate categories may include "Less than a high school education, a high school education, some college education, a college degree." We call these "discrete" categories because they are "closed-ended" variables—supplying respondents with a specific, discrete set of responses (we look at these types of variables more closely in Chapter 14 when discussing survey research). For variables that have a limited, or smaller, number of categories, we refer to them as **categorical variables** or variables that have these discrete, limited number of category options.

For the example measuring socioeconomic status and health in a graveyard, the gravestones used to measure socioeconomic status could have the attributes of small stone, medium stone, or large stone—in effect measuring the size of the stone to denote status (bigger stones indicating higher socioeconomic status, smaller stones lower status). In terms of measuring age, the dates of birth and death would denote age (ages being the attributes) to measure level of health. For any given gravestone, the age will vary into distinct age categories (e.g., 88 years old, 76 years old, 93 years old, 67 years old, 49 years old, etc.). Because each age represents its own category, possible categories range from age = 0 to age = 101 (for example). This age range could then be more than 100 categories. When variables have a broad range of categories (more than 20, for example) we refer to them as **continuous variables**.

Apart from having attributes that are categorical or continuous, variable categories must also be *exhaustive* and *mutually exclusive*. **Exhaustive** attributes mean that all possible categories are included in the variable. You have probably taken a survey where you know your answer almost before you get to the answer categories. Yet sometimes you find that your answer is not included. What do you do? Do you pick the next best answer or do you skip the question? Do you get frustrated because your answer is not there and stop filling out the survey altogether? Making sure that each variable has exhaustive categories is important. In religion research, it is not uncommon to see a question asking for someone's religious preference. Look at the following variable:

What is your religious preference?
1. Protestant
2. Catholic
3. Jewish

Given a particular cultural context, these categories may make sense. For example, in the United States, these are the largest religious groups. You probably recognize, however, that the categories are not exhaustive—not everyone fits into these categories. A first reaction might include thinking of all possible religious groups:

What is your religious preference?
1. Protestant
2. Catholic
3. Jewish
4. Muslim
5. Church of Jesus Christ of Latter Day Saints
6. Reformed Church of Jesus Christ of Latter Day Saints
7. Fundamentalist Church of Jesus Christ of Latter Day Saints
8. Church of Christ Scientist
9. Jehovah's Witnesses
10. Hindu
11. Buddhist
12. Baha'i
13. Theosophy
14. Zoroastrianism

We hope you recognize that there are so many varieties of religious preferences that we can never come up with a comprehensive list of all possible religious preferences in order to make the categories exhaustive. There is an easier way to make a question exhaustive:

What is your religious preference?
1. Protestant
2. Catholic

3. Jewish
4. None
5. Other

By adding the categories of "None" and "Other," the question is now exhaustive. Anyone who does not fit into Protestant, Catholic, or Jewish is some other religious preference or has no religious preference; thus the categories now encompass anyone and everyone.

As well as having exhaustive categories, variables also must have categories that are mutually exclusive. Having **mutually exclusive** categories means that while everyone can answer one category (Protestant, Catholic, Jewish, None, Other), people should be able to choose *only* one category. Several years ago a survey associated with a medical school asked respondents the following question:

What is your religious preference?
1. Catholic
2. Christian
3. Jewish
4. Muslim
5. Other

First of all, without a category for "None," the question attributes are not exhaustive. The category for "Christian," however, is problematic because Catholicism is one of two branches of Christianity—they are the same thing. Protestant Christians have only one option here ("Christian"), but what does a Catholic Christian choose? These attributes are not mutually exclusive: Catholics fit into two categories. When attributes are not mutually exclusive, respondents are **misclassified**; that is, the categories do not reflect the respondent's true response/classification.

Some questions are designed specifically to have respondents select "all that apply." We call these "cafeteria" questions, and expect that each category represents a distinctive answer, for which there are multiple categories. When you have a question that is expected to be one answer per respondent, variable attributes should be both exhaustive (everyone can answer at least one) and mutually exclusive (everyone can answer only one).

VARIATION

We are concerned with *what* variables measure and *how* variables are measured. When variables have exhaustive and mutually exclusive attributes, they should vary appropriately. We want variables to vary because it is through their *variation* that we can measure change in another variable. **Variation** is the amount of change included in a variable; each variable has 100 percent variation.

Let's look at the variance—distribution—of the following two GSS variables. The variable EVCRACK in Figure 5.1 measures whether or not respondents have ever used crack cocaine (with answer categories of "yes" or "no"). The variable GRASS in Figure 5.2 measures whether respondents think that marijuana should be made legal (with answer categories of "should" [be made legal] or "should not" [be made legal]).

Both of the variables in Figures 5.1 and 5.2 are considered **dichotomous variables** because they each have two category options (yes/no and should/should not,

Have you ever, even once, used 'crack' cocaine in chunk or rock form? (EVCRACK)

■	1) Yes	115	6.3
▥	2) No	1710	93.7
	Total (N)	1825	100.0
	Missing	219	

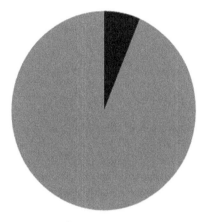

Figure 5.1. The variation in EVCRACK.[1]

Do you think the use of marijuana should be made legal or not? (GRASS)

		Freq.	%
■	1) Should	603	47.9
▥	2) Should not	656	52.1
	TOTAL (N)	1259	100.0
	Missing	785	

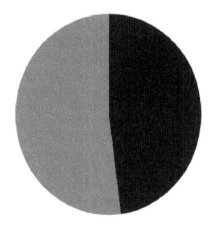

Figure 5.2. The variation in GRASS.[1]

[1] Pie chart derived using MicroCase Statistical Software.

Do you believe there is a life after death? (POSTLIFE)

	Freq.	%
■ 1) Yes	1462	81.1
▦ 2) No	341	18.9
Total (N)	1803	100.0
Missing	241	

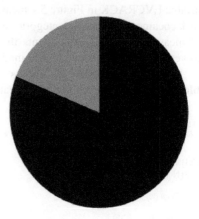

Figure 5.3. The variation in POSTLIFE.[1]

respectively). Think about each of these variables as having 100 percent variation. In each of these examples, variation is split into two attributes. For EVCRACK, 6.3 percent of people report having ever tried crack cocaine; 93.7 percent of people report not having tried crack cocaine. The closer a variable gets to having any one category reach 100 percent, the less variance there is in the variable. When most people select the same response, a variable has little variance.

In contrast, the variable GRASS has a nearly maximum amount of variation. Those responding that marijuana should be made legal versus should not be made legal are 47.9 percent to 52.1 percent, respectively. For a dichotomous variable (a variable with two answer categories), the maximum amount of variation is 50 percent for each category. Unlike EVCRACK, GRASS has almost as much variation as possible.

Here are two additional dichotomous variables. Look at Figure 5.3 for the variation in the variable for POSTLIFE, which measures whether or not a person believes there is a life after death, and Figure 5.4 for the variation in ASTROLGY, which measures whether respondents ever read a horoscope or their personal astrology report.

Which of these two variables has the most variation? Figure 5.3 shows the univariate distribution for POSTLIFE with 81.1 percent of people reporting that they believe in a life after death while 18.9 percent of people report they do not believe in a life after death. The results in Figure 5.4 for ASTROLGY show that 57.6 percent of respondents read a horoscope or a personal astrology report, and 42.4 percent of respondents do not. Remember that the maximum variation for a dichotomous variable is for each of the two categories to be closer to 50 percent. That makes ASTROLGY the variable with the most variation (as opposed to POSTLIFE).

When using a dichotomous variable (a variable with only two attributes/answer categories), variation is important to gauge. A widely used rule for dichotomous variables is that they need to have at least 20 percent or more variation between categories. If one category encompasses less than 20 percent of the respondents, there's often not

Do you ever read a horoscope or your personal astrology report? (ASTROLGY)

	Freq.	%
■ 1) Yes	1073	57.6
▨ 2) No	789	42.4
Total (N)	1862	100.0
Missing	2648	

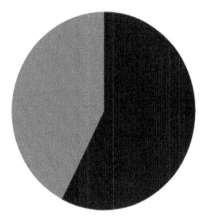

Figure 5.4. The variation in ASTROLGY.[1]

Fundamentalism/liberalism of respondent's religion (FUND2)

	Freq.	%
■ 1) Fundamentalist	521	26.9
▨ 2) Moderate	795	41.0
■ 3) Liberal	623	32.1
TOTAL (N)	1939	100.0
Missing	105	

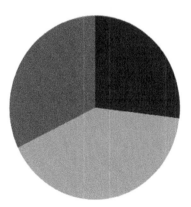

Figure 5.5. The variation in FUND2.[1]

enough variation within the variable to use it to explain the change/variation in another variable. The more variance within a variable, the better able that variable is to explain variance in another variable.

One way to increase variation in a variable is to increase the number of answer categories. By adding a third answer category to the variable FUND 2 in Figure 5.5, we have created more variation. What would maximum variation be for FUND2, which measures the fundamentalism/liberalism of respondent's religion? If each variable contains 100 percent variance, to attain maximum variance a variable with three categories would have categories that approach 33 percent (100 percent variation/3 categories). FUND2 is fairly evenly distributed between categories.

We hear a lot of talk these days about liberals and conservatives. I'm going to show you a seven-point scale
on which the political views that people might hold are arranged from extremely liberal--point 1--
to extremely conservative--point 7. Where would you place yourself on this scale? (POLVIEWS)

	Freq.	%
■ 1) Extremely lib	76	3.9
▨ 2) Liberal	259	13.1
■ 3) Slightly lib	232	11.8
▨ 4) Moderate	746	37.8
■ 5) Slightly con	265	13.4
▨ 6) Conservative	315	16.0
■ 7) Extremely con	80	4.1
TOTAL (N)	1973	100.0
Missing	71	

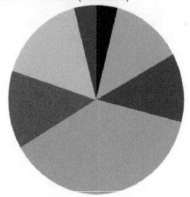

Figure 5.6. The variation in POLVIEWS.[1]

We hear a lot of talk these days about liberals and conservatives. I'm going to show you a seven-point scale on which
the political views that people might hold are arranged from extremely liberal--point 1--to extremely conservative--point 7
Where would you place yourself on this scale? (POLVIEWS2, collapsed categories from POLVIEWS)

	Freq.	%
■ 1) Liberal	567	28.7
▨ 2) Moderate	746	37.8
■ 3) Conservative	660	33.5
TOTAL (N)	1939	100.0
Missing	71	

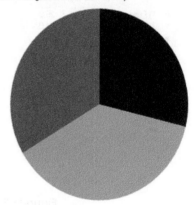

Figure 5.7. The variation within POLVIEWS2.[1]

What about the distribution of POLVIEWS? With seven category answers, we have
a lot of variation. Notice the largest part of the pie chart has "Moderates" representing
37.8 percent of the variance. With more categories of somewhat even distribution
between options for "liberal" and "conservative," there seems to be ample variation.
Of course, we can collapse the POLVIEWS variable to get a better sense of how the
distribution works by including all three categories that measure conservative and
liberal.

The POLVIEWS2 variable shown in Figure 5.7 has slightly more variation than
the FUND2 variable in Figure 5.6, since each category of POLVIEWS2 is closer to the
33 percent of maximum variation expected in a variable with three categories. Social

researchers are interested in variation because higher variation in one variable helps to explain variation in other variables. Sometimes, however, variables that appear to have variation have no variation at all; these are called constants.

Maximizing Variation

It is good practice, whenever possible, to begin with an operationalization that maximizes the number of categories. When you begin with a continuous variable—a broad range of categories—you see the full variation of a variable. It is generally useful to begin with a variable that elicits full variation. This allows you to create additional variables that are useful in multiple analyses. The question we've been using to operationalize education ("What is the highest level of education that you have completed?") does not begin as a question that is measured using only four categories. The original General Social Survey (GSS) 2010 variable categories run from 0, meaning "no formal schooling," through "8 years of college." These categories represent 20 distinctive levels of education.

Beginning with these continuous categories allows us to pinpoint year-by-year differences in education but also allows us to appropriately *collapse* the categories for other uses. When we describe four categories that encompass these 20 categories, the data are still meaningful and can be analyzed using cross-tabular analysis—four categories to read from a screen are much easier than 20. The fewer categories also allow us to generalize to meaningful groups (people will less than a high school degree, people with more than four years of college, etc.). You want to begin with maximum variation because if you begin with truncated—fewer—categories, you can never go back to see the full variation.

CONSTANTS

When looking at the variance of a variable, you want to make sure that you do not accidentally confuse a constant for a variable. A **constant** has a value that does not change—it has no variation. The GSS has been conducted most years since 1972, and then biennially beginning in 1994. One of the items contained in each GSS is the variable YEAR, or "GSS year for this respondent." Since the GSS uses multiple survey years to conduct time-trend studies, it is important to have a variable that allows us to control for year of survey given. When a codebook is created, each year that the survey has been conducted is included as an answer category, so the descriptor and attributes might appear in the codebook with a list of attributes for each GSS year like this:

GSS Year for the Respondent

1. 1972	7. 1978	13. 1986	19. 1993	25. 2004
2. 1973	8. 1980	14. 1987	20. 1994	26. 2006
3. 1974	9. 1982	15. 1988	21. 1996	27. 2008
4. 1975	10. 1983	16. 1989	22. 1998	28. 2010
5. 1976	11. 1984	17. 1990	23. 2000	
6. 1977	12. 1985	18. 1991	24. 2002	

GSS Year for this respondent (YEAR)

	Freq.	%
■ 1) 2010	2044	100.0
TOTAL (N)	2044	100.0

Figure 5.8. The variation in YEAR for GSS 2010.

The number of categories for the variable indicates that the variable has a good amount of variation. Yet when you run a univariate statistic on the variable for the GSS 2010, 100 percent of the respondents give the same answer, shown in Figure 5.8.

While YEAR may appear to be a variable because the categories are coded for each possible year the GSS has been administered, the actual value for a given year will not change because all of the respondents will have the same value (same year). YEAR is therefore a constant and not a variable when used in the GSS 2010. When looking at more complex variables, you need to make sure enough variation (change) exists within one variable to explain the change within another variable.

LEVELS OF MEASUREMENT

Levels of measurement help us think about what variables tell us. Depending on how a variable is measured, we can also tell what statistical analysis is best for the type of information contained in a variable. For example, we've been using cross-tabulation analysis to look at research questions. The variables that are most appropriate to cross-tabular analysis are those that have seven or fewer categories—the categorical variables we discussed in a previous section. Categorical variables tend to be *nominal-* or *ordinal-level* variables, but they can be created from any level of data. For example, we noted earlier that in eliciting variation for a variable, it is better to get the most variation. Most surveys ask respondents about their age. It is much better to begin with a question such as, "What is your age? _____," or "In what year were you born?" to allow respondents to give you their actual age. Asking their exact age or year of birth gives us the maximum variation. Once you have an actual age (which is a ratio-level variable), you can create a categorical variable that groups the age variable into smaller categories (e.g., "younger than 18 years old, 18 to 29 years old, 30 to 39 years old, 40 to 49 years old, 50 to 59 years old, 60 to 69 years old, 70 years old or older). This categorical variable is now an ordinal-level variable.

There are four levels of measurement: nominal, ordinal, interval, and ratio. Each subsequent level of measurement builds on the previous level, adding a unique quality for that level (while each of the four levels of measurement distinctly describe what is being measured by a variable, as we move from nominal to ordinal to interval to ratio, each next level takes on an additional quality, like having an inherent order). The first level of measurement is the *nominal* level. The categories/attributes for a nominal-level variable *distinctly* describe that variable. For example, look at the variable REG16:

In what state or foreign country were you living when you were 16?

0. Foreign
1. New England
2. Middle Atlantic
3. East North Central
4. West North Central
5. South Atlantic
6. East South Central
7. West South Central
8. Mountain
9. Pacific

A **nominal measure** distinctively describes the variable by the answer categories/attributes of the variable (as we mentioned before, the descriptor tells you what the variable measures—"In what state or foreign country were you living when you were 16?"—while the categories tell you how the variable is measured). Does each category of REG16 give you a clear, distinctive description of which category of the country you lived in at 16? Each of the answer categories distinctly describes in what region of the county a respondent lived at 16. There is no inherent order to the list of attributes (at least from a scientific perspective—our students generally try to explain why the Pacific region is better than the others); each attribute simply describes a region of the country. None is higher, bigger, or faster than the other. Each is simply distinct.

Ordinal measures are the next level of measurement. Like nominal measures, ordinal measures have answer categories that distinctively describe the variable. Ordinal measures, however, can also be ordered along some kind of continuum, allowing us to rank them logically. Look at the variable for SOCBAR.

How often do you do the following things: Go to a bar or tavern?
1. Almost every day
2. Once or twice a week
3. Several times a month
4. About once a month
5. Several ties a year
6. Once a year
7. Never

Notice that the categories have a logical order that tell us how more or less frequently respondents spend time in a bar or tavern. Ordinal measures are useful because we can compare respondents' frequency of going to a bar, making clearer generalizations, like "the more often one goes to a bar or tavern, the less likely they are to attend church."

Frequency of religious attendance would be another example of an ordinal measure:

How often do you attend religious services?
0. Never
1. Less than once a year
2. Once a year
3. Several times a year
4. Once a month
5. Two to three times a month
6. Nearly every week
7. Every week
8. More than once a week

As long as the categorical attributes of a variable can be ordered along some sort of a continuum, the minimum level of measurement would be ordinal.

The next level of measurement is the interval level. **Interval measures** build on the previous two levels; interval measures have answer categories that distinctly describe the variable like nominal and ordinal measures, and they can also be ranked logically in some kind of order like ordinal measures. The unique quality that interval-level measures add to data is that interval measures also have *equal distance between units*. Whereas an ordinal-level variable ranks the categories within groupings, interval-level measures give a more precise degree of difference between categories by telling us how much more one category is than another.

For example, if I asked, "What is the current temperature?" Depending on where you live and what season it is, you could give answers like, "72 degrees" or "37 degrees." The temperature would be a good example of an interval-level variable, where the assumption is that each degree is an equal measure of temperature. So the difference between 72 and 37 degrees is 35 degrees, just as the difference between 10 and 45 is 35 degrees. We can say that because we know that each degree measures the same unit—degrees. Interval measures are generally considered *continuous variables* because they are structured in such a way as to maximize variation (creating many attributes).

Interval measures, however, are made somewhat more unique because they contain no meaningful zero point. What is a meaningful zero point? We expect that when a variable is measured at "zero" that zero is the absence of the variable altogether. So when you answer "zero" to a question like, "How many classes did you skip last semester?" we expect that no classes—zero classes—were skipped last semester. Interval-level measures, however, are unique. When we say it's zero degrees outside, that does not mean there is no more temperature. Temperatures are measured below zero, into the negatives. Thus temperature has no meaningful zero point. This makes interval measures somewhat unique.

Using the same building blocks within levels of measurement—distinctively describing a variable, being logically ordered along some continuum, and having equal distance between units—are **ratio** measures. Ratio-level measures expand interval-level measures by having a meaningful zero point—at the measure of zero, there is no more of the variable. Because ratio measures have a meaningful zero point, we can talk about ratios (2 times, half as much, fourfold.). For example, if Ernie weighs 220 pounds and Bert weighs 110 pounds, we can say that Ernie weighs twice as much as Bert. Weight would be an example of a ratio measure.

Another example of a ratio measure would be:

How many children do you have?

  0. 0
  1. 1
  2. 2
  3. 3
  4. 4
  5. 5
  6. 6
  7. 7
  8. 8
  9. 9
  10. 10
  . . .

In this variable, each category represents the actual number of children a person has (equal distance between units); therefore, the variable is a ratio measure. Having "zero" children means that the respondent has no children.

We can take the same concept—even the same variable—and measure it using multiple levels of measurement depending on what we want to learn about the variables or glean from a particular analysis. As we've explained, if variables initially elicit the maximum variation, different analyses can be used based on the interval- or ratio-level measure while the measure can be collapsed as an ordinal or nominal measure to use in cross-tabular analyses. For example, the variable that measures "Number of hours worked last week" can be measured in two ways. The first variable describes "Number of hours worked last week" and is asked as an open-ended question (one where respondents fill in their own answers), eliciting the maximum variation (with a range of working 0 to 168 hours per week)[2]. What level of measurement is this? Does each hour given describe how many hours the respondent worked? Yes, so minimally the variable is nominal. Are the categories ordered on any kind of a continuum? Yes, lower categories indicate fewer working hours and vice versa, which would mean we've moved to the ordinal level. Does each category have equal distance between units? Yes, each category represents an hour; for one respondent who worked 40 hours, he or she would have worked 20 hours less than someone who worked 60 hours or 128 hours less than someone who worked 24 hours a day, 7 days a week. Therefore we've moved up to the interval level of measurement. Last question: do the categories have a meaningful zero point? In other words, do those who report working zero hours last week mean that they did no work at all? Yes. Thus the categories for this variable are ratio-level measures.

Of course, if we want to do a cross-tabular analysis, having 169 possible categories is daunting (168 possible hours per week plus the category of zero hours is 169 total category possibilities). How would you be able to represent or read this data in a 2×2 cross-tab? In order to do a cross-tabular analysis, we need a categorical variable. Taking the original variable, we can collapse the categories into meaningful groups that denote those who work less than full time, full time, and more than full time. We do this by combining all the hours between 0 and 39 to represent less than full time, keep 40 hours as the category for full time, and take categories 41 to 168 as working more than full time. These then become "less than 40 hours," "40 hours," and "more than 40 hours" worked last week. What level of measurement do we now have? Does each category describe how many hours the respondent worked? Yes, so minimally the variable is nominal. Are the categories ordered on any kind of a continuum? Yes, lower categories indicate fewer working hours and vice versa, which would mean we've moved to the ordinal level. Does each category have equal distance between units? No. Of the three categories, 1, 2, and 3, each is composed of unequal units; that is, the category for working less than 40 hours has at least 40 possible hours in it; the category for worked 40 hours has only one hour, the category for 40 to 168 hours worked has many more. The categories—1, 2, and 3—are not measured using the same units. Therefore the variable does not meet the criteria for an interval measure and must then be an ordinal measure.

---

[2] It is not uncommon for some respondents to write in "168" when asked how many hours per week they work. This response indicates that they believe they work 24 hours a day, 7 days a week.

Exercise: Levels of Measurement
Look at the variable descriptions here. What are the stated or implied response categories? Based on these variable descriptions and response categories, what level of measurement does each variable represent? Circle the level of measurement that each variable represents.

1. About how many hours do you spend on social networking sites in a given 24-hour period? _____
   Nominal   Ordinal   Interval/Ratio
2. What is your favorite music genre? _____
   Nominal   Ordinal   Interval/Ratio
3. What is your home state? _____
   Nominal   Ordinal   Interval/Ratio
4. Do you strongly agree, agree, disagree, or strongly disagree that you love country music?
   Nominal   Ordinal   Interval/Ratio
5. Not counting e-mail, about how many hours of the week do you use the Web? ____
   Nominal   Ordinal   Interval/Ratio
6. Have you ever been widowed?
   Nominal   Ordinal   Interval/Ratio
7. Do you accept others, even when you thing they do things that are wrong, most of the time, some of the time, seldom, never?
   Nominal   Ordinal   Interval/Ratio
8. How many organizations would you say you belong to (school, community, place of worship, etc.)? _____
9. How often do you spend a social evening with relatives, almost daily, once or twice a week, several times a month, once a month, several times a year, once a year, or never?
   Nominal   Ordinal   Interval/Ratio
10. Have you ever felt discriminated against because of your age?
    Nominal   Ordinal   Interval/Ratio

UNITS OF ANALYSIS

Now that we've looked at how variables are measured, we address **units of analysis**, the *things* a hypothesis directs us *to observe*, or the cases in a data set (the thing being studied). Two general categories of units include individual-level units (people) and aggregate-level units (geographic places—states, nations or things—and social units— churches, schools, clubs, etc.). Aggregate units are single units (like states) that are composed of individuals. Units of analysis are the things that have the variables. Let's go back to the relationship between education and income. If we hypothesize that higher levels of education lead to higher levels of income, we want to identify the independent and dependent variables. Independent variables are our starting point. We assume (infer) that the independent variable is causing changes to a second variable, the dependent variable. Our hypothesis implies that if we know a person's level of education, we can predict that person's level of income. So education is the independent variable (the causal variable) and income is the dependent variable (the effect).

To understand units of analysis, you want to ask the question, "Who or what can have an education level and an income level?" The only thing that can get an education and have income are people/individuals; thus people are the units of analysis. What if our hypothesis stated that countries with a high percentage of the population with college degrees will have a lower percentage of hate crimes? In this hypothesis, who or what has a percentage of the population with college degrees and a percentage of hate crimes? Countries.

Exercise: Units of Analysis

Here are some hypotheses. For each hypothesis, identify the independent variable, the dependent variable, and the unit of analysis:

Example: <u>People</u> with higher <u>socioeconomic status</u> are less likely to <u>be arrested</u>.
UA   IV   DV

1. Churches with higher levels of participation will have higher levels of financial giving.
2. Counties with fewer urban centers will have higher birthrates.
3. Magazines that are geared toward popular culture will have higher readership.
4. Older people will be more likely to hold conservative political views.
5. Those living in higher income neighborhoods are more likely to feel safe at night.
6. Country song lyrics will be more depressing than pop song lyrics.
7. Cities that have seatbelt laws will have lower rates of vehicle deaths.
8. Parents' with higher socioeconomic status will have children with higher levels of education.
9. Higher levels of mental health are a result of being married.
10. States with higher percentages of youth will have higher property crime rates.

RELIABILITY AND VALIDITY OF MEASURES

When we engage in research, we trust that the measures we use to define a variable actually capture the meaning of the concept and that they do so consistently. These are the considerations of *validity* in the former instance and *reliability* in the latter. For purposes of our discussion, we can use the following standard definitions of the concepts:

- Validity is the extent to which a research measure *actually captures the meaning of the concept it is intended to measure*. For example, if we are interested in job satisfaction, does our questionnaire or survey of job satisfaction actually measure the extent to which workers are content with their work, or do the questionnaire items really reflect other concepts like "acceptance of work responsibilities"?
- Reliability is the extent to which a research measure provides a *consistent evaluation of a concept*. Thus we can create a questionnaire that contains items focused on a person's attitude toward their manager, but would the resulting attitude measure be the same if we asked the same worker(s) on a different occasion? The extent to which the questionnaire items are free of measurement error (e.g., incorrectly worded items, etc.) is the consideration of reliability.

A research measure can be reliable, but not necessarily valid, as, for example, using height as a measure of athletic prowess. Height can be measured very accurately, but it does not necessarily indicate how much of an athlete a short or tall person may be.

However, a measure may be valid, but it may not have strong reliability as is the case in some glucometers. When people with diabetes check their blood sample for sugar content, they will have a valid measure of blood glucose, but the instrument may give varying readings from day to day (and therefore not a reliable measure) due to a variety of factors (e.g., quality of test strips, etc.).

Types of Validity and Reliability

Validity

- Face validity: The extent to which a measure has a ring of plausibility about it. In other words, does the measure appear to measure what it is intended to measure? (Does a measure of religious orientation contain items that mention religion?)
- Content validity: Does the measure adequately represent all the content domains of the concept it is intended to measure? (Does a measure of academic achievement contain measures of math, reading, writing, science, etc., or merely reading?)
- Predictive validity: Does the measure provide predictions that can later be confirmed by examining the behavior or content from the same individuals? (Does a Success in College questionnaire predict whether a student attains a certain academic record?)
- Concurrent validity: Does the measure correlate with other measures attempting to measure the same thing? (Does my job satisfaction instrument correlate highly with your job satisfaction instrument if we administered both to the same individuals?).

(Continued)

- Construct validity: Does a measure of a concept relate strongly with another measure that it *should* correlate strongly with (converging measures), and negatively with measures it should not agree with (diverging measures)? (In the first instance, does my measure of happiness correlate strongly and positively to a measure of job satisfaction? In the second instance, does my measure of happiness correlate strongly and negatively with a measure of depression?)

Statistically measuring validity is a subject beyond the scope of this book, but the procedures are straightforward. Researchers use correlation (which we do discuss in this book) and factor analysis techniques (a more complex procedure we do not discuss in this book) primarily to establish the strength of validity.

Reliability

- Interrater reliability: The extent to which there is agreement between two or more expert judges classifying the presence of some measure. (Do three trained raters agree on the presence of a new teaching technique among the math teachers of a study school?)
- Retest reliability: The extent to which the responses of a measure correlate highly with the same measure administered at another time or with a similar form of the measure. (Do my job satisfaction scores of a study group correlate highly with the scores we receive from the same measure we obtain a week later; do the scores from the original different job satisfaction measure correlate highly to a slightly different job satisfaction measure?)
- Internal consistency measures of reliability: The extent to which the items of a single instrument correlate with one another either in separate halves of the test (split-half reliability) or to which the items correlate among themselves if there is a single theme or content in the instrument.

Statistically measuring reliability typically involves some variety of correlation, although special formulas exist for measures of internal consistency (e.g., Kuder-Richardson formulas, etc.). Interrater reliability is measured by a variety of methods of varying sophistication, from percentage agreement to intraclass correlation coefficients.

6

USING SPSS IN RESEARCH

This is an introduction to using SPSS, a very powerful statistical software program. Ordinarily, research methods and design courses do not teach readers to use this software. Most such courses typically discuss the nature of research designs without "getting their hands dirty in data." This book does not take this approach. We want to examine the key issues of research design and then use real-world data to show you how to create your own analyses and interpret research findings.

By using statistical software, we can place more attention on understanding how to *interpret findings*. Our approach encourages you to focus on how to understand and make applications of the results of research findings. SPSS and other statistical programs are efficient at performing the analyses; the key issue in our approach is how to interpret research results in the context of research questions.[1]

REAL-WORLD DATA

As mentioned, we focus on using real-world data in this book. One reason is that you need to be grounded in approaches you can use with "gritty" data. We want to make

---

[1] Parts of this section are adapted from M. L. Abbott, *Understanding Educational Statistics Using Microsoft Excel® and SPSS®* (Wiley, 2011), by permission of the publisher.

Understanding and Applying Research Design, First Edition. Martin Lee Abbott and Jennifer McKinney.
© 2013 John Wiley & Sons, Inc. Published 2013 by John Wiley & Sons, Inc.

sure you are prepared to encounter the little nuances that characterize every research project.

Another reason we use real-world data is to familiarize you with contemporary research questions in social science and public health. Classroom data often are contrived to make a certain point or demonstrate a specific research design, which are both helpful. But we believe it is important to draw the focus away from the procedures per se and understand how the procedures will help the researcher resolve a research question. This is an active rather than a passive learning approach to understanding research design and analysis.

COVERAGE OF STATISTICAL PROCEDURES

The statistical applications we include in this book are "workhorses." We present introductory treatments of key statistical procedures and then review the basic procedures that allow you to understand more sophisticated procedures. We will not be able to examine advanced statistical procedures in much detail. As you learn the capability of SPSS, we hope you can explore more advanced procedures on your own, beyond the end of our discussions.

Some readers may have taken statistics coursework previously. If that is you, we hope the book enables you to enrich what you learned previously and to develop a more nuanced understanding of how to address research design problems using SPSS. Our intention is to help you become "research literate" in order to recognize what statistical processes should be used with different research designs, how to use the procedures correctly, and how to make appropriate research conclusions.

SPSS BASICS

This book explores the use of SPSS with research design problems. Therefore, we have included the following sections to provide some familiarity with the basic functions of SPSS. We will introduce more advanced uses in later sections that correspond to the statistical procedures connected to specific research designs we discuss.

Several statistical software packages are available for managing and analyzing data; however, in our experience SPSS is the most versatile and responsive program. Because it is designed for a great many statistical procedures, we cannot hope to cover the full range of tools within SPSS in our treatment. We will cover, in as much depth as possible, the general procedures of SPSS, especially those that provide analyses for the research design processes we discuss in this book.

The calculations and examples in this book require a basic familiarity with SPSS. Generations of social science students and evaluators have used this statistical software, making it somewhat a standard in the field of statistical analyses. In the following sections, we make use of SPSS output with actual data in order to teach you how to use research findings that shed light on research problems stemming from specific research designs.

In this section we illustrate the SPSS menus, so it is easier for you to negotiate the program. The best preparation for the procedures we discuss, and for research in general, are to become acquainted with the SPSS data managing functions and menus. Once you have a familiarity with these processes, you can use the analysis menus to help you with more complex methods.

GENERAL FEATURES

Generally, SPSS is a large spreadsheet that allows the researcher to enter, manage, and analyze data of various types through a series of drop-down menus. The screen in Figure 6.1 shows the opening page where data can be entered. The tab on the bottom left of the screen identifies this as the "Data View" so you can see the data as they are entered.

A second view of the opening spreadsheet is available by choosing the "Variable View," also located in the bottom left of the screen. When you choose Variable View, SPSS switches to a different screen showing all the variables that are included in the set of data shown in the Data View. As shown in Figure 6.2, the Variable View allows you to see how variables are named, the width of the column, number of decimals, variable labels, any values assigned to data, missing number identifiers, and so on. This information can be edited within the cells or by the use of the drop-down menus, especially the "Data" menu at the top of the screen. One of the important features on the page in Figure 6.2 is the "Type" column, which allows the user to specify whether the variable is "Numeric" (i.e., a number), "String" (a letter, for example), or some other form (a date, currency, etc.). This information is important because it will help you to

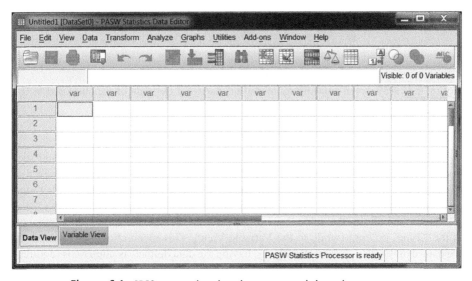

Figure 6.1. SPSS screen showing data page and drop-down menus.

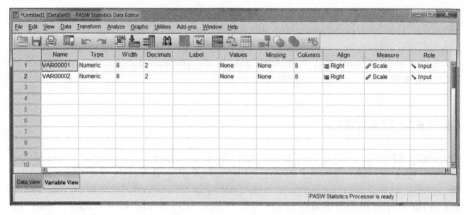

Figure 6.2. SPSS screen showing the Variable View and variable attributes.

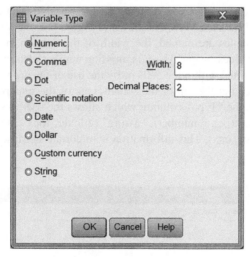

Figure 6.3. SPSS screen showing submenu for specifying the type of variable used in the data field.

identify the specific kind of data you choose to include in your research. In later sections, we discuss different types of data that we might collect; SPSS makes sure that the data you gather is appropriately identified as a number, a letter, a date, or other kind of information.

Figure 6.3 shows the submenu available if you click on the right side of any cell in the "Type" column (while in the Variable View). This menu allows you to specify the nature of the data. For most analyses, having the data defined as numeric is required, since most analyses require a number format (you can't add, subtract, multiply, and divide letters). The "String" designation, shown at the bottom of the choices, allows you to enter data as letters and words, such as quotes from research subjects, names of subject groups, and so on. If you use a statistical procedure that requires numbers, make sure the variable is entered as a "numeric" variable, or you will receive an error message and your requested procedure will not be executed.

USING SPSS WITH GENERAL SOCIAL SURVEY DATA

Now that we have explored some of the SPSS basics, we can get a better sense of how to use it by looking at a data file available from the Internet. The General Social Survey (GSS) is a biannual survey describing the structure and nature of American society. It is funded through the National Science Foundation and has been administered by the National Opinion Research Center since 1972. According to the GSS, it is a "full-probability, personal-interview survey designed to monitor changes in both social characteristics and attitudes currently being conducted in the United States."

Figure 6.4 shows an SPSS screen shot of selected data from the 2010 GSS. We selected five items of information from the overall survey: age, education, income, socioeconomic indicator, and sex.[2] These are **variables**, or sets of information, that can take a range of values. For example, "age" is a number that the interviewers obtained from survey respondents who provided their year of birth.

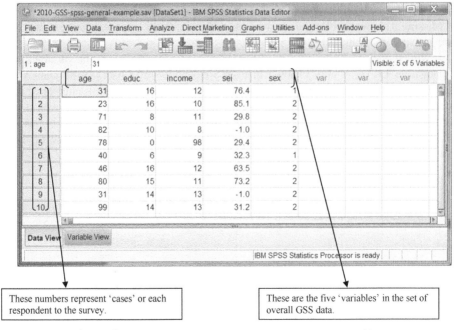

These numbers represent 'cases' or each respondent to the survey.

These are the five 'variables' in the set of overall GSS data.

Figure 6.4. The example GSS data shown in an SPSS data file.

[2] There are over 2000 of these items in the total survey.

Since the interviewers could only interview adults, the range of possible ages in this variable is from 18 to "89 OR OLDER." You can see the first 10 cases in Figure 6.4. (**Cases** are individuals who provided information, or "respondents.") The first case (number 1, which is highlighted) is a 31-year-old person. This person happens to be a male since the interviewers identified the respondent as a male and assigned the code of 1, which they used for males.

The other variables in our example GSS data file in figure 6.4 are as follows:

* Education is the "highest year of school completed"
* Income is "total family income"
* Sei is the "respondent socioeconomic index"
* Sex is the "respondent's sex"

By the way, the values in the data file also show another feature of SPSS, ways of coding data that are missing. If you notice in Figure 6.4, case 10 appears to be 99 years old! While this is possible, it is not so with GSS. The value of 99 is used as a place-holder for missing information. Thus, when the interviewer is not able to obtain the year of birth of the respondent, he or she must put a *placeholder* in the set of data for that respondent. Figure 6.5 shows the Variable View of this set of information in which we have highlighted the "Missing" column.

As we explained earlier, the Variable View is simply a list of the variables in our data file shown with several pieces of descriptive information. The highlighted field ("Missing") shows that for the age variable, values of 0, 98, and 99 indicate values that are missing in the interviewer's records.

Notice that one of the missing value placeholders for another of the variables, sei, is negative 1.0 (or, "−1.0") shown in Figure 6.5. Now refer back to the Data View in Figure 6.4 and you will see that cases number 4 and 9 are both missing, since they show values of −1.0. SPSS can still use the nonmissing values for a case even if one or more of the values are missing. Placing a list of the missing values in the "Missing" field (in Variable View) simply *informs* SPSS that when it encounters one of these numbers not to include it in a specific operation.

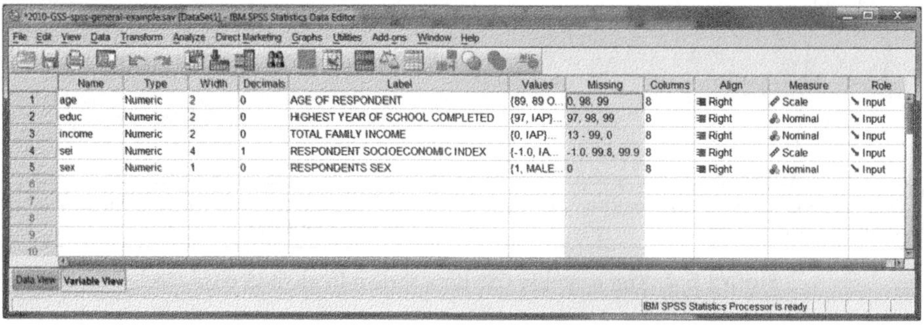

Figure 6.5. The Variable View in SPSS highlighting the "Missing" values for each variable.

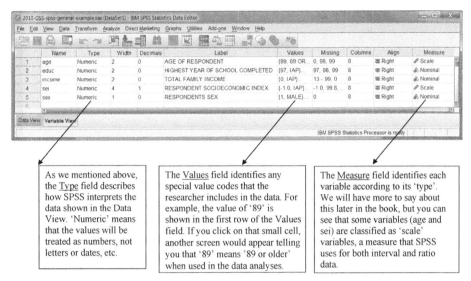

Figure 6.6. The Variable View in SPSS highlighting several categories for each variable.

The other categories in the Variable View are also very helpful. Figure 6.6 shows the Variable View noting additional information about the variables in our example GSS data set. Figure 6.6 is a copy of Figure 6.5 but without highlighting the "Missing" category.

That is enough about SPSS for now. As we encounter different statistical procedures, we will introduce more information about how to use SPSS to assist with data analysis. We also have prepared several book sections that give you more information if you are interested.

7

CHI-SQUARE AND CONTINGENCY TABLE ANALYSIS[1]

In a previous section, we introduced and discussed some examples of using General Social Survey (GSS) data to help test whether or not hypotheses about relationships among variables are supported. Figure 2.2 shows how voting behavior differs by social class identification, for example. This analysis made use of a **contingency table,** which displays data in rows and columns so that you can identify potential patterns in the relationships. As we saw in Figure 2.2, the analysis revealed a pattern whereby a greater percentage of those in middle and upper classes voted versus those in lower and working classes.

In this section, we take a closer look at contingency tables and some statistical measures that can help us to have a clearer picture of the extent of the relationships among study variables. First, we learn how to set up a contingency table, and then we use SPSS to see whether or not the data reveal meaningful patterns of relationships.

CONTINGENCY TABLES

Contingency tables are created by presenting data in rows and columns according to the independent and dependent variables of a study. In this way, the researcher can

---

[1] Parts of this section are adapted from M. L. Abbott, *Understanding Educational Statistics Using Microsoft Excel® and SPSS®* (Wiley, 2011), by permission of the publisher.

Understanding and Applying Research Design, First Edition. Martin Lee Abbott and Jennifer McKinney.
© 2013 John Wiley & Sons, Inc. Published 2013 by John Wiley & Sons, Inc.

easily see how the data are arrayed across the categories of the variables. These tables of data containing the frequencies are called contingency tables because the data in the row cells are *contingent upon or are connected to* the data in the column cells (independent variable). Statisticians and researchers often refer to the analysis of contingency tables as "cross-tabulation," or simply, **crosstabs.**

Having the data displayed in rows and columns according to the categories of the variables making up the contingency table is helpful to the researcher. Simple visual inspection may help to detect patterns not ordinarily apparent when the data are not placed in tables. Typically, contingency tables are used with nominal or categorical data in which we simply identify the frequencies, or numbers of observations, in each category. For example, in Figure 2.2, you can see that there were 45 individuals (81.8%) who reported that they were in the upper class and that they voted in the 2008 election. By contrast, 284 individuals (32.2%) identified themselves as belonging to the working class and that they did not vote in the 2008 election.

When researchers wish to present the results of their analyses, or to simply list the data in the tables, *they must also use percentages along with frequencies.* This is because the frequencies in cells are often different across rows and columns. Therefore, percentages are a way to present the frequency data in a standardized way. Raw frequency differences are thereby transformed to a common expression across the entire contingency table. For example, Figure 2.2 shows that 598 working-class individuals voted versus only 45 upper-class individuals. However, the percentages of these cells are 67.8 percent and 81.8 percent, respectively. Thus the frequencies alone are not a reliable indicator of the magnitude of a finding; percentages provide a common base, or standardized interpretation, since the raw numbers are based on different totals.

One convention some researchers use when presenting data in contingency tables is to present the independent variable categories in columns and the dependent variable categories in rows.[2] In this way, the column data percentages are created to total 100 percent and the researcher can *compare values of the independent variable across the column categories and within rows of the dependent variable categories.* This provides a common way of interpreting the data from visual inspection and from chi-square analyses.

Work Autonomy and Personal Health: An Example of a Contingency Table

Melvin Kohn (1977) discussed the connection between work autonomy and social class in studies showing how conditions at work affected an individual's values (self-direction and conformity, for example). We can extend these ideas by looking at how work conditions might affect other factors like personal health. We might ask the question, "Do the conditions at work affect people's perceptions of their general health?" This

---

[2] Other researchers organize contingency tables in the opposite fashion, with the independent variable categories in rows. As you will see in the analyses to follow, it does not matter which convention you use as long as you remember how to create the interpretation.

Healthcond * workautonomy crosstabulation

Health condition	Work autonomy		Total
	High	Low	
Excellent or good	572	64	636
	56.3%	44.1%	54.8%
Fair	291	54	345
	28.6%	37.2%	29.7%
Poor	153	27	180
	15.1%	18.6%	15.5%
Total	1016	145	1161
	100.0%	100.0%	100.0%

Figure 7.1. GSS contingency table relating work autonomy to general health.

is a very compelling question because work is central in most people's lives. If there are dynamics at work that affect how we feel about our physical and mental health, it seems reasonable to try to understand what they may be and to what extent they dominate our perceptions. We can use the GSS data to help respond to this research question.

Figure 7.1 shows an example of a contingency table reflecting these concepts. As you can see, the table presents GSS data for two variables relating to the relationship between work life and individual health. The independent variable, presented in columns, is "Work autonomy," which represents the number and percentage of respondents who reported different attitudes regarding the following statement: "[I have] a lot of freedom to decide how to do [my] job." The respondents are categorized in the table as having high autonomy at work if they responded either "Very true," or "Somewhat true" to the statement. The respondents categorized as having low autonomy were those who responded with "Not too true" or "Not at all true" to the same statement about freedom to do their job.

The dependent variable, health condition, represents the respondents to the GSS item identifying their "Condition of Health." These responses are placed in rows, since they are measures of the dependent variable. The categories of the responses are "Excellent or Good," "Fair," and "Poor."

As you can see from Figure 7.1, we created percentages of the raw numbers according to our rule noted earlier. Both the work autonomy categories (in the columns) are reported as percentages that total 100 percent. (You can see this by looking at the bottom of the columns where 100 percent is listed for each column category.) This allows the researcher to compare how categories of the independent variable (work autonomy) vary *across categories of the dependent variable* (health condition). Thus 56.3 percent of those reporting high work autonomy felt that they had excellent or good health versus 44.1 percent of those reporting low work autonomy. By contrast, those with different levels of work autonomy reported the opposite pattern with respect to judging their health to be fair. Only 28.6 percent of those reporting high autonomy believed they had fair health versus 37.2 percent reporting low autonomy. The results for those with poor health are less discrepant, but most closely resemble those with fair health.

Viewing these percentages, one would conclude that the conditions of work (high or low autonomy) appear to have an effect on perceptions of general health. Those with more work autonomy are more likely to judge their health as better than those with low work autonomy.

Describing versus Analyzing Contingency Table Data
Earlier we *described* frequency data by the way they were displayed in rows and columns. Percentages, along with the frequencies, allow the researcher to understand how each of the categories of one variable (e.g., subjective class identification in Figure 2.2) are compared within categories of the other variable (voted or not in Figure 2.2). Thus, in Figure 2.2, it appears that the respondents of the different subjective classes reported quite different voting patterns (with those in the middle and upper classes reporting more voting behavior).

In the next section, we use statistical analyses to *analyze* the data to determine whether these differences between categories can be said to be expected by chance or if they reflect a relationship between the variables that cause them to display a meaningful pattern. Thus, our statistical procedures will help us to understand whether some dynamic of subjective class identification results in different voting behavior. Using statistical analyses to shed light on the overall differences we observe will give us confidence in the meaningfulness of the patterns.

USING CHI SQUARE TO DETERMINE THE SIGNIFICANCE OF RESEARCH FINDINGS

The differences in the percentages that we observed from Figure 7.1 indicated that the categories of work autonomy had a *differential impact* on the categories of general health perceptions. The question that follows such an observation is "How different do the data (in row and column cells) have to be before we could conclude that the data patterns are meaningfully different?" The answer to this question leads us to the topic of statistical significance. We deal with this question more comprehensively in a later chapter on inferential statistics, but for now, we can examine how researchers use SPSS to help assess whether, and to what extent, work autonomy affects general health perceptions.

Statistical Significance

A quick definition of statistical significance may be of use here until we can develop the topic more comprehensively in later sections. This is a concept that requires a bit of groundwork to make sense, but we can suggest here that **statistical significance** *refers to whether the measure of a variable can be said to be greater or lesser than what would be expected by chance alone.*

For example, if we took a group of students (randomly) from a school and measured their math knowledge, it would most likely be very similar to the math knowledge of all the students in the school (i.e., we could say the sample of students were *representative* of the entire population of students at the school). If we then taught the sample students a new way to understand math, they might then perform better than everyone else at the school (assuming the new method is better than the math curriculum normally used in the school).

Statisticians have suggested criteria that would help researchers recognize whether this math improvement could be said to just occur by *chance* (i.e., perhaps the possibility that we simply ended up with a bunch of high math performers among our sample) or whether the new math instruction method pushed the math achievement average of the sample so far away from the average of the entire population of students that the sample students no longer are the same as the students in the population. In this example, we would conclude that the math performance of the sample group was now *statistically significantly different* (greater) than that of the population of students.

Statistical significance is different than "practical significance" or "effect size," which is discussed in the following sections. Effect size refers to the magnitude of a finding, not whether or not it is a chance or nonchance finding. We develop both of these matters in the following sections.

Researchers use the statistical procedure of **chi square** to analyze the differences among the data in contingency tables to determine whether the patterns of difference are different enough to be considered statistically meaningful. Chi-square results can be interpreted in different ways in research, but we can use it in the present example to indicate *the strength of the relationships* among our study variables.

Chi square is a statistical procedure that primarily uses nominal (or categorical) and ordinal data. It works by examining "frequency counts" or simply the number (frequency) of people or observations that fit into different categories. As we noted earlier, our example uses this level of data, since the categories of work autonomy and health perception are frequency counts of the number of respondents who indicate different perceptions of work autonomy and health.

USING SPSS FOR THE CHI-SQUARE TEST OF INDEPENDENCE

You can use the "Crosstabs" command in SPSS to create a **chi-square test of independence.** I will demonstrate the procedure with the same data I used in the work autonomy–health example in Figure 7.1, and then I will show an alternative procedure that is best when you input the data table directly into SPSS.

THE CROSSTABS PROCEDURE

SPSS prepares a contingency table with the data you specify and provides a range of findings. Starting with the GSS table in which we code the variables as we did in Figure 7.1 (e.g., with the GSS frequencies placed in "high" or "low" work autonomy), we can choose the Crosstabs procedure through the main SPSS menu of choices: "Analyze—Descriptive—Crosstabs." Figure 7.2 shows the Crosstabs menu that results from this choice.

As you can see in Figure 7.2, we called for the column variable (work autonomy) to be the independent variable and the row variable (health condition) to be the dependent variable for this analysis according to the protocol we discussed earlier. We chose these variables from among the total list of variables in the 2010 GSS database. Using the specification shown in Figure 7.2 will result in a table similar to the one in Figure 7.1. (We changed the SPSS output headings slightly in creating Figure 7.1 so that the example is clearer.)

The "Statistics . . ." button shown in Figure 7.2 (top right corner of the Crosstabs window) allows the researcher to choose which statistical analyses are desired. Figure 7.3 shows this window in which we call for the overall chi-square statistics and a series of measures for nominal data: Contingency Coefficient, Phi, and Cramer's V.

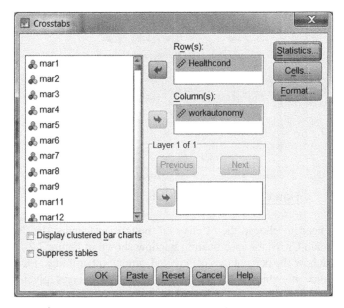

Figure 7.2. The SPSS Crosstabs specification window.

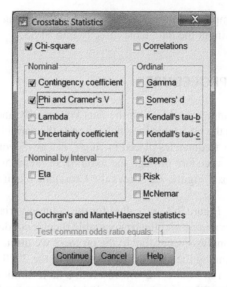

Figure 7.3. The Crosstabs: Statistics menu in SPSS.

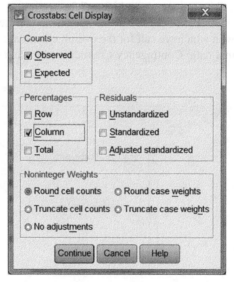

Figure 7.4. The SPSS Crosstabs: Cell Display menu.

You also need to choose the "Cells . . ." button (just under the "Statistics . . ." button in the main Crosstabs specification window shown in Figure 7.2). In this menu, you can choose how the percentages are created, among other things. Figure 7.4 shows this menu choice in which we called for column percentages given the protocol we discussed earlier where the independent variable is represented in the column variable.

Symmetric measures

		Value	Approx. Sig.
Nominal by	Phi	.081	.022
Nominal	Cramer's V	.081	.022
	Contingency Coefficient	.081	.022
N of valid cases		1161	

Figure 7.5. The Crosstabs output showing effect size findings.

This series of choices results in SPSS output showing the chi-square analyses needed for a research analysis. (We are not including all the output in the present example; we explain the other output in later sections.) If we were to finalize the cross-tabs procedure, SPSS would return a table that looks like Figure 7.1. The output would also include other information relevant to the interpretation of the data. We examine one of these findings here, the **effect size.** We discuss effect sizes much more in the following chapters, but for now, we can understand effect sizes to be a number that represents the *magnitude* of a finding.

In chi-square analyses (using Crosstabs procedures and output from SPSS), the researcher would get the output table shown as Figure 7.5. As you can see, there are three effect size measures provided. Essentially, these effect size measures are like correlations in that they represent the *degree of relatedness* of the study variables. Since crosstabs makes use of nominal (or categorical) data as is the case in our example, we are limited to which we can use. Each of these three measures (Phi, Cramer's V, and Contingency Coefficient) is interpreted differently. One of the factors that helps drive the interpretation is the size of the contingency table. In our example (Figure 7.1), the contingency table has three rows and two columns of data. This makes it a 3 × 2 chi square, since we identify the table by the number of rows (3) times (×) the number of columns (2).

EFFECT SIZE: CONTINGENCY COEFFICIENT

We can start with the most common effect size measure for chi square, **contingency coefficient** (symbolized as C), since it is appropriate to use with any size table. Generally, as the number reported (0.081 in Figure 7.5) increases, the relationship between the study variables is stronger. As we discuss later in the book, effect sizes can be classified as small, medium, and large, according to statistical tables that have been created for this purpose. In the case of C, the maximum possible value cannot reach 1.000.[3] Therefore, the effect size values for small, medium, and large vary with the conditions of the study (e.g., number of cells). Cohen's (1988) suggestions are generally 0.100 (small), 0.300 (medium), and 0.500 (large), depending on the factors affecting C.

[3] The maximum value for C increases as the number of cells increases, up to a certain point.

In our example, Figure 7.5 shows a contingency coefficient of 0.081, which, according to our criteria, would allow us to interpret this as a small finding. That is, work autonomy conditions do affect perceptions of health, but the relationship is small. One ballpark interpretation method is to look at the percentage difference between the column values on each of the categories of the rows. Thus, for example, there is a 12.2 percent difference (56.3% minus 44.1%) in those who perceive themselves as having good (or better) health between those who have high and low work autonomy. There are no established criteria for how big the percentages have to be before we would judge them to be big or small, so we can use the effect size criteria to help make this determination.

EFFECT SIZE: PHI COEFFICIENT

One effect size measure, **phi coefficient** (symbolized as φ), is designed specifically for contingency tables with two rows and two columns (i.e., a 2×2 table). Phi values can reach a maximum of 1.000, so the interpretation is a bit closer to other effect size measures we talk about later in the book. Generally, values of 0.000 indicate no strength of relationship, and values of 1.000 indicate the strongest possible strength of relationship. As you can see from Figure 7.5, the phi value is 0.081 (in our example, all three measures of effect size are the same, which is not always the case, since the results depend on the size of the contingency table).

The same criteria for judging C apply also to phi. Cohen's (1988) suggestions are generally 0.100 (small), 0.300 (medium), and 0.500 (large). Thus, in our example, as with C, the effect size is judged to be small.

EFFECT SIZE: CRAMER'S V

Contingency tables larger than the 2×2 table (i.e., that have more than two rows and/ or two categories) typically use another measure of effect size, Cramer's V. Like phi values, **Cramer's V** calculates effect size values that range between 0.000 and 1.000.

The judgment for the magnitude of Cramer's V does not always use the guidelines we discussed for the other effect size measures (0.100 for small, 0.300 for medium, and 0.500 for large) that pertain to the other tests. This is because there are adjustments made to the formula caused by the shape of the table. Cohen (1988) provides an adjusted set of guidelines in a series of tables. For the 2×3 table in our example, we can use the same 0.100, 0.300, and 0.500 guidelines for judging the effect size for Cramer's V as we did for C. Larger tables will have reduced magnitude effect size criteria that determine small, medium, and large effects.[4] (For example, if our table had been 3×4, the effect size criteria for a large effect would be 0.354, not 0.500.)

---

[4] We do not have the space in this book to develop this matter given its technical nature. However, SPSS reports the actual significance appropriate to any size of table, so in a practical sense, we do not need to delve more deeply into the subject at this point.

CREATING AND ANALYZING THE CONTINGENCY TABLE DATA DIRECTLY

Another method of using SPSS to calculate chi square from a contingency table is to *create the summary table directly* into the SPSS spreadsheet rather than using the Crosstabs menus with raw data, like GSS data as we did earlier. We will still end up using Crosstabs and interpreting the output in the same way, but this alternative procedure allows you to simplify the data file prior to conducting the analyses by using a "weight cases" specification. Figure 7.6 shows the spreadsheet that we can create directly into SPSS if we know the data. This allows us to skip the process of reading the entire database (e.g., the entire GSS file) into SPSS before we use the Crosstabs procedure.

As you can see in Figure 7.6, the contingency table cells are identified in the SPSS spreadsheet by numbers corresponding to column and row position in a separate variable that we have called "Number." Thus, in the Workautonomy variable, 1 identifies the first column and 2 signifies the second column; the rows are identified by 1, 2, and 3 in the Health variable. This arrangement identifies that, for example, 572 respondents indicated high work autonomy and excellent or good health (first column and first row). In the same fashion, 54 respondents identified low work autonomy and fair health (second column, second row). The other values are similarly identified by column and row numbers.

Once created in the SPSS spreadsheet format, you can choose the "Data—Weight Cases" menu, found in the main SPSS menu ribbon at the top of the spreadsheet. Figure 7.7 show how this appears.

When you select "weight cases," the separate menu window appears that allows you to direct SPSS to weight the cases appropriately. Figure 7.8 shows this separate menu window.

This procedure calls for SPSS to consider 1161 cases (i.e., respondents) corresponding to the breakdown of the table cells as we earlier specified (with Workautonomy as 1 and 2, and Health as 1, 2, and 3). This is simply a shortcut method to avoid entering all 1161 cases of data or reading the entire data file into SPSS. As you can

	Workautonomy	Health	Number	var	var	var	var	var	var
1	1	1	572						
2	1	2	291						
3	1	3	153						
4	2	1	64						
5	2	2	54						
6	2	3	27						

Figure 7.6. The SPSS data table for the Crosstabs procedure.

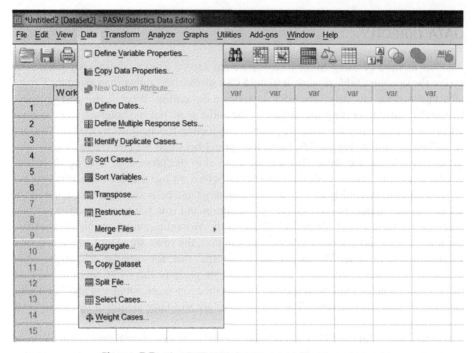

Figure 7.7. The SPSS Weight Cases specification menu.

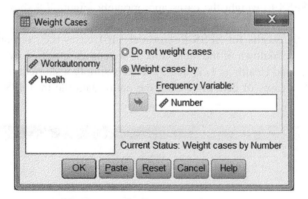

Figure 7.8. The Weight Cases menu window.

see, we called for the program to use the "Number" variable as a way to virtually recreate the raw score data file.

Once you make this choice, you can use the Crosstabs menus to create the same chi-square output described earlier (in Figures 7.2 through 7.4). This method is much easier if you have a large sample of data to input.

CONCLUDING COMMENTS

This concludes our introduction to chi-square and contingency analysis. You will see contingency tables and chi-square results used in examples throughout the remaining chapters of the book because they are among the most versatile statistical procedures available. Chi square can be used with any level of data, both in descriptive and inferential analyses, and is a straightforward way to represent patterns in data.

8

LEARNING FROM POPULATIONS: CENSUSES AND SAMPLES

Social scientists are often concerned with *populations*. One of the hallmarks of social research is precision; researchers must be very precise when defining concepts. Unfortunately, for many of the concepts we use there is a popular usage of terms that mistakes the very precise definition scientists must use. Generally when using the word *population,* what comes to mind is *the* population, meaning the population of a nation (and more specifically *our* nation). When social scientists use the word *population,* the population may not include any people at all. A scientific **population** contains all units of a set (sometimes referred to as a universe). So if we're interested in the population of the United States, every single person in the United States is part of the set/universe. If we are interested in how fourth grade teachers in the state of Washington prepare their students for the state exam, all fourth grade teachers in the state of Washington are the population. If we are interested in how themes in pop songs from the 1960s differ from pop songs of the 2000s, then all pop songs from both decades would be the population of interest (of course, we would also need to define what counts as a "pop" song before we would want to collect songs from either decade). Researchers determine a population based on their research question and design. Depending on their goals or the size of the population, researchers can choose to collect data from a *census* or a *sample*.

Understanding and Applying Research Design, First Edition. Martin Lee Abbott and Jennifer McKinney.
© 2013 John Wiley & Sons, Inc. Published 2013 by John Wiley & Sons, Inc.

CENSUSES

A **census** is an official count of a whole population (all units in the set) and recording of certain information about each unit. For example, the most famous American census is the U.S. Census, which attempts to enumerate every resident in the United States every 10 years (a decennial census). The 2010 U.S. Census counted 308,745,538 Americans and cost $13 billion (U.S. Census Bureau 2010). As shown in Figure 8.1, the 2010 Census asked just 10 questions.

As you might guess, it is difficult to count more than 300 million people. In the previous national census, the 2000 U.S. Census, Americans were undercounted by more than 3 million people (PricewaterhouseCoopers 2001). Unfortunately, those who do not get counted are not missed randomly; those who are undercounted are not representative of the nation as a whole. The poor, as well as people of color and immigrants, are often undercounted. To correct for this bias, the 2010 Census recruited places of worship, charities, and community organizations to try to get marginalized populations to fill out the census form (*The Economist* 2010). The Census Bureau also embarked on a marketing campaign using celebrities and high-profile events like the Super Bowl to inform the American public of the arrival of the census form on April 1, 2010. These efforts helped the Census Bureau to obtain a 74 percent response rate by mail, saving $1.6 billion of the operational budget. Another 22 percent of forms not returned by mail were collected through interviews, an impressive return/response rate (U.S. Census Bureau 2010).

Censuses do not have to encompass a nation's entire population and can be quite small. College professors have a census of their classes by way of their class rosters. A class roster enumerates all the members of the class, giving the professor information on name, class standing, major, e-mail address, and so on. Any enumeration of all the units in a set results in a census.

There are some limitations to doing a census. As the set of units in your population increases, it becomes more difficult and more costly to count all the units. The 2000 U.S. Census cost $3 billion, counting 281,421,906. Adding 10 years and 27,323,632 cost an additional $10 billion. Social scientists have been lobbying Congress for several years to move to a more cost-effective and more accurate count, using a random *sample* of Americans, rather than a census. Unfortunately, many people—even congressional representatives—do not fully understand the logic of counting a smaller *sample* of the population and how that can actually correct for those who are undercounted and give an accurate picture of the whole population. Like the word *population,* the words *random* and *sample* have taken on popularized meanings apart from the precise definitions used by social scientists. How often have you said to someone, "The most random thing happened to me"? Usually what we mean is that something odd or weird happened; in fact, whatever happened that you thought "random" was probably anything but random.

SAMPLES

In contrast to a census, a **sample** is a subset of units taken from the population. Samples are typically done in studies for practical reasons. It is unusual for researchers to have

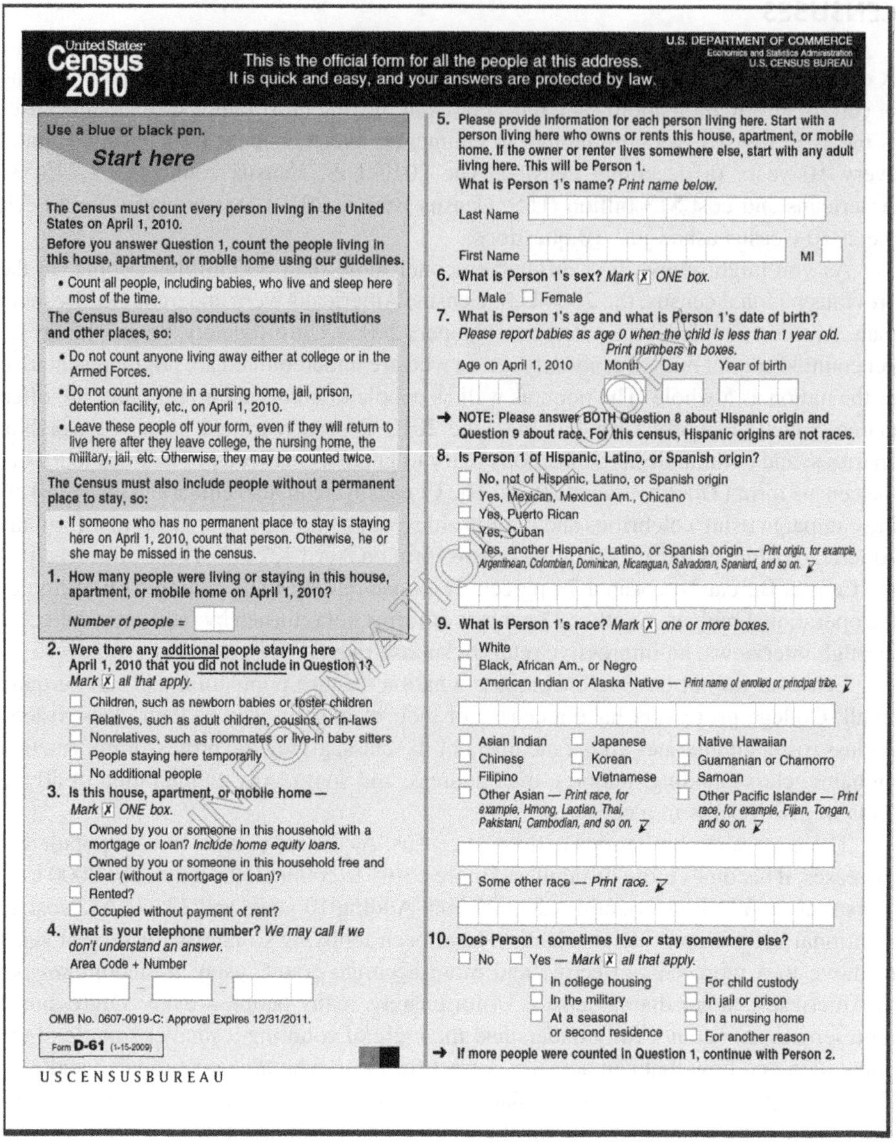

United States Census 2010

This is the official form for all the people at this address.
It is quick and easy, and your answers are protected by law.

U.S. DEPARTMENT OF COMMERCE
Economics and Statistics Administration
U.S. CENSUS BUREAU

Use a blue or black pen.

Start here

The Census must count every person living in the United States on April 1, 2010.

Before you answer Question 1, count the people living in this house, apartment, or mobile home using our guidelines.

- Count all people, including babies, who live and sleep here most of the time.

The Census Bureau also conducts counts in institutions and other places, so:

- Do not count anyone living away either at college or in the Armed Forces.
- Do not count anyone in a nursing home, jail, prison, detention facility, etc., on April 1, 2010.
- Leave these people off your form, even if they will return to live here after they leave college, the nursing home, the military, jail, etc. Otherwise, they may be counted twice.

The Census must also include people without a permanent place to stay, so:

- If someone who has no permanent place to stay is staying here on April 1, 2010, count that person. Otherwise, he or she may be missed in the census.

1. How many people were living or staying in this house, apartment, or mobile home on April 1, 2010?

Number of people = ☐

2. Were there any additional people staying here April 1, 2010 that you did not include in Question 1? *Mark ☒ all that apply.*

☐ Children, such as newborn babies or foster children
☐ Relatives, such as adult children, cousins, or in-laws
☐ Nonrelatives, such as roommates or live-in baby sitters
☐ People staying here temporarily
☐ No additional people

3. Is this house, apartment, or mobile home — *Mark ☒ ONE box.*

☐ Owned by you or someone in this household with a mortgage or loan? *Include home equity loans.*
☐ Owned by you or someone in this household free and clear (without a mortgage or loan)?
☐ Rented?
☐ Occupied without payment of rent?

4. What is your telephone number? *We may call if we don't understand an answer.*

Area Code + Number

☐ - ☐ - ☐

OMB No. 0607-0919-C: Approval Expires 12/31/2011.

Form D-61 (1-15-2009)

5. Please provide information for each person living here. Start with a person living here who owns or rents this house, apartment, or mobile home. If the owner or renter lives somewhere else, start with any adult living here. This will be Person 1.

What is Person 1's name? *Print name below.*

Last Name ☐

First Name ☐ MI ☐

6. What is Person 1's sex? *Mark ☒ ONE box.*

☐ Male ☐ Female

7. What is Person 1's age and what is Person 1's date of birth? *Please report babies as age 0 when the child is less than 1 year old.*

Print numbers in boxes.

Age on April 1, 2010 | Month | Day | Year of birth

→ NOTE: Please answer BOTH Question 8 about Hispanic origin and Question 9 about race. For this census, Hispanic origins are not races.

8. Is Person 1 of Hispanic, Latino, or Spanish origin?

☐ No, not of Hispanic, Latino, or Spanish origin
☐ Yes, Mexican, Mexican Am., Chicano
☐ Yes, Puerto Rican
☐ Yes, Cuban
☐ Yes, another Hispanic, Latino, or Spanish origin — *Print origin, for example, Argentinean, Colombian, Dominican, Nicaraguan, Salvadoran, Spaniard, and so on.*

9. What is Person 1's race? *Mark ☒ one or more boxes.*

☐ White
☐ Black, African Am., or Negro
☐ American Indian or Alaska Native — *Print name of enrolled or principal tribe.*

☐ Asian Indian ☐ Japanese ☐ Native Hawaiian
☐ Chinese ☐ Korean ☐ Guamanian or Chamorro
☐ Filipino ☐ Vietnamese ☐ Samoan
☐ Other Asian — *Print race, for example, Hmong, Laotian, Thai, Pakistani, Cambodian, and so on.*
☐ Other Pacific Islander — *Print race, for example, Fijian, Tongan, and so on.*

☐ Some other race — *Print race.*

10. Does Person 1 sometimes live or stay somewhere else?

☐ No ☐ Yes — *Mark ☒ all that apply.*

☐ In college housing ☐ For child custody
☐ In the military ☐ In jail or prison
☐ At a seasonal or second residence ☐ In a nursing home
☐ For another reason

→ If more people were counted in Question 1, continue with Person 2.

USCENSUSBUREAU

Figure 8.1. The Census 2010 U.S. census form.[1]

the time and money to measure every unit of a population (census). Therefore, researchers collect a random proportion of the population to serve as the study group and assume that the findings from this sample reflect the same values in the population.

Samples can be drawn using procedures that ensure they are **representative** of the population—representative means that the characteristics of the sample *represent*

[1] The U.S. 2010 Census form is reproduced here by permission.

the overall characteristics of the whole population. By using random (representative) samples, researchers can reach conclusions about the entire population based on the sample study results (this is a probability sample drawn in such a way that each unit has an equal chance of being chosen). For example, we may not be able to measure every student's academic achievement in the state of Washington, but we can select a smaller group of students in Washington (a sample) to measure their academic achievement and then generalize the results to all students in Washington based on those in the study.

It goes without saying (but we are going to say it anyway) that the way in which the sample is created has a huge impact on whether the sample values reflect the same values as those in the population (representativeness). In some cases, samples are drawn using procedures that do not result in representativeness. These "sampling" procedures, while sometimes necessary due to the nature of the study, may distort study findings because they do not reflect everyone in the population. In the following sections, we examine several ways in which samples can be created in the attempt to ensure representativeness.

PROBABILITY SAMPLING

To provide a useful description of the total population, a sample of individuals from a population must contain essentially the same variations that exists in the population. What does that mean? For example, if the population you're interested in is composed of 50 percent women and 50 percent men, we would expect a random sample to closely match those percentages. The sample must be selected using **probability sampling,** *where every unit in the set (population) has an equal probability to be selected for the sample,* or the sample cannot be used as a representative sample (any data gleaned from a nonprobability sample is not generalizable to the population being studied). Taking nonprobability "samples" can be problematic if the intent is to generalize to a larger population. There are some reasons why you would not take a random sample, for example when doing research on unusual populations (members of new religious movements), populations that would be difficult or impossible to randomly sample (street youth), or when you are doing exploratory research (talking with people before you have a sense of any population). But generalizing from a nonprobability "sample" violates the assumptions of probability statistics and can mislead us about the nature of a population.

A probability sample is representative of the population from which it is selected because within a probability sample all members of the population have an equal chance of being chosen for the sample. Because probability samples rest on mathematical probability theory, we are able to calculate the accuracy of the sample. Even though a sample may not exactly mirror the characteristics of the population (for example, we may get 53 percent women and 47 percent men), probability sampling ensures that samples are representative of the population we are studying as a whole. Probability samples also eliminate conscious and unconscious biases because a representative probability sample approximates the aggregate characteristics within the population.

TYPES OF PROBABILITY SAMPLES

Sampling is the process of selecting observations. How these observations are chosen is important in social scientific research. A probability sample means that all units in the set (the population) have an equal chance (or at least a known probability) of being selected for the sample. As random/probability sampling is the backbone for scientific research dealing with describing populations (statistical generalization), it is important to think about the research question and design in choosing a random sampling technique. We present six probability samples, each contributing uniquely to selecting samples that make the most sense for the type of population being used and the type of data being collected.

Simple Random Sample

The **simple random sample** (SRS) is the basic probability sample. If you've ever seen someone pull a name from a hat, or watched a late-night lottery drawing, or played Bingo, then you've seen a simple random sample: all of the units from the population are included and have an equal chance of being selected. In Bingo, each letter and number is represented on balls that are then put into a container. As each ball is pulled out of the container and called, each ball has been selected at random—meaning that each had an equal chance of being chosen.

For smaller populations it is easy to select units/observations from a simple random sample. As populations get larger, however, this technique becomes more onerous. Imagine taking a 10 percent sample of a 10,000-student university population. In order to take a sample of 1000 students, you would first need to find a **sampling frame,** or a comprehensive list of the units that make up the university's student population. Generally the registrar or the financial aid offices keep current lists of all enrolled students. These lists would be examples of sampling frames. If you take a printed list (sampling frame) and cut out all 10,000 student names in exactly the same way (so they are all equal in size), put them all in one big hat and draw out 1000 names (a 10 percent sample), you will have your random sample of university students and possibly a lot of paper cuts.

To avoid the paper cuts, you can assume that with the appropriate sampling frame, each student would coincide with a number (students 1 to 10,000). Another way to take a random sample would be to employ an electronic random number generator, like Random.org. Random.org uses true random number generation and is easy (and free) to use for most purposes. If you use the integer generator from Random.org to select a 10 percent (1000 people/units) sample from 10,000, what do you need to know? First, you need to plug in the total number of numbers you need (1000). Next, the program asks for the range of numbers needed. If the university has 10,000 students, the range would be 1 to 10,000. Figure 8.2 shows a screen shot of Random.org.

Click on "Get Numbers," and the screen will fill with 1000 integers all between the range of 1 to 10,000. Figure 8.3 shows the screen shot results for this specification. This is much simpler than drawing 10,000 names out of a hat.

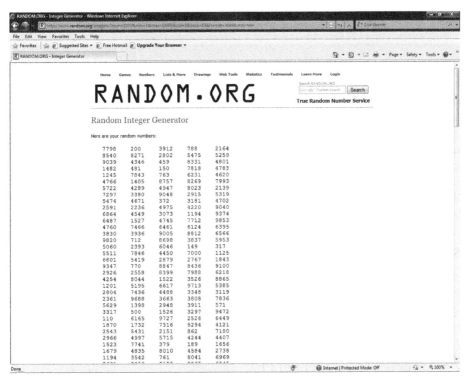

Figure 8.2. The Web site screen for Random.org.

Figure 8.3. The list of random numbers chosen from Random.org.

Figure 8.4. Selection of a smaller sample using Random.org.

There is one problematic issue with a random integer generator, however. We change our example to select 50 numbers from a range of 1 to 100. Figure 8.4 shows this result.

Do you notice anything about the distribution of numbers in Figure 8.4? It is likely that when choosing *random* numbers from a random number generator that integers will be repeated (like the numbers 6, 67, and 26 shown in Figure 8.4). That is problematic because you've set the parameters of the generator to match your needs. Since you will never know how many numbers will be repeated and if it will negatively impact your random selection, there is another way to use probability sampling that is helpful as populations increase.

Systematic Random Sampling

All probability samples use randomness as the basis for selecting observations. A **systematic random sample** does the same using a two-step process. Once you have a sampling frame—a list with all of the units in the population—the very first case/unit is selected randomly. Refer back to our example of taking a 10 percent sample of a 10,000-member university student body. Our sampling frame from the registrar's office lists all currently enrolled students. Thinking that each name on the list corresponds to a number (from 1 to 10,000), we can use a random number generator like Random.org

Figure 8.5a. Using Random.org to specify the first case of a sample.

to select the first case. This is shown in Figure 8.5a. Using the integer generator (because we are selecting only one random number, there will be no problem with duplicate integers) to select one random number between 1 and 10,000, say we get the number 5398.

Person 5398 on our list shown in Figure 8.5b becomes our random start. The second step in the process is to calculate a *sampling fraction*. A **sampling fraction** is calculated by taking the total number of the population and dividing it by the number of units/people needed for the sample: *n*.

$$\frac{\text{Total population}}{\text{Number needed in sample}} = n \quad \text{or} \quad \frac{10,000 \text{ university students}}{1000 \text{ students for the sample}} = 10$$

By dividing the total population (10,000) by our sample size (1000), we compute 10 as our sampling fraction. Since a random number (5398) has been generated for the starting number, each subsequent number is random. We begin by selecting person 5398 from the list for our sample and then take the sampling fraction to select every *n*th (or *10*th) case until the end of the list is reached. Once 5398 is the selected person, 5408 is selected next, then persons 5418, 5428, 5438, and so on. Once we've hit person 10,000, we go back to the beginning of the list to select every *10*th person until we get

Figure 8.5b. The first case in a systematic sample.

back to person 5398. (Since we started with number 5398, we have to make sure to systematically sample the cases "in front of" number 5398.)

Using SPSS to Obtain a Random Sample. Another way to obtain a random sample is to use the random selection process in SPSS. The procedure uses a process of nonreplacement so that the same case is not chosen twice. This means that once a case has been chosen, the remaining cases are selected from the group that remains (i.e., the number in the population minus the chosen case). Figure 8.6 shows the specification window.

Figure 8.7 shows the additional specification windows that appear once the SPSS user has requested "Select cases" from the "Data Analysis" window. As you can see in the example, we asked for "Exactly 50 cases" from the total number of cases in the population (1221).

Telephone Polls (Random-Digit Dialing)

One of the ways social scientists conduct efficient probability samples is through the use of *random-digit dialing*. Most households have one telephone line given a unique 10-digit code (we called these telephone numbers). The 10-digit codes used as tele-

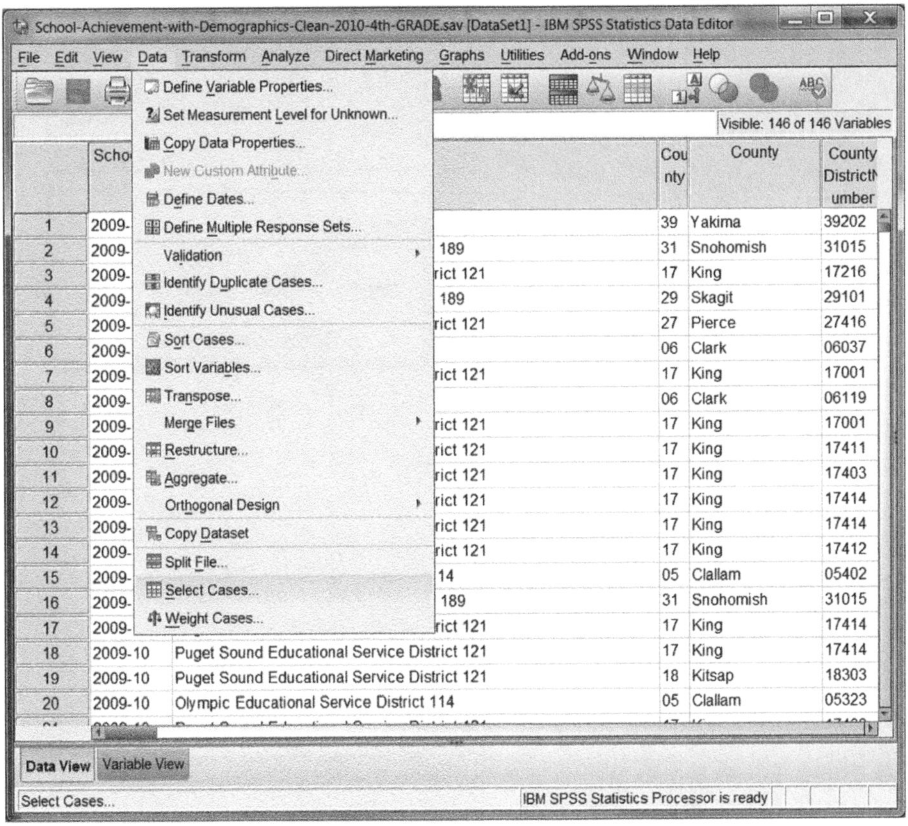

Figure 8.6. The Selection window in SPSS that provides sampling.

phone numbers are more complex than just assigning households a unique identifier. The phone number is based in three parts: the area code, the exchange, and the four-digit random number. Together, these three parts create a number identifying roughly 96 percent of all American households. These numbers are also useful because both the area code and the exchange code are geographically based. Random-digit dialing allows scientists to program into software the area codes and exchanges that were needed to reach the appropriate sample populations for their research. For example, if we wanted to survey Seattle residents on their attitudes toward green energy initiatives, we could program the first three digits of the phone number to Seattle's 206 area code. We could even program specific neighborhood exchanges to pinpoint areas even more precisely.

Using random-digit dialing gives researchers access to any household/number. Unlisted telephone numbers are equally likely to be dialed; therefore biases related to who has listed and unlisted numbers are eliminated. With the increasing use of cell phones, however, this process becomes more complicated. When researchers employ probability sampling, they have to make sure that every unit in the population has

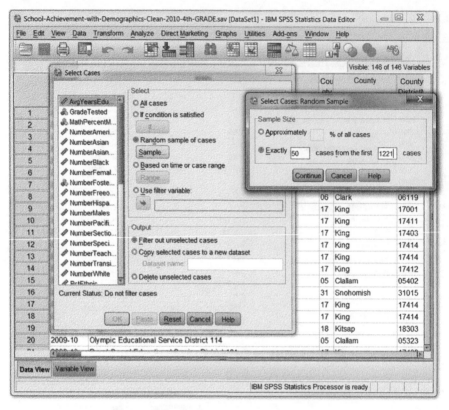

Figure 8.7. The Random Sample menus in SPSS.

an equal chance of being chosen for the representative sample. In households that have multiple land lines and/or cell phones, the risk of duplicating people or households increases, impacting the representativeness of the sample.

Stratified Random Samples

As populations become larger and research questions more complex, probability sampling has evolved to take population characteristics into account. **Stratified random sampling** uses probability theory in a somewhat more complex way. The word *strata* means "layer." You may think of stratification, or the rankings/layers of a society (e.g., richest to poorest, oldest to youngest, etc.). Depending on your research question, you may need to take this more complex approach. For example, if we are interested in how physics gets taught at the secondary level, we could do a probability sample of all secondary physics teachers in the United States. As we think about physics as a discipline, however, we might realize that it has an unbalanced gender ratio—men are more likely to teach physics than women. If we do a systematic random sample (or a simple random sample), we will have a representative sample of physics teachers. But what if our theory or previous research tells us that male physics teachers teach differently from female physics teachers? If physics teachers are 68 percent male and 32 percent female, we may not have enough variation between male and female to use sex as a variable. What if our random sample underestimates the percentage of female physics teachers? If sex is an integral variable to our research question, we need to make sure that we have enough variation between female and male to be able to use sex as a variable. Since significantly more physics teachers are male, we run the risk of getting more male respondents in our random sample and not enough female respondents.

Knowing that sex in an important piece of the research, we may want to employ a stratified random sample. A stratified random sample evaluates among the population the categories of a characteristic (stratum) that is important to the research. For our example, sex is a variable given two categories—female and male. In order to stratify by sex, we would need to know the proportion of the stratum in the population. Based on a report by the American Institute of Physics (White and Tesfaye 2010), we know that 68 percent of those teaching physics are male and 32 percent are female. When we use stratified random sampling, we sample with the same proportion of strata as the true value (parameter), eliminating statistical (random) error.

Rather than sampling the entire population, we would first create a sampling frame for each category of the variable for which we are stratifying. In our example, the variable for sex has two categories, so we would need two sampling frames—one for female physics teachers and one for male physics teachers. Once we have each group, we would randomly sample from each. Since each physics teacher has an equal probability of being chosen for the sample (because all are accounted for within the sampling frames), the sample is representative of physics teachers as a whole and can be generalized to the larger population of physics teachers. Stratifying a sample ensures that the sample is representative of the population as a whole while at the same time ensuring that the sample contains the appropriate variation for important variables.

Oversampling

In the stratified sample discussed in the text, representativeness is generated through using appropriate proportions within layers of the population. Sometimes, there are reasons why researchers may be concerned about certain segments of the population being represented in a sample even if it is representative. For example, suppose a researcher is conducting a worker discrimination study but is only able to create a small sample size. Suppose further that there may be a very small segment of the population with known discrimination experience and concern (i.e., those with specific physical disabilities). Gathering a stratified sample would yield a representative sample of the population, but it might only capture a very few of the respondents who would be able to speak specifically to the reason for the study. In this case, the researcher might *deliberately sample a greater proportion of this type of respondent to ensure their contribution to the study*. It is important for the researcher to describe this decision in the study, and especially to discuss the impact of the oversample on any statistical conclusions drawn from the analyses.

Cluster Sampling

Cluster sampling is another two-step sampling process. What is unique about **cluster sampling** is that rather than beginning with individual units, we begin with aggregate units. By randomly selecting units, we can then survey the individual units within the randomly selected clusters and still have a random sample of individuals. Since each cluster has an equal chance of being chosen, by default, the individual units within the clusters also have an equal chance of being chosen, ensuring randomness.

For example, say that the United Methodist Church wanted to survey members on their attitudes toward politics. It is very difficult for the national denomination to maintain an accurate list of all United Methodists in the country, since people join congregations that are affiliated with the denomination, instead of signing up with the denomination. What the national denomination does have, however, is an up-to-date list of all of their church congregations. Using the sampling frame made up of all United Methodist churches in the country, researchers can randomly sample each church congregation. Once the churches have been randomly selected, it is easier to obtain the membership lists from each selected church, getting updated addresses for members and then sending each member a survey. Even though you are surveying people from the same congregation, as long as the congregation was chosen as a random cluster, the results of the survey will be representative of United Methodists as a whole—all members, via their congregation, have an equal chance of being selected. That is the key. There is an issue, however, of representativeness if your sample of churches includes several megachurches. Because megachurches have a disproportionate number of members (and they are not distributed randomly), the sample could be skewed. In order to correct for this, there is a sampling procedure that takes cluster size into account: probability proportional to size.

Probability Proportional to Size

When drawing from clusters, we cannot assume all clusters are the same size. This will impact the random sampling. Methodologists have taken this into account with probability proportional to size sampling. **Probability proportional to size** (PPS) sampling allows researchers to equalize clusters, making sure each cluster has approximately the same number of individual units, ensuring true random sampling. With PPS, a sampling fraction is calculated based on the size of the cluster relative to the size of the population. As with general cluster sampling, each cluster has the same probability of being included in the sample.

A good example of PPS sampling comes from the General Social Survey (GSS), which we have been using. Conducted by the National Opinion Research Center (NORC) in Chicago (see GSS 2010), the GSS is a semiannual survey that samples noninstitutionalized people living in the United States who are 18 years old and older. Based on U.S. Census data, PPS sampling draws from the full national population using probability statistics to obtain a sample of approximately 3000 people using geographic clusters (since it is nearly impossible to create a sampling frame of 308 million individual units). NORC uses a multistage PPS procedure as an advanced cluster sample (in effect, the United States is broken into geographic blocks that represent similar-size clusters). NORC divides the United States into primary sampling units (PSUs) based on counties. Each county has a total number of households, as determined by the most recent census. Counties with less than 2000 households are merged with adjacent counties to create similar-size PSUs. PSUs are randomly sampled, and then interviewers travel from the first dwelling unit in the block beginning at the northwest corner of the block and proceed in a specified direction until they have completed the block quota (canvassing the block finding the appropriate proportion of people, as designated by the census tract data). Therefore, a region, nation, and other very large areas can be broken into clusters with similar numbers of individuals so each cluster—and thus individual—has an equal probability to be chosen for the sample.

SAMPLING AND STATISTICS

Probability sampling not only allows us to collect representative data of populations, it also allows us to calculate how close we are to the actual (true) characteristics of the entire population. When we measure the "true" value of a variable within the population, we are measuring the parameter. The **parameter** is a "fixed" value, meaning that in the population there is one true value for that variable at that moment in time. Currently, the percentage of women to men in the United States is 51 percent to 49 percent, respectively. These percentages would be the true, or fixed value of women to men; they represent the parameters of the variable for gender. When taking a representative sample of the population, we are likely to slightly overestimate or underestimate the percentages of women and men (in other words, get a slightly higher or lower percentage of women and men than the parameter). We expect this since sampling deals with what we observe, or statistics.

Statistics are the observed values of a variable within a sample of the population. For example, the 2010 Census confirms that the consistent parameter (true value) of men and women in the United States is 49 and 51 percent, respectively. Looking at two probability samples of Americans gives us slightly different statistics—or observed values. The 2010 GSS sampled Americans finding 56.5 percent of their sample to be women and 43.5 percent men. The 2008–9 American National Election Study (ANES) counted 42.5 percent men and 57.5 percent women. Both surveys overestimate the percentage of women while underestimating the percentage of men in the population.

Again, getting statistics that slightly overestimate or underestimate the parameters of a variable is normal. When a sample includes a substantial proportion of the population, the statistic should be quite accurate: the larger the sample, the closer the statistic should be to the parameter. Sometimes we know the parameters, making it easy to compare to the statistics. Sometimes, however, we do not. For example, in national elections probability samples are taken regularly to determine the outcome of an election before the election happens. In the months leading up to a presidential campaign, a host of news agencies use random-digit dialing to collect data about voting preferences. Targeting approximately 1000 to 2500 adults over age 18, polls ask who you would vote for if the election were held that day.

Three organizations that polled Americans on their voting preferences on November 2, 2008, included Gallup, Diageo/Hotline Daily Tracker, and Reuters/C-SPAN/

	Gallup	Diageo/Hotline Daily Tracker	Reuters/C-SPAN/Zogby
McCain	44%	45%	44%
Obama	55%	50%	51%

Figure 8.8. 2008 preelection presidential poll statistics.

	Gallup +/- 2	Diageo/Hotline Daily Tracker +/- 3.5	Reuters/C-SPAN/Zogby +/- 2.8
McCain	44% (42%-46%)	45% (41.5%-48.5%)	44% (41.2%-46.8%)
Obama	55% (53%-57%)	50% (46.5%-53.5%)	51% (48.2%-53.8%)

Figure 8.9. 2008 preelection presidential poll statistics with confidence intervals.

Zogby. These three organizations reported the statistics (observed values) of the percentage of the popular vote expected to go to McCain and Obama for November 2, 2008, and found in Figure 8.8.

What's nice about the observed values (statistics) listed in Figure 8.8 is that we will find out how accurate they are when the poll results (parameter) come in on that first Tuesday of November (designated as Election Day). Before that day, however, we can calculate how close the statistics are to the parameters—even though we do not know yet what the true value/parameter will be. The magic of mathematical probability theory allows us to calculate a confidence interval and a confidence level.

The **confidence interval** is an estimated range of values, based on the sample that should contain the actual population value a certain proportion of the time. A **confidence level** is the probability that the parameter (i.e., the "true" value) falls in that range (the confidence interval). For example, whenever political pollsters report their results, they give a range of percentage points (e.g., "plus or minus 2 percentage points"). The 2008 preelection results given in Figure 8.8 were reported with various margins of error. The percentage points given help us to create the range (confidence interval) for which we expect the parameter to fall. We take the statistic and we add the margin of error points to the statistic (e.g., "plus 2 percentage points") and subtract the margin of error points (e.g., "minus 2 percentage points") to create the range within which we expect the parameter to fall. Figure 8.9 calculates the confidence intervals for each of the three organizational polls.

Thus, according to the example in Figure 8.9, a "95 percent confidence interval" would indicate which two percentage values that would contain the actual population

	Gallup +/- 2	Diageo/Hotline Daily Tracker +/- 3.5	Reuters/C- SPAN/Zogby +/- 2.8	**Parameter (actual percent of the popular vote received)**
McCain	44% (42%-46%)	45% (41.5%-48.5%)	44% (41.2%-46.8%)	**45.7%**
Obama	55% (53%-57%)	50% (46.5%-53.5%)	51% (48.2%-53.8%)	**52.9%**

Figure 8.10. 2008 preelection presidential poll statistics with confidence intervals and parameter.

opinion value 95 times out of 100. If the poll *predicted* McCain getting 44 percent of the popular vote, we would be 95 percent confident that the *actual* popular vote would fall between 42 percent and 46 percent. These values are shown in Figure 8.9 by the ±2 indicated as the margin of error in the Gallup Poll.

After Election Day we can evaluate how close the statistics (observed values) come to the parameter (the actual percentage of those who voted for McCain or Obama). Figure 8.10 reports the confidence intervals, as well as the *actual* percentage of the popular vote each candidate received.

Clearly the parameter (true value) of the percentage of the popular vote for McCain and Obama fall within each of the three confidence intervals calculated on the preelection poll statistics.

Sample Size

One of the most frequent questions we hear from researchers and students as they consider conducting a study is "How big should the sample be?" This sounds like a fairly straightforward question, and it is indeed a very important one. But the answer is complex.

The size of a sample should primarily be about its representativeness. As we have seen in our discussion, the purpose of a sample is to represent a population from which it was drawn. Even very small samples can do that if they are drawn using the appropriate procedures we have discussed (especially random sampling using probabilistic methods). If sampling follows these rules, then the answer to the size question is easy: the sample should be as large as possible! Although even small samples drawn with appropriate procedures can be representative, larger samples provide the additional advantage of making sure that even narrow segments of the population are chosen in greater number (on this point, see our discussion of oversampling).

We still do not offer a target number for size under even the best of conditions. Part of the consideration may be the limitations of time and cost. For

(Continued)

example, in some of our evaluation research work, we are called upon to sample large organizations (for membership information, etc.). Large samples may be best (if they are representative), but the costs of sending out materials (question-naires, etc.) may be prohibitive; thus smaller samples are better. The bottom line still applies: *if representative, the sample can be quite small and still be useful and valid.*

The size of a sample is related to how it is used in statistical analyses. This is the part of the answer to the original size question that gives people the most confusion. However, these considerations are the most important for determining sample size. Here are some of the relevant parts of these considerations.

Sample size can be used to improve the chances for a statistically significant finding. The best example of this is correlation. If a researcher uses a correlation analysis, they can successfully observe a significant finding if they have a correla-tion of 0.482 (at the 0.05 level) with a sample size of 15. However, if they have a sample size of 100, they only need a correlation of 0.195 to achieve statistical significance! The same dynamics are true for other statistical tests as well. (The reason why this is the case includes the simple matter of the formulas for these processes that have the N in the denominators.)

Sample size is related to statistical power. As we have stated elsewhere (Abbott 2011), statistical power represents the ability of a statistical analysis to detect a "true" finding. Therefore, the greater the power, the greater the likeli-hood that the research will provide meaningful results. Power is usually expressed as a proportion. For example, researchers want to observe a power level of 0.800 rather than 0.500, since the former indicates that the findings will more likely lead to a rejection of the null hypothesis (and therefore indicate a meaningful finding).

Sample size is related to "effect size." We have discussed effect size through-out this book in the statistical sections. Essentially, effect size is a measure of the impact of a finding, independent of its statistical significance. Thus, if we perform a correlation analysis, the effect size (r^2) indicates how much of the variance in the outcome variable is attributed to the predictor variable. In this case, the stronger the correlation (r), the larger the effect size, since more variance is explained in the outcome variable.

Sample size is related to the nature of the statistical procedure. As we dis-cussed earlier about correlation, the formulas for statistical procedures indi-cate the importance of varying sample sizes. Since each statistical procedure (e.g., correlation, t test, regression, chi square, etc.) uses different formulas to determine significance, the magnitude of sample sizes varies by the procedure a researcher uses.

The statistical determination of sample size involves sample size, effect size, and power considerations. The reason the overall question of sample size is so complex, therefore, is because it cannot be considered in isolation of the other elements. In effect, sample size is one part of a negotiation involving power and effect size. A formal consideration of sample size might proceed something like this: if I desire an outcome with 0.800 power, and a medium effect size (e.g., of 0.500 with a t test), we would need a sample size of about 80.

Cohen (1988) provides the most comprehensive discussion and tables for determining sample sizes using all the pertinent elements. In that work, you will find separate tables showing sample size according to (1) the statistical procedure, (2) desired effect size, and (3) power. If, for example, we only anticipate a low effect size at the same power level, we would need a much greater sample size to achieve statistical significance (e.g., with an effect size of 0.100 with the *t* test, and a power level of 0.800, we would need a sample size of over 2000.)

Actually, there is an easier way to check on power and sample size. There are several online sites that do it for you for free! Simply search for "Online power and sample size calculators," and you will find several that may be helpful.

So the answer to the original question is "it depends!" Sample size is a simple, but somewhat fuzzy concept when applied to real-world research. As you can see, it is related to the representativeness of the sample, as well as such considerations as power, effect size, and the statistical procedure that you anticipate using.

POTENTIAL BIASES IN PROBABILITY SAMPLES

We have tried to make clear that for scientific research to be done on large populations (small ones too), representative data are useful. The only way to obtain representative data is to use random sampling techniques—probability sampling; otherwise, regardless of how well a survey is written, the data obtained cannot be generalized to the population. Once a probability sample has been diligently selected, there are a few things that can impact the representativeness, and thus the generalizability, of the sample. These issues include nonresponse, selective availability, and areal bias.

Nonresponse

We've outlined the steps needed to select a representative sample of a population. Even the most diligent researchers, however, must monitor the amount of **nonresponse**—people who do not respond—within the sample. In any data collection there will be people who refuse to participate, people who are unavailable to participate, and/or people who are too ill to participate. These people are the "nonresponders." Researchers expect some people selected by probability sampling to simply not answer their phones, not return a survey, and so on. Unfortunately, if there is significant nonresponse, we cannot generalize the results of the data collection to the population because people who do not respond are not distributed randomly. Thus a high rate of nonresponse results in a *sample bias*. There are actually patterns or groups who are less likely to respond—younger people, males, residents of cities, conservatives, the very poor, and/ or the very wealthy.

Selective Availability

Somewhat akin to nonresponse is the problem of selective availability. Although they may be willing to respond to a survey, some people are just difficult to find, including those who are institutionalized and those who are simply less available, resulting in **selective availability.** Institutionalized populations—people in the military, people who are incarcerated, and people who are in college—are less available and therefore less likely to participate, which again, can negatively impact our ability to generalize findings to a whole population.

Areal Bias

Another type of bias that can hurt the generalizability of a probability sample is **areal bias,** a geographic bias that occurs because sampling is based on geographic units. Stark and Roberts (2002) outline some of the issues regarding areal bias in the GSS's primary sampling units (PSUs) from the 1990s. In selecting random units, none of them happened to fall in two regions: the western region or the Miami region. This means that the results of the GSS can be generalized to residents in America but cannot be used to generalize about certain subgroups including Mormons, cattle ranchers, and Cuban Americans. These specific groups are underrepresented in the data.

Response Rates

Researchers are constantly asked, "What is an appropriate *response rate* when I administer a questionnaire?" (a **response rate** is the rate at which respondents complete the survey). The short answer is "100 percent." Of course, no one expects to be able to capture the questionnaires from all the sample subjects to whom the questionnaire was sent. For a variety of reasons (e.g., disinterest, distrust, method of delivery, etc.), researchers must be content with as large a sample return as possible.

As we have repeatedly emphasized, a sample, when it is chosen with probabilistic methods, will yield a representative picture of the population from which it is selected. This means that, in order to capitalize on this representativeness, researchers must obtain information from each of the individuals in the sample group if they are conducting a study, the results of which they intend to generalize to the population.

The upshot of not obtaining information from all sample subjects is that the results will not be representative of the population. Thus, *the conclusions of a study will be compromised to the extent that not all subjects participate.* Response rate is therefore directly related to representativeness and to the confidence that one has in the meaningfulness of the study results.

Some researchers have suggested rules for return rate in order to bolster our confidence in study findings that are not based on 100 percent return. And these rules vary by the type of study (i.e., mailed questionnaires, phone interviews, online surveys, etc.). For example, self-administered (mailed) questionnaires might be expected to have a lower return rate than the completion rate for personal interviews. Babbie (2007) suggested 50 percent as adequate for survey researchers; Dillman (1999) suggests a structured method for obtaining higher response rates in mailed surveys. Recent experience with online surveys suggests response rates are much lower than these rates. Irrespective of the type of survey or the method by which it is delivered, if response rates are less than 100 percent, there will be an impact on the confidence of one's conclusions and the generalizability of the findings.

This does not mean that studies with less than 100 percent return are meaningless. Some studies (i.e., program evaluations) are undertaken to shed light on the nature of a company work group or a school-based initiative. Even if generalizability to wider populations is restricted, the study still may provide a look at a majority segment of the specific study group. For example, if we are interested in what our company employees think about a new policy, the results of a questionnaire (with less than full return) may not be generalizable to similar work settings, but the results may provide greater insight into our own workforce.

This latter circumstance is related to a case study method in which we confine the focus of our attention to a single instance or group rather than attempt to make a statement about the larger population on the basis of studying a sample. Our case study conclusions still may be compromised if we do not obtain a full return of survey materials, but it will provide a body of findings that can be combined with other findings to yield a better picture (the latter process is known as *triangulation*).

NONPROBABILITY "SAMPLES"

The previous examples of probability sampling are the only sampling techniques that allow researchers to *generalize* to a population. Unfortunately, we sometimes run into data that are collected using **nonprobability "samples"** to generalize to some population (often by well-meaning but ill-informed people). If random sampling—or probability sampling—is not used, then the data taken cannot be generalized to any population because the sample is not representative of any population. Just because we have data gathered from people within the population does not mean it is representative, and thus generalizable, to that population.

What if we are interested in how students think about the alcohol policy at our university? We develop a survey about the alcohol policy and then ask the students in our classrooms to complete it. We then take the data, analyze it, and report to our colleagues that our university's students believe that the alcohol policy is a good policy. What is the problem with doing this? Do the students in our classrooms represent all the university's students? They are students at our university. Do not they represent our students? The question to ask is "Did every student at our university have an equal opportunity to be chosen to take our survey?" If we only took our own students (even if we asked our faculty friends to give the survey to their students, too), then they are not representative of the university's students as a whole, and any data we get from them cannot be considered representative. The data may tell us something about the students who responded to the survey but does not represent the students at the university and therefore cannot be generalized to the university student population. In such cases, the findings of our research could only be generalized to the set of students who took the survey. Data that cannot be generalized to the target population are less meaningful than representative data because we do not know who is represented by the results of the data.

A classic example of this problem lies in the popular prediction of a winner for the 1936 presidential election between Alf Landon and Franklin Delano Roosevelt (FDR). In several earlier elections, the largest periodical of the day, *Literary Digest,* had accurately predicted the winners of presidential elections. *Literary Digest* created an ambitious method to determine the presidential election outcome. Sending 10 million ballots to Americans, the periodical received 2 million responses—a fair response rate. The results of the poll showed Alf Landon with 57 percent of the popular vote and FDR with 43 percent of the popular vote. It looked as if Landon would win in a landslide over FDR. When Election Day rolled around, however, Landon carried only two states and FDR won 61 percent of the popular vote. How could *Literary Digest*'s statistics have been so far off the parameters?

The first question to ask is how did *Literary Digest* get a list of addresses for 10 million Americans? What was their sampling frame? *Literary Digest* got lists of all automobile owners and all telephone subscribers in order to send out their ballots. Are these two groups representative of all Americans? Think about what you know about the country in 1936. The United States was in the midst of the Great Depression. Did most Americans have their own cars and/or telephones? In the midst of the Depression in the United States, only those considerably more wealthy than average received ballots. There is a correlation between wealth and voting Republican, which may explain the inflated statistics favoring Alf Landon. Obviously these groups were not representative of the population as a whole; thus the results of the survey were meaningless.

In their defense, *Literary Digest* did not use probability sampling. They were using some of the most up-to-date techniques of taking data, but probability sampling was not in common use at the time. What we know now is that even with a much smaller sample (for example, national election polls draw on just 1000 people from a population well over 300 million), as long as each unit in the population has an equal chance of being selected for the sample, a probability sample can accurately predict/describe the entire population.

Convenience Samples

As we've noted previously, social researchers use words very precisely. When discussing samples, there is an expectation that a sample is equal to a representative group of units from some larger population. In popular usage, however—sometimes creeping into the vocabulary of the sciences—the word *sample* is often used to refer simply to a group of units (most often a group of people who may or may not be members of the population of interest, but who are not necessarily representative of that population). Technically, groups of units or people from the population who have not been randomly selected are not a sample; they are just a group of people. Data taken from this type of group are generally not useful because the results of the data cannot be generalized to any population—even the one to which they may belong.

The most egregious type of a nonsample "sample" is the **convenience sample.** Just as the name implies, units are selected conveniently, drawing on those that are most available. Even when some of the units are found within a common population, if they are not chosen randomly, the units are not representative of the population (or what we call a relevant population). Because the units are not representative of a relevant population, the data taken from them is not generalizable to any relevant population.

A great example of this would be the late-night talk show host who walks outside of his studio to ask people questions. A couple of years ago one of these hosts asked for graduates in his audience to answer the question, "How many planets are there?" In the seven or eight clips that were shown on television, none of the graduates could accurately identify the number of planets (this example predates Pluto's demotion from a planet). The graduates' answers caused great hilarity for the audience. When the host came back into the studio, his response (and that of the audience) highlights the problem with a convenience sample. The late-night host noted, "Look at the state of American education!" The audience clapped raucously in agreement with him that graduates from American schools—including high school, college, and one graduate student—could not identify something simple like the number of planets in our solar system. The host was generalizing to all people with educations from American institutions. Did the people he interviewed on camera represent all people with an American education? No, but both he and the audience (without too much thought) generalized his findings to the whole population of everyone who has an American education. We do not expect professional comedians and television hosts to know or use probability sampling. We must also recognize that journalists and even some scientists not trained in taking reliable sampling data from people may not understand the principles behind probability sampling and generalizing from a group of people/units to any larger relevant population.

Quota Sampling

Another type of a so-called sample that has a veneer of scientific accuracy is quota sampling. **Quota sampling** is based on a *matrix* that describes certain characteristics of a target population. The key to quota sampling is to find a current and accurate accounting of these characteristics. Like stratified random sampling, quota matrices rely on up-to-date information regarding the proportion of demographic characteristics in a target population. For example, several malls have video stations where you may be asked to participate in a study showing movie trailers to get your reaction to the films. These stations are usually staffed by people with a clipboard or electronic tablet and a quota matrix. The matrix will have a breakdown of the characteristics that are important to the organization collecting data. For films, perhaps market researchers want to find out how a general population responds, so a matrix will target perhaps 100 people, broken into several categories. If 51 percent of a population is female, the data collection will ask for 51 females to view the film trailers (and then 49 males). Each sex category may also be further broken down into age groups. So of the 51 women who are supposed to view the film trailers, perhaps 20 percent should be under 18, 30 percent between 18 and 25, 40 percent between 26 and 45, and 20 percent over 45, each percentage representing the proportion of those age groups who attend movies. So the person stationed at the video booth would need to have 10.2 women who were under 18, 15.3 women who were 18 to 25 years old, 20.4 women 26 to 45 years old, and 10.2 percent of women who are over 45 years old. Of course getting those extra "point 2s" may be tricky.

From gender and age, the matrix could be further broken down into more categories, for example race. Once each category has been filled, the data are collated and taken back to their respective organizations, where decisions are made based on the results (which are not random, even though the data are from people who represent each category). Although quota sampling resembles probability sampling, it still is not a scientific substitute for probability sampling. It is difficult to find accurate and current information on the population proportions of characteristics, and again, *not everyone in those categories has an equal chance of being selected*, so they cannot represent any type of relevant population, and thus the findings cannot be generalized to any relevant population.

Snowball Sampling

The remaining type of nonprobability "sample" can be helpful when doing exploratory research—when you do not know enough about a population to undertake a full-scale data collection using a probability sample. Snowball sampling is also useful for learning from special populations that may be difficult to locate (e.g., drug users, illegal aliens, sex workers, etc.). Like the "snowball effect," **snowball sampling** occurs when researchers begin with one person of interest and ask that person to refer others from the population or who share the characteristics to be studied to the researcher. Although this type of group can give valuable insights, the data collected cannot be generalized to others—even those who share the same characteristics being studied.

For example, how do people decide to place their loved ones in a nursing home? Having no type of sampling frame that would identify people facing the decision to place loved ones in a nursing care facility, snowball sampling would be a good way to begin to collect data on this special population. Identifying support groups through hospitals or nursing care facilities may be a good way to begin to find people willing to be interviewed about making this decision. Once interviewed, those in the process of placing a loved one in nursing care mayknow of others experiencing a similar situation and choose to refer them to the researcher. Snowball sampling can be quite helpful in this type of research context.

Weighted Sample Data

There are several situations that present challenges to the researcher's ability to ensure representativeness. One such situation applies primarily to those studies in which data are taken from a group and used to represent an individual. For example, in large national studies (like the GSS), interviewers sample household units in their attempt to locate appropriate respondents. The difficulty is that there may be multiple people in the household unit, leading to the necessity to select only one for an interview. Technically, each of the household units are given an equal probability of being chosen, but within the household, if there are more than one potential respondent, each person has a lower probability of being chosen. The **weighted sample data** procedure may include an adjustment of this respondent's interview, since it is representative of more than one respondent in the unit.

Formulas for weighting are quite complex, but the idea is fairly straightforward: make the one interview you collected count for more than one when you conduct the statistical analysis of all respondents. In effect, we use weights to compensate for the unequal probabilities.

Exercise: Populations and Sampling

Now that you have had a chance to think about how researchers collect data by drawing samples of populations, let's explore the ARDA again to look at the ways data are sampled.

Go the www.theARDA.com and click on the tab for the Data Archive. The archive is broken into several categories of data including International Surveys and Data, U.S. Church Membership Data, U.S. Surveys, and Other Data. Click on the link for the Baylor Religion Survey. The first survey listed for Baylor is from 2005. Click on the link to that survey to go to the Summary page. There is a variety of information available from the summary page of the survey, including the number of cases and variables, Data Collection and Collection Procedures (summary pages often include a section for Sampling Procedures, as well).

As you read through the summary of the data collection, what does it tell you about the population targeted for the survey and how that population was sampled? How were the data collected? According to the summary, the target population was "a national sample" of adults age 18 and older (in the continental United States). These adults were sampled using random-digit telephone sampling (random-digit dialing). The data were collected using telephone interviews and self-administered mail surveys.

Find five additional data files in the ARDA's Data Archive. List the name of the data set, the target population, sampling procedure, and how the data were collected.

1. Data File: _____
   Population: _____
   _____

(Continued)

Sampling: _____

_____

Data Collection: _____

_____

2. Data File: _____
   Population: _____

   Sampling: _____

   _____

   Data Collection: _____

   _____

3. Data File: _____
   Population: _____

   Sampling: _____

   _____

   Data Collection: _____

   _____

4. Data File: _____
   Population: _____

   Sampling: _____

   _____

   Data Collection: _____

   _____

5. Data File: _____
   Population: _____

   Sampling: _____

   _____

   Data Collection: _____

   _____

9

CORRELATION[1]

In previous chapters, we discussed descriptive statistics as ways of illustrating patterns embedded in data. We continue that discussion by examining correlation, the statistical process that measures the extent to which the values of two (or more) variables are related or linked. Correlation processes also allow the researcher to understand spuriousness more clearly.

Most everyone has an understanding of the basic principles of correlation; it is somewhat intuitive. Technically, **correlation** is the statistical process of measuring how changes in two variables are related to one another. Thus, for example, we may observe that students who get the highest reading achievement test scores are also the ones who read the most. Or, stated another way, as the amount of time spent reading increases, so do the achievement scores (and vice versa). Thus, in this example, the values of both variables increase together, or covary.

Of course, as we have already learned, not every relationship is what it seems! There may be additional variables not taken into account in the analysis that give the original two variables the appearance of covarying. The examples showed how this

[1] Parts of this section are adapted from M. L. Abbott, *Understanding Educational Statistics Using Microsoft Excel® and SPSS®* (Wiley, 2011), by permission of the publisher.

Understanding and Applying Research Design, First Edition. Martin Lee Abbott and Jennifer McKinney.
© 2013 John Wiley & Sons, Inc. Published 2013 by John Wiley & Sons, Inc.

might happen on a theoretical level. By studying correlation statistically, we can document to the extent to which *spuriousness* is present among a set of study variables and help to debunk mistaken theoretical assumptions.

THE NATURE OF CORRELATION: EXPLORE AND PREDICT

Researchers use correlation to *explore* the relationships among a series of variables they suspect may be important to a research question. Other evaluators may use correlation to help *predict* an outcome knowing that the predictor and the outcome variables are related. Explanation and prediction are two important uses of correlation.

DIFFERENT MEASUREMENT VALUES

Correlation is somewhat unusual in that the researcher can measure the relationship between two variables that are operationally measured differently. For example, in the opening paragraph, we mentioned the relationship between reading achievement and amount of reading. If we operationally define *reading achievement* as the score on a state achievement test and *amount of reading* as the number of books read per month, we will end up with two different sorts of scores. The state reading test may represent a standardized score that ranges from 100 to 600, whereas the amount of reading may simply be the number of books a student reads in a month (the number will not be 100 to 600 but more typically 0 to 15, given the length of the book). Despite the difference in the scales of the variables, correlation can accurately detect whether *changes in one variable are linked to changes in the other variable*.

Correlation analyses are so powerful that a researcher can also calculate the correlation of two variables measured with different *levels of data* (i.e., interval, ordinal, or nominal). Thus a researcher might correlate reading achievement scale scores (interval level) with students' subjective appraisals of the extent of their reading (ordinal level) as their rating of "A lot, Some, A little, or None."

CORRELATION MEASURES

There are many ways to calculate correlation, since the procedures are customized to the nature of data from a researcher's study question. Thus, for example, there are correlation calculations used when variables are nominal, as we saw in Chapter 7 on chi square (i.e., Cramer's V and phi). In this section, we discuss the most common method of calculating correlation with interval-level variables.

Named after Karl Pearson, the **Pearson's correlation coefficient**, symbolized by r, is used to measure the relationship between two interval-level variables. We used the previous example of reading achievement and number of books read to show how this method can be quite versatile and helpful.

INTERPRETING THE PEARSON'S CORRELATION

Pearson's *r* is a calculated number that varies from −1.000 to +1.000. The closer the *r* value is to 0, the less the two variables are related to one another. Here are the two primary dimensions of Pearson's *r* that are helpful for interpreting the relationship:

* Strength: *The closer the* r *value gets to either −1 or +1, the stronger the correlation between two variables.* An *r* value of 1.000 would indicate that every time one variable increased by one unit, the second variable increases by one unit. It is also the case that a value of 1.000 would indicate that each time a variable decreases by one unit, the second variable also decreases by one unit. An example might be our achievement/time spent reading example.
* Direction: *When the variables change their values in the same direction, the* r *is a **positive correlation**. Whenever the variables change in opposite directions, the* r *value is negative.* Positive and negative do not mean good and bad; they simply indicate the direction of change in both variables. **Negative correlations** are also called *inverse correlations,* since one variable is going up as the other is going down in value. An example of this might be the relationship between achievement and the time spent watching TV.

An Example of Correlation

In this section, we use two interval-level variables to demonstrate how to calculate a correlation and how to use SPSS to analyze correlation data. We use a fictitious example to demonstrate how to calculate Pearson's *r*, and then we introduce an example using real data. The real-world data example uses aggregate data following from the discussion of aggregate data in a previous section.

Table 9.1 shows the fictitious data we will use to show how to calculate Pearson's correlation. The Reading Achievement variable consists of test scores from eight students on a reading test with 150 points possible. The SES variable is the socioeconomic measure (from a hypothetical index of family income and education, etc.) of each student on a scale from 0 (lowest) to 100 (highest measure). The research question is

TABLE 9.1. Fictitious Data for Correlation Example

Reading Achievement	SES
120	100
20	33
50	71
90	52
110	90
100	84
40	66
10	30

whether there is a correlation between the two variables. Do reading achievement levels change in the same direction as the SES student measures?

ASSUMPTIONS FOR CORRELATION

As with the other statistical procedures, there are assumptions that must be met before we can use Pearson's r correlation. The *primary* assumptions (there are others we discuss in later chapters) are as follows:

- Randomly chosen sample
- Variables are interval level (for Pearson's r)
- Variables are normally distributed
- Variances are equal. Pearson's r is robust for these violations unless one or both variables are significantly skewed.
- **Linear relationship**. The two variables must display a straight line when plotting their values on a scattergram. (Violations of this assumption might include **curvilinear** relationships in which plotted data appear to be in the form of a U.) Formally, we can detect these curvilinear relationships using SPSS.

We should note that correlation is a robust test, which means that it can provide meaningful results even if there are some slight violations of these assumptions. However, some are more important than others in this regard as we will see.

PLOTTING THE CORRELATION: THE SCATTERGRAM

In an earlier section discussing visual representations of data in descriptive statistics, we showed how to create histograms. These are useful graphs for showing the distribution of single variables used in research. When we use a correlation design that includes two variables, we can use another procedure variously known as a scatter diagram, scatterplot, scatter graph, or simply **scattergram**. These are visual graphs that show the relationship between two variables. In essence, they are like combining the histograms of two study variables.

Figure 9.1 shows the scattergram between our two example variables, reading achievement and SES. As you can see, the dots are displayed from the lower left side of the plot to the upper right side. This pattern indicates a positive correlation, since as one variable increases in value, the other value also increases.

Reading the plot is straightforward. The values in the table of data are presented in pairs, with each pair representing a single student's scores. Thus the top pair of values in Table 9.1 indicate that this student scored 120 on the reading achievement test and had an SES measure of 100. We filled in the dot in Figure 9.1 to show the pair of scores for the next-to-last student in Table 9.1 (the student scoring 40 on the reading test with SES = 66).

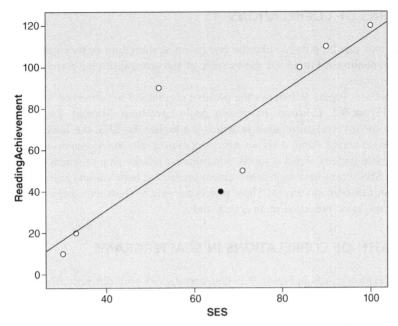

Figure 9.1. The scattergram between reading achievement and SES.

As you can see, the pairs of scores are entered into the plot simultaneously so that each dot represents the pair of scores of an individual student. Typically, the outcome variable is placed on the y axis and the predictor variable is placed on the x axis. The fact that there is a correlation between the two variables does not mean that SES *causes* achievement levels to be high. There may be many other variables (not included in the analysis) that would influence reading achievement. We discuss this further in the following sections.

The Direction of Correlation
In correlation designs it is not always apparent which variable is the outcome and which is the predictor. You simply have to understand which is which from the research question. For example, suppose you were studying the relationship between wages and work performance. Are higher wages related to better work performance (i.e., due to wages motivating workers to work harder), or does work performance lead to changes in salary levels (i.e., because the amount of work drives how much workers get paid)? It really depends on what you, as a researcher, are hypothesizing and what other, related, research has been conducted and reported in this area.

The research question in our hypothetical study is whether the SES of the student will influence their reading achievement (i.e., better prepare the student academically resulting in better performance). Therefore, SES is the predictor variable (X) in this study. However, could not the case be made that a student's academic performance, over time, results in higher social standing (as measured by SES)?

PATTERNS OF CORRELATIONS

Correlations can be positive like the one in our scattergram, or they can be negative and even nonlinear. Figure 9.2 shows some of the possibilities for correlation patterns in scattergrams.

Panel a in Figure 9.2 shows the positive correlation we observed with the actual data in Figure 9.1. However, the results could have been different. Panel b shows a negative (or inverse) correlation in which the higher the SES, the lower the reading achievement score! Panel c shows no correlation at all; the scores do not fall into a recognizable pattern. Panel d shows a curvilinear relationship in which students with medium SES scores have high achievement scores, but both low and high SES students have poor achievement scores! These panels are only to show the correlation possibilities that might be produced in an actual study.

STRENGTH OF CORRELATIONS IN SCATTERGRAMS

The correlation panels in Figure 9.2 show correlations with different directions or patterns. The dots could extend upward to the right (positive correlation), downward to

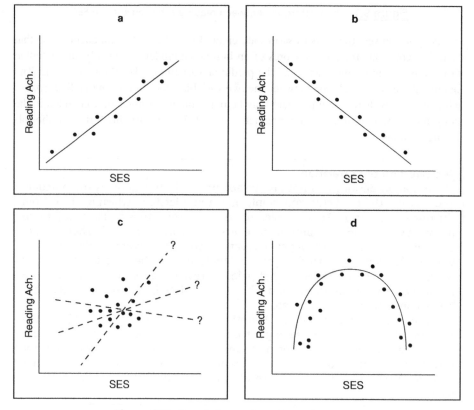

Figure 9.2. Correlation patterns in scattergrams.

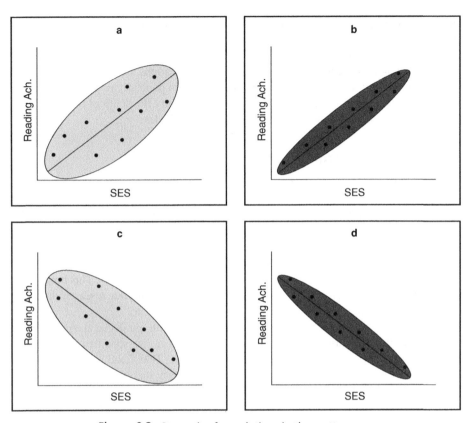

Figure 9.3. Strength of correlations in the scattergram.

the right (negative correlation), or in other patterns. Figure 9.3 shows scattergrams that indicate the *strength* of the correlation.

The top two panels (a and b) represent positive correlations. The dots are arrayed from bottom left to upper right. In scattergrams such as this, we use a line to represent the scatter and pattern of the dots. For now, we can look at the lines as a way of helping to show the direction of the scatter as well as how close each of the dots are to each other. We discuss this line, known as the **"line of best fit"** or "regression line," further in Chapter 10. It is created by a formula that is helpful for establishing predictions of outcome values with known predictor values.

Notice that the *extent of the scatter around the line* is different. When the dots have a wide scatter, like the scattergram in panel a, the correlation is weaker. This would indicate that as values of one variable (SES) increase, the values of the other also increase, but not one for one. Panel b shows a positive correlation with a tighter pattern of dots that indicate more of a one-for-one increase in the values of both variables. In this case (panel b), the correlation would be stronger.

Panels c and d of Figure 9.3 show negative (or *inverse*) correlations. Panel c shows that as one variable increases in value, the other decreases; however, the values do not change in a one-for-one relationship. The negative correlation in panel d shows a much stronger correlation, since the dots are very close to the line, indicating more of a one-to-one change in values.

The panels underscore the fact that negative correlations are not necessarily bad, since negative only refers to direction. As you can see in panel d, this negative correlation is very strong. We indicated strength of correlation by the depth of the shading in the panels, with stronger correlations showing darker shades.

Creating the Scattergram with SPSS

You can easily draw a scattergram freehand with pairs of data. However, SPSS provides a simple procedure to create the graphs.

SPSS has a straightforward way to create scattergrams using the main Graphs menu. Figure 9.4 shows how to create the scattergram through the Graph, Legacy Dialogs path of menu choices. Near the bottom of the list is Scatter/Dot, which will produce a specification box that allows you to design your scattergram.

When you choose Scatter/Dot, the menu box in Figure 9.5 appears. This box allows you to choose the scattergram that matches the complexity of your research data. For our fictitious example, the Simple Scatter choice is appropriate.

Choosing the Simple Scatter design from the menu produces a specification window like the one shown in Figure 9.6. As you can see, we specified that the SES variable should be placed on the x axis. We can select the "Titles" and "Options" buttons in the upper right corner of the window to further specify the graph, but we prefer to edit the scattergram once it is produced.

When you choose OK from the window in Figure 9.6, SPSS produces a graph in the output file like the one shown in Figure 9.1. If you double-click on the graph, you can make a series of edits using the available menu screens. (We added the line to the basic graph to produce the graph in Figure 9.1.)

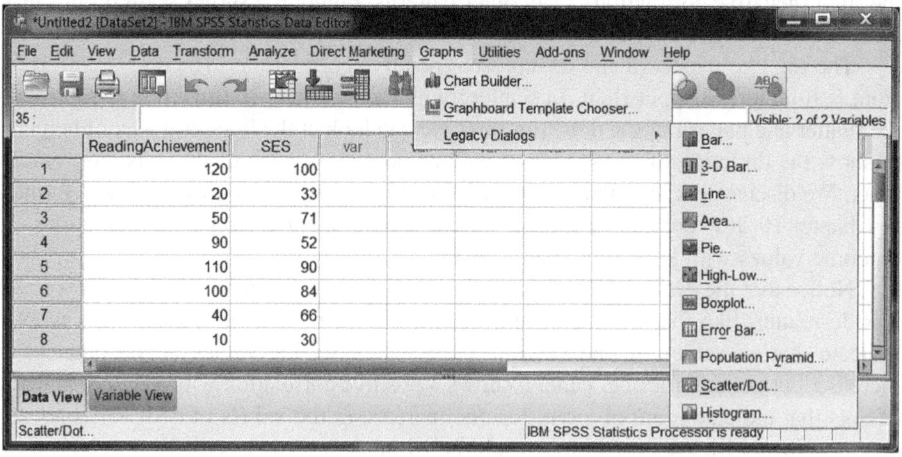

Figure 9.4. The SPSS Graph menu for creating scattergrams.

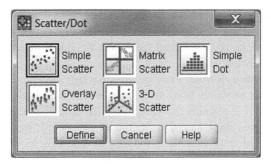

Figure 9.5. The Scatter/Dot menus.

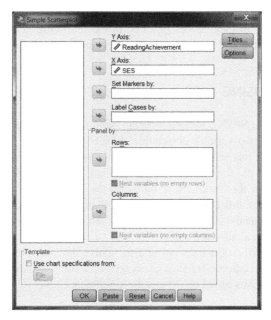

Figure 9.6. The scattergram specification window in SPSS.

EVALUATING PEARSON'S *r*

If a calculated Pearson's *r* is a number between −1.000 and +1.000, how can a researcher evaluate how strong it is? At what point does it become strong? For example, if we have an *r* = 0.400, is that considered large? How can we judge the strength of the relationship simply by looking at the scattergram?

There are several ways to answer these questions, but we focus on two: statistical evaluations and practical evaluations. *Statistical evaluations* typically rely on hypotheses tests and comparing calculated *r* values with table values of *r* that have been determined by statisticians to indicate various levels of meaningfulness. Thus we might determine, given a certain N size and with a certain degrees of freedom (which we discuss in later sections), our calculated value of 0.400 might be considered significant or not a result of chance alone.

Practical evaluation methods examine how much of an explanation of an outcome variable a correlation provides from the predictor variable. Typically, this is accomplished on the basis of explaining the variance in the outcome variable. We know that most all study variables show a certain amount of variance. If we know how much of the variance in the outcome variable is reduced by knowing its relationship with the predictor variable, we can speak of its "practical significance" or impact. This is the case because we are removing uncertainty (the extent of variation) as a result of understanding the outcome variable's relationship to the predictor variable.

To get an idea of how to evaluate r, we can first calculate the value of r using SPSS. Introductory statistics textbooks will show you how to calculate r from a set of raw data if you wish to explore this process. In the following example, we use SPSS to analyze the data in Table 9.1 on reading achievement and SES.

CORRELATION USING SPSS

Figure 9.7 shows the SPSS correlation menu. Note that it is called Bivariate because we are specifying a two-variable correlation (bi = two).

When you make this selection, you will get the menu shown in Figure 9.8. As you can see, we have moved the variables in the database (which normally are listed in the left-most window) to the Variables: window on the right side of the menu using the arrow button in the middle. Although there are many specifications a researcher can use to customize the analysis, you can take the default settings (Pearson shown checked, etc.) to arrive at the calculated r value.

Choosing OK will produce the correlation matrix shown in Figure 9.9. As you can see, the Pearson Correlation is 0.859. The correlation table also indicated N, the number of cases the analysis was based on (8) and the significance level (0.006) that we discuss in a later section.

INTERPRETING r: EFFECT SIZE

Returning to the matter of evaluating r, we can discuss whether 0.859 is strong or not by examining its explained variance. We refer to this as the **effect size** of the correlation. Thus, if the correlation is large, it explains more of the variation in the outcome variable, and therefore it has a greater effect size stemming from its relationship to the predictor variable.

With correlation, this effect size is known as the *coefficient of determination*, or r^2. *This value is simply the square of* r *and refers to the amount of variance in one variable explained by the other*. What this means is this: we can consider the fact that a distribution of scores (e.g., reading achievement test scores) vary a certain amount or

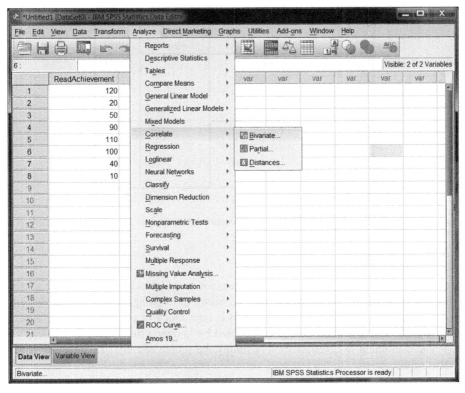

Figure 9.7. The SPSS correlation menu.

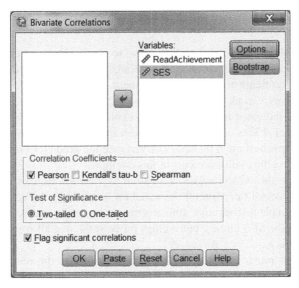

Figure 9.8. The correlation specification window.

Correlations

		ReadAchieve-ment	SES
ReadAchieve-ment	Pearson Correlation	1	.859**
	Sig. (2-tailed)		.006
	N	8	8
SES	Pearson Correlation	.859**	1
	Sig. (2-tailed)	.006	
	N	8	8

**. Correlation is significant at the 0.01 level (2-tailed).

Figure 9.9. The SPSS correlation matrix.

are spread out around a mean score. The question is why reading achievement scores vary. There are many potential reasons (e.g., nature of teaching, natural ability, reading experience, etc.). What we are interested in is how much of an impact on the understanding of reading achievement is contributed by knowing the SES of the student.

The coefficient of determination is simply the square of r (r^2). This amount is the extent of the variance in the outcome variable explained by the predictor. Thus, in our example where $r = 0.859$, 73.8 percent (0.859^2) of the variance in reading achievement is explained by SES. Stated differently, knowing the correlation between reading achievement and SES helps researchers to understand almost three fourths of the reason why there is so much variation in reading achievement. Obviously, this is an example with a limited number of cases, but it shows the process of how to interpret the strength of r.

Figure 9.10 shows how the variables relate to one another in terms of explaining variance. If you consider the variance of a variable visualized as the area of a circle, the top circle in the figure is reading achievement. If we could perfectly explain why reading achievement varied, we could say that we explained all the area in the circle (perhaps visually indicated by shading in the explained portion). The effect size of a correlation in effect shades in the amount of the variance in the outcome variable. That is, it shows how much of an explanation of variance the predictor variable contributes to understanding reading achievement, in this example.

Thus knowing the student's SES is a partial explanation of their reading achievement scores. SES doesn't explain all the variation in reading achievement, since there is still a lot of unexplained variance (the amount of the reading achievement circle not overlapping with the SES circle), but it chips away at the overall spread of the scores. The r^2 value is this *explained variance*.

Knowing how much variance is explained is helpful to the researcher, but what guidelines do we have that will help us judge the extent of the explained variance? Cohen (1988) provides the following conventions for the r^2:

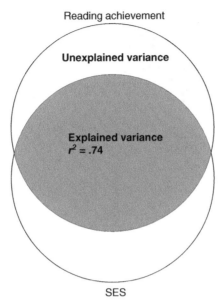

Figure 9.10. The effect size of correlation: explaining variance.

$$\text{Small Effect Size:} \quad r^2 = 0.010 \, (r = 0.100)$$
$$\text{Medium Effect Size:} \quad r^2 = 0.090 \, (r = 0.300)$$
$$\text{Large Effect Size:} \quad r^2 = 0.250 \, (r = 0.500)$$

These r^2 size conventions may appear to be small and for some studies they may not be appropriate. The researcher is ultimately in charge of establishing the meaningfulness of the correlation effect size, since it is tied to the nature of the research question. Thus, we might not be excited about an r^2 of 0.090 in a small exploratory study of job satisfaction (outcome variable) and job tenure, but we might be very excited about the same effect size in a study of the relationship between a new drug and the incidence of Alzheimer's disease among a segment of an urban population.

CORRELATION INFLUENCES

Several factors can affect the size of a correlation and therefore its effect size. These are things that are not related to the meaning of the variables, but to the dynamics of the research design and other such factors. We mention some of the big factors:

1. *Sample size:* The larger the sample size, the more likely it is to be judged statistically meaningful.

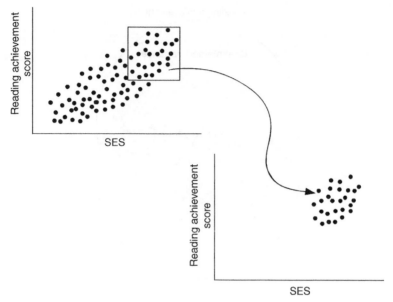

Figure 9.11. The correlation problem of restricted range.

2. *Restricted range:* The strength and size of a correlation is dramatically affected by "restricted range," or the selection of scores that do not display full variability. Figure 9.11 shows this problem with a scattergram. As you can see, if we restrict a study to only members of high SES (the bottom panel of the figure), then we will not see the overall pattern of relationships that are in the data when all the cases are included (the top panel).

3. *Extreme scores:* Outlier scores can have a dramatic effect on the calculated r, especially in studies with fewer cases. We have shown this in Figure 9.12. This figure represents the data in Table 9.1 in which we have changed the SES value of the first case from 100 to 10 (the black dot in the upper left corner of the figure). This might happen as an accident while entering the data, for example. When we calculate r from this altered data set, we find that r is now 0.225 (down from 0.859)! The r^2 is reduced (from 0.74) to 0.05! Changing one value thus results in reducing the effect size by 69 percent (from 75 to 5 percent).

4. *Curvilinear relationship:* We showed an example of a curvilinear relationship in Figure 9.2 (panel d). In these types of relationships, the pattern of the data may extend in one direction and then break in a different direction creating a nonlinear path. These are often not easy to detect by simply looking at a scattergram, but SPSS has a straightforward procedure that helps to detect curvilinear relationships. In any case, you can see that if the data are not related in a linear fashion, calculating a Pearson's r will yield a much smaller r (and therefore a smaller effect size).

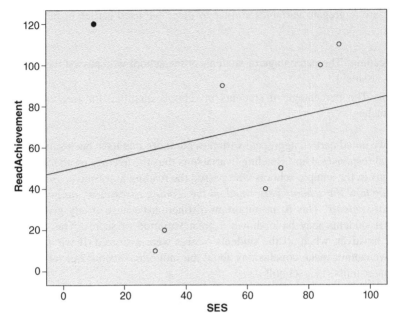

Figure 9.12. The problem of outliers.

CORRELATION IS NOT CAUSATION

We noted this injunction previously. The problem with discovering correlations is the temptation to assume that the variables are *causally* related. Thus we might be tempted to assume that a higher SES causes reading achievement scores to increase. This conclusion does not logically follow from our research findings of a meaningful correlation. There are many things that are related to reading achievement test scores other than SES, and some of them may be better explanations of increases in scores. Researchers use specific statistical measures (e.g., regression) to understand causal relationships. We examine these in a later section.

We now turn to an example of correlation using real-world data to show how to use SPSS to create an *r* value and then how to evaluate the results. The example, as we mentioned earlier, will use aggregate data. Furthermore, we will use this example to discuss the relationship between correlation and causation.

AN EXAMPLE OF CORRELATION USING SPSS

The example data we will use is a random sample of 40 schools taken from the Washington State school database.[2] The variables we consider for the correlation

---

[2] The data are used courtesy of the Office of the Superintendent of Public Instruction, Olympia, Washington. The Web site address is http://www.k12.wa.us/.

example are aggregate variables similar to those we used earlier in the hypothetical example:

- Reading: The percentage of students at the school who passed the state reading assessment.
- FR: The percentage of students in schools qualified for free and/or reduced lunches.

As we noted earlier, aggregate variables are those that have been summed up from individual responses. Our "Reading" variable is thus the average of all the percentages of students in the sample schools who passed the reading assessment. *We therefore are measuring how the schools performed on the reading assessment, on average, not the individual students.* This is an important distinction because at any given school the individual students may have shown a great variation of scores. This variation was masked, however, when all the students' scores were averaged. If we use aggregated scores, we cannot make conclusions about the individual scores, but rather the scores of the larger units (i.e., schools).

The other variable, "FR," is also an aggregate variable measuring the average percentage of students eligible for free and reduced lunch across the sample schools. This variable is typically used in education research, since it is one of the only variables that indicates SES level. You must remember when you examine your findings that high values of FR indicate low SES, since FR measures the percentage of students at a school with low enough family income to qualify for reduced meal prices.

Table 9.2 shows the sample data in two columns (each). We will use this database to provide a real-world example of calculating correlation and making appropriate conclusions with aggregate data. For this example, create an SPSS file and see if you can calculate Pearson's *r* for the relationship between (school level) reading and FR. We provide the steps and answers below.

Example Worksheet: Correlation

1. What is the correlation?
2. What is the effect size?
3. Show the scattergram.
4. Are there potentially problematic outliers?
5. What conclusions can you draw? State your findings.

Example Worksheet Answers: Correlation
Before you begin the exercise, you might look over the data carefully and perform some of the descriptive statistics exercises you learned in former sections. You will find that the reading variable is not exactly normally distributed, whereas the FR variable is closer to normally distributed. (In both cases, the numerical values for skewness and kurtosis indicate they are within normal bounds.)

TABLE 9.2. The School Database Sample

Reading	FR		Reading	FR
	14		67	23
41	97		75	19
65	87		44	67
72	32		66	51
71	69		70	53
71	65		57	39
76	31		45	71
33	97		94	7
73	50		44	63
48	44		40	0
73	36		63	30
68	43		73	0
66	35		61	51
51	68		54	49
51	70		48	61
59	64		48	87
71	51			0
83	31		74	52
67	41		89	15
76	54			51
Cases 1–20			**Cases 21–40**	

You will also notice that there are several missing cases (in both variables). SPSS makes adjustments for these cases, so you do not have to enter values unless there is a reason to do so (which we have not yet discussed). Notice that there are several FR values of 0 *that are not missing values* but rather indications that there were no students at the study school qualified to receive free or reduced-price meals.

Refer back to the procedures for using SPSS to calculate the correlation and create the scattergram. Here are the answers. How did you do?

1. What is the correlation? Figure 9.13 shows the SPSS correlation matrix, indicating a negative (or inverse) correlation between Reading and FR. Thus

(*Continued*)

Correlations

		ReadingPercent MetStandard	PercentFreeorRe ducedPricedMeals
ReadingPercentMetStandard	Pearson Correlation	1	$-.567^{**}$
	Sig. (2-tailed)		.000
	N	37	37
PercentFreeorReducedPriced Meals	Pearson Correlation	$-.567^{**}$	1
	Sig. (2-tailed)	.000	
	N	37	40

**. Correlation is significant at the 0.01 level (2-tailed).

Figure 9.13. The correlation results for the example problem.

schools with larger percentages of low-income students show higher average reading achievement. As you can see, the correlation is (–0.567), based on 37 schools (three schools with missing reading data were not included in the analysis).

2. What is the effect size?

The effect size is 0.320 (–0.567^2). According to the criteria for judging the effect size we discussed earlier, this finding would represent a large effect size. Another way to express the effect size is to note that the FR percentage of the schools explains 32 percent of the variation of the aggregated reading percentages of the schools.

3. Show the scattergram.

Figure 9.14 shows the scattergram between reading and FR (using the actual names of the variables in the database). Note several aspects of the graph:

- There is a downward trend to the scatter consistent with a negative correlation.
- The scatter is fairly evenly distributed around the line of best fit and the scatter is not widely dispersed (with a couple of exceptions).
- There is one outlier indicated by the black dot on the bottom left of the graph.

4. Are there potentially problematic outliers?

We highlighted the outlier value in the scattergram to point out how much of an impact they make on the results. In this example, excluding this school (reading = 40 and FR = 0) increases the correlation value to –0.722 (from –0.567) and the effect size to 0.520 (from 0.320). This is a dramatic change in the findings. The question to the researcher is whether dropping the case is justified according to some research rationale. When we look at the data closely (in the original database), we conclude that the 0 FR value is probably an error, possibly because the school did not submit the proper data to the state. It is therefore

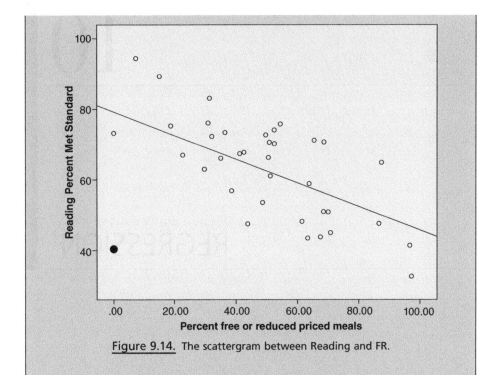

Figure 9.14. The scattergram between Reading and FR.

probably justified to drop this case and report the stronger findings. You as a researcher ultimately control the decision regarding outliers.

5. What conclusions can you draw? State your findings.

On the basis of these findings, we would conclude that the FR percentage of schools has a strong relationship to the reading achievement of the study schools. Of course, we must remember that we cannot conclude a *causal* relationship, only a *strong relationship*. There may be other reasons why schools' reading percentages vary.

NONPARAMETRIC CORRELATION

Nonparametric correlation procedures are those that do not make reference to theoretical distributions. As we saw in the contingency table section, we can measure the relatedness (correlation) of two variables that use frequency. One procedure we discussed is Cramer's V, a nonparametric procedure that does not rely on a theoretical distribution. Typically, statistical procedures using nominal and ordinal data are considered nonparametric. With correlation, ordinal data can be analyzed using Spearman's rho correlation that uses ranked data (ordinal). The procedure is available in SPSS through the Analyze–Correlate–Bivariate menus that we used earlier to create Pearson's *r*. You can simply check a box in the Bivariate Correlations menu screen to obtain Spearman results.

10

REGRESSION[1]

Researchers attempt to understand social problems using specialized statistical procedures. When we explored the use of contingency tables, for example, we saw that we could use chi square to help us determine whether there were meaningful patterns among categorical data. We also saw that contingency tables could be elaborated by adding more independent variables. Social problems are complex and therefore require methods that allow us to recognize and understand this complexity. Adding more variables to the contingency table is one way to see how *multiple* influences on a dependent variable measure can provide a deeper understanding of the phenomena.

Multiple linear regression (MLR) procedures follow the same logic. This statistical procedure assesses the influence on an outcome variable of more than one predictor variable. When we use correlation procedures, we can understand how one variable influences or is related to another variable. For example, job stress is a pervasive problem in some companies. Not only does it have an impact on productivity and worker satisfaction, it may also influence what happens at home. As Melvin Kohn (1977) demonstrated, the nature of work influences how people respond to one another within their families.

[1] Parts of this section are adapted from M. L. Abbott, *Understanding Educational Statistics Using Microsoft Excel® and SPSS®* (Wiley, 2011), by permission of the publisher.

Understanding and Applying Research Design, First Edition. Martin Lee Abbott and Jennifer McKinney.
© 2013 John Wiley & Sons, Inc. Published 2013 by John Wiley & Sons, Inc.

In the real-world context of research, understanding a social phenomenon is rarely achieved only by examining one potential influence. Thus, for example, if we wanted to understand worker stress in more depth, we would probably be able to create a more comprehensive view than simply finding a correlation between stress and health. We may want to examine several possible influences (e.g., social class, social stress, age, exercise, dietary habits, etc.), not just one (perceived state of health).

Crosstab analyses of contingency table data (categorical data) have a practical limitation of three or four variables. As the number of variables increases, the combinations increase, making it very hard to analyze results.

In this chapter, we take a brief look at multiple regression, which has great capacity to examine many separate influences on a single (interval variable) outcome. As you might imagine, this is a complex procedure, so we cannot hope to cover it extensively here. However, we can discuss some of the practical features of the program and show how to use it with a real-world example. Several sources develop the procedure in much greater depth if you would like to explore it further (e.g., see Abbott 2010, 2011).

MLR is a process that creates a model for (1) *predicting* values of an outcome variable from a set of predictor variables, and (2) a way of *explaining* the variance in an outcome variable based on the influence of a set of predictor variables.

UNDERSTANDING REGRESSION THROUGH CORRELATION

The best way to understand regression is to recall what we learned in the chapter on correlation. As you recall, correlation allows you to consider how two variables are related to one another. Do the values of one variable change along with the values of another variable? We examined a research question when we considered correlation: Is socioeconomic status (SES) related to reading achievement? We found in that (fictitious) example that SES is strongly and positively related to reading achievement; as SES increases, reading achievement improves. Figure 10.1 shows how the variance in reading achievement is explained by SES.

As you can see, the shaded portion indicates that about 74 percent of the variance in reading achievement is explained by SES. That leaves only a small portion of the variance in reading achievement unexplained. But what could account for that 26 percent of the unexplained variance? Could we add to our understanding of the relationship between SES and reading achievement by adding variables to the research study that could possibly account for more of the unexplained variance? Perhaps gender or previous reading experience, or other such variables could increase our understanding.

REGRESSION MODELS

Regression models are built in such a way that they explain as much of the variance in the outcome variable as possible. In effect, we add additional correlations and fit

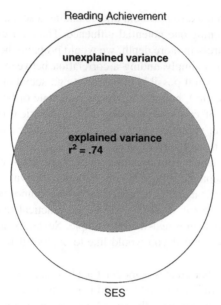

Figure 10.1. The Venn diagram of SES and Reading Achievement indicating explained variance.

them together in a statistical model that is the best explanation of the most variance in the outcome. Another way of stating this is that we are using a procedure to extend the insights provided by correlation to help us further understand a given research outcome variable. By using regression, we create several helpful analyses:

1. We can examine individual correlation pairs of all the relevant variables in a research study.
2. We can see whether the entire set of predictor variables is a significant explanation of the variance in the outcome variable. This explanation shows the *combined* or *joint effect* of all the predictor variables.
3. We can identify the pieces of unique outcome variance explained by each of the predictor variables when the other predictors are not allowed to influence the analysis (i.e., when the other variables are controlled). This explanation shows the *net* or *unique effects* of specific predictor variables.
4. We can look at the relationship among the predictor variables in order to detect possible spuriousness and to specify whether a given predictor is likely to be an antecedent or intervening variable.

Consider Figure 10.2, which shows the impact of adding another correlation to the relationship between reading achievement and SES. In Figure 10.1, one variable (SES) explains about 74 percent of the variance in an outcome (Reading achievement). In Figure 10.2, we have added a second predictor variable (designated ??? because it could

READING ACHIEVEMENT

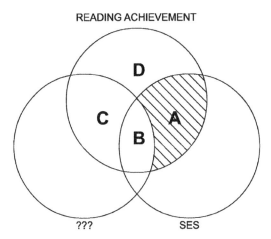

Figure 10.2. Examining all the sources of influence on the dependent variable.

be any of several possibilities). In this figure, section A represents the amount of variance explained by SES. But now you can see that the second predictor not only has increased the explained variance in reading achievement (section C), but it is also correlated with the first predictor. Thus section B represents the overlap of the two predictors in their explanation of the outcome.

The sections of Figure 10.2 illustrate how regression can use the intercorrelations of several predictor variables to explain an outcome variable. Additional predictor variables (potentially) add to the explanation of variance in the outcome variable. However, the intercorrelation of the predictor variables are problematic to explain because we cannot identify the source of the explanation between the predictors. Regression analyses allow the researcher the ability to see these relationships and measure each of the pieces of influence in the relationships. Thus we could quantify the amount of variance in the outcome variable that originated in each of the sections of the figure: A, B, C, and D.

This explanation glossed over a great deal of detail! If you are interested in understanding these relationships in greater depth, you might explore the procedures of partial correlation and semipartial (or part) correlation. For our purposes, it is enough to understand that when you add predictor variables to an analysis, it helps to explain variance in the outcome variable, but it introduces complexity that must be simplified. We can do all of these using SPSS procedures, some of which we have already introduced.

USING SPSS TO UNDERSTAND REGRESSION

In this section, we present a multiple linear regression (MLR) study to show some of the ways to interpret the findings using real-world data. SPSS is specifically designed for these kinds of complex analytical procedures, so we review the output in some detail.

In Chapter 9 we discussed how to use correlation to understand the relationship between reading achievement and FR, an aggregate measure of family income. In what follows, we extend this example to include a third variable, ethnicity. We can create a research study question with these variables such as, "Do ethnicity and family income affect reading achievement?" This question implies that we are attempting to understand whether (1) ethnicity and family income *combined* explain a great deal of the variance in reading achievement, and (2) how much unique (*net*) variance in reading achievement do each of the predictors explain (i.e., the independent impact of a predictor on the outcome when the other predictor is added)?

Table 10.1 shows the data for this example. These are fourth grade student data in 50 schools randomly selected from the Washington state database for 2011.[2] The following are the study variables:

TABLE 10.1. The Data for the Regression Example

Math	Ethnicity	FR	Math	Ethnicity	FR
44.2	2.2	84.0	45.2	57.8	62.5
9.0	3.7	86.3	58.3	58.4	46.5
50.0	15.3	68.3	76.7	58.7	27.9
47.1	23.0	87.0	69.0	61.2	30.4
41.7	26.0	47.2	75.0	63.0	54.0
45.3	28.4	73.9	71.8	68.7	54.6
41.9	29.0	73.4	75.0	71.1	50.3
91.8	33.3	80.4	84.2	71.7	12.7
59.6	33.7	70.6	78.7	77.1	10.5
51.0	34.1	78.2	80.8	77.6	57.7
90.5	34.2	7.7	62.1	78.4	32.1
25.8	34.8	79.8	89.4	80.2	3.2
26.5	35.2	71.5	32.8	80.3	78.0
38.3	35.3	76.0	68.6	81.3	27.9
33.9	40.2	75.1	89.5	82.0	14.8
53.3	43.4	29.6	82.8	82.1	31.3
54.3	45.9	48.8	40.0	83.1	75.4
75.4	47.3	21.5	78.7	84.4	47.0
55.3	48.1	64.3	50.0	85.2	46.2
54.5	51.7	54.5	96.0	85.2	17.5
50.0	51.8	89.3	62.9	85.8	36.5
79.3	51.8	38.9	77.5	86.6	6.2
40.4	52.3	38.3	64.8	90.1	38.3
71.9	54.3	46.0	53.7	90.2	41.3
34.1	55.8	56.8	18.2	90.6	28.1

[2] The data are used courtesy of the Office of the Superintendent of Public Instruction, Olympia, Washington. The website address is: http://www.k12.wa.us/. Data represent a sample of schools from the total list of schools with data on all study variables.

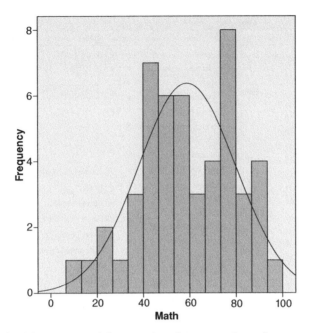

<u>Figure 10.3.</u> The histogram used for assessing the assumption of a normally distributed variable.

- "Math" is the percentage of fourth grade students in the sample schools who passed the math assessment (without a previous pass).
- "Ethnicity" is the percentage of students at a school who are classified as white or nonethnic.[3]
- "FR," as we have seen earlier, is the percentage of students by school who qualify for free or reduced price meals due to family income.

Before we begin with the analysis, it is important to note that the outcome variable (math) meets the normal distribution requirement for using regression. As you can see in Figure 10.3, the graph appears normally distributed. It does not appear to be perfectly normal, but a series of statistical tests suggest that the data are close enough to be considered normally distributed for this analysis.[4]

Specifying the Analyses in SPSS

The first step in using SPSS for a regression analysis is to use the Analyze pull-down menu. Figure 10.4 shows what this looks like for this example. As you can see, we

[3] The Washington school database provides the percentage of students in various ethnic categories at schools in the state. Studies commonly use the percentage of white students at a school as an indicator of "white/ nonwhite" proportion. That is, if you report the percentage of white students, this is also a way of reporting the percentage of students who are nonwhite.

[4] We assessed skewness and kurtosis along with a series of other statistical tests, which confirmed that the data meet the assumption.

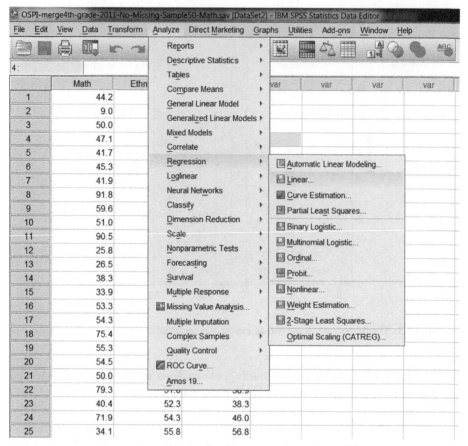

Figure 10.4. The SPSS specification menu window for multiple linear regression.

specified a "Linear" "Regression" analysis. The list of additional tests shows how complex the procedure can become, but many analyses can be performed using the straightforward procedure of linear regression where we assume that the variables are related in a straight line fashion when plotted on a line graph.

Figure 10.5 shows that we are calling for a regression analysis in which the math variable is the outcome and the first predictor variable is ethnicity. You can see that Ethnicity is placed in the "Independent(s)" box using the arrow buttons. The next step is to enter the additional (second) predictor FR. We can do this in a couple of ways, but it is instructive to enter it separately so that we can see what happens to the explained variance in math as a result of adding each predictor independently.

We do this by adding FR in a separate "Block." Figure 10.6 shows how we speci-fied this procedure. When we entered Ethnicity, as in Figure 10.5, we chose the "Next" button (immediately above the "Independent(s):" box). This placed the second predictor in a separate space. The results will then show both blocks of results.

The next step is to identify which outputs we want from the analysis. To do this, we select the "Statistics" button on the upper-right column of choices. Figure 10.7

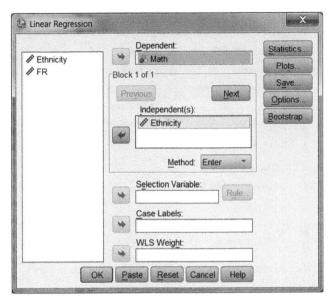

Figure 10.5. The specification window for analytical procedures in multiple regression.

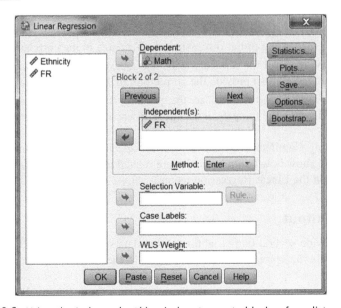

Figure 10.6. Using the Independent(s): window to create blocks of predictor variables.

shows the set of choices we made for this example analysis. We asked for the following:

- "Model fit" is the statistical likelihood that the model, or use of two predictors, provides a better than chance explanation of variation in school math achievement.

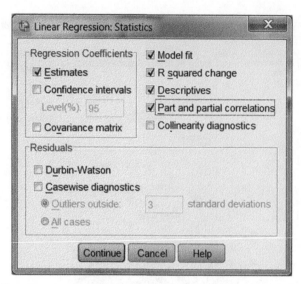

Figure 10.7. Choosing procedures from the Statistics menu in multiple regression.

- "R squared change" is the combined effect size of both predictors on the outcome variable (according to our discussion of effect size in the correlation chapter).
- "Descriptives" call for summary statistical measures for the variables involved. This choice will also provide the correlation matrix of all the study variables.
- "Part and partial correlations" are the measures that allow us to test for spuriousness and assess the unique contributions of each of the predictor variables.

By choosing Continue, we can complete the specification and examine the SPSS output that will show how all the variables are related to each other. (Choose Continue and then OK on the Linear Regression menu.)

The SPSS Output

SPSS will provide several output tables for this data set, but we show only the main tables to use in interpretation. The first table is the correlation matrix shown in Figure 10.8.

Because it is a matrix, you can see the results as a symmetrical table in the upper panel of the output. We show this with different shades. In either set of values, you see the following:

- There is a statistically significant *positive* correlation between Ethnicity and Math. The higher the percent of the student population in the school categorized as white, and consequently, the lower the percentage of nonwhite students at a school, the higher the math percentage passing rate ($r = 0.418$, $p < 0.001$).

Correlations

		Math	Ethnicity	FR
Pearson Correlation	Math	1.000	.418	−.635
	Ethnicity	.418	1.000	−.595
	FR	−.635	−.595	1.000
Sig. (1-tailed)	Math	.	.001	.000
	Ethnicity	.001	.	.000
	FR	.000	.000	.

Figure 10.8. The correlation matrix from the multiple regression output.

- There is a statistically significant *negative* correlation between FR and math. The greater the proportion of students eligible for free/reduced lunches in a school, the lower the math achievement score of the school (−0.635; $p < 0.0001$)[5].
- There is a statistically significant *negative* correlation between ethnicity and FR. The higher the ethnicity (white) measure in a school, the lower the proportion the student's eligible for free/reduced lunches in a school (−0.595; $p < 0.0001$).

On the basis of the correlations alone, we draw the inference that a school's math percentage passing is associated with (1) greater percentages of white students per school, and (2) fewer proportions of students qualified for free and reduced lunches. How can regression analyses help us to look deeper into these relationships?

INTERPRETING MULTIPLE REGRESSION: THE COMBINED, OMNIBUS FINDINGS

As we noted earlier, regression helps us to understand many things, but two are noteworthy. The first of these is the ***omnibus test***, *which reveals the combined effect of all the predictor variables on the dependent variable.* Thus, for example, if we were analyzing the results for our previous example of job stress and health, the omnibus test would show the extent of the relationship between the outcome variable (stress) and the set of predictor variables (for example, social class + age + exercise + dietary habits + . . . , etc.).

The numerical way multiple regression indicates the omnibus, or combined, effect of the predictor variables on the outcome variable is by using R^2. You may recall that this is the effect size for correlation. It is also the effect size for multiple regression except that it represents the percentage of the variance of the outcome variable explained by the set of all the predictor variables.

---

[5] When SPSS reports the significance level as 0.000, it means that the significance level is very small. The figure .000, when shown with many decimal places, will be non-zero significance, but tiny. In the present example, the significance of the correlation between math and FR is p < 0.000000371139 . . . '! In this case, it is convention to report the significance simply as p < 0.0001.

Model Summary

Model	R	R Square	Adjusted R Square	Std. Error of the Estimate	Change Statistics				
					R Square Change	F Change	df1	df2	Sig. F Change
1	.418ᵃ	.175	.158	19.1434	.175	10.187	1	48	.002
2	.637ᵇ	.405	.380	16.4248	.230	18.205	1	47	.000

a. Predictors: (Constant), Ethnicity
b. Predictors: (Constant), Ethnicity, FR

Figure 10.9. The model summary from the SPSS output.

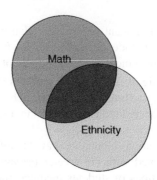

Figure 10.10. The overlap of the two study variables.

Figure 10.9 helps us to gain insight into the combined effect of the predictors.[6] Although there are many sources of information in the SPSS output, let us concentrate on the shaded portions first. As you can see, the output table is presented in two "models" (see first column). These models correspond to the *blocks of variables that we specified when we created the analysis*, as shown in Figures 10.5 and 10.6.

The first model includes only the correlation between ethnicity and math. The effect size is 0.175 (i.e., $R^2 = 0.175$). This indicates that ethnicity accounts for 17.5 percent of the variance in math in schools *by itself*. Figure 10.10 shows how this might look as a Venn diagram. As you can see, there is an overlap between the two circles indicating the amount of explained variance (17.5 percent). The remainder of the math circle is unexplained variance.

The results in the second model show how the effect size (R^2) *changes as a result of adding the second predictor* to the analysis. The effect size (shown in the second row) is now 0.405, indicating that 40.5 percent of the variance in math is explained by the *combination of ethnicity and FR*. These effect size changes in the R^2 is shown in the "R Square Change" column of Figure 10.9. As you can see, where there is only one predictor (model 1), the effect size changes from 0 to 0.175. When we add the second predictor, the effect size changes by 0.230, or increasing the R^2 from 0.175 to 0.405 (i.e., $0.405 - 0.175 = 0.230$). Therefore, ethnicity and FR together account for 40.5

[6] For a more complete explanation of the values and categories of Figure 10.9, consult Abbott (2010, 2011).

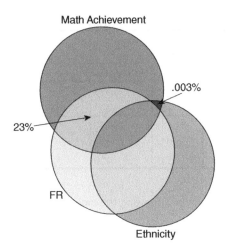

Figure 10.11. The omnibus, or combined, effects of ethnicity and FR on math achievement.

percent of the variance in math in schools. Figure 10.11 shows these proportions in a Venn diagram.

Figure 10.11 shows that FR is a much larger contributor to math variance than ethnicity by itself. When FR is added to the MLR, the unique contribution of ethnicity to math achievement drops to less than 1 percent![7] This is a powerful and provocative finding. Are we really saying that ethnicity has nothing to do with school-level math achievement? No, but we are saying that the relationship among the variables is complex, and MLR can help us to understand it in a new way.

A full explanation of how we derived the proportions of variance in Figure 10.11 is beyond the scope of this book, but you can see that MLR can be very helpful in bringing clarity to a complex situation. MLR also allows you to add more predictors to the analysis, using the same SPSS procedures we explained earlier.

INTERPRETING MULTIPLE REGRESSION: THE INDIVIDUAL PREDICTOR FINDINGS

The second set of information that SPSS provides is the information on the importance of each individual predictor on the dependent variable. It is good to have an idea of the combined impact of all the predictors on the outcome variable (omnibus effects), but it is also important to know how much each predictor contributes to an understanding of the outcome variable. SPSS does the latter by providing an output showing the coefficients of each predictor variable. *These coefficients generally indicate the importance of each predictor.* We can see how this works by examining the output for our example.

---

[7] We determined the unique proportions of variance in math due to the predictor variables using squared part correlations. This approach allows the researcher to view the independent contributions of each predictor *irrespective of the order in which the predictors were added to the analysis.*

Coefficients[a]

Model		Unstandardized Coefficients		Standardized Coefficients		
		B	Std. Error	Beta	t	Sig.
1	(Constant)	38.415	6.976		5.506	.000
	Ethnicity	.361	.113	.418	3.192	.002
2	(Constant)	81.158	11.670		6.954	.000
	Ethnicity	.054	.121	.063	.448	.656
	FR	−.511	.120	−.597	−4.267	.000

a. Dependent Variable: Math

Figure 10.12. The SPSS output showing the individual predictor, or unique effects.

Figure 10.12 is the next table produced in the SPSS output. This table explains the coefficients in the regression analysis. When we conduct an MLR analysis, we create a linear equation that expresses how the predictor variables relate to the outcome variable. We can determine the unique impact of each predictor variable by examining the standardized (**Beta**) and unstandardized coefficients of the equation.

The Beta coefficient is an expression in standard deviation units (like the z scores we will examine later in the Statistical Procedures Unit C section). It is interpreted as the standard deviation amount change in the outcome variable that results from a one standard deviation increase in that particular predictor variable. *Generally*, the larger this coefficient, the more impact that predictor has on the outcome variable when the other predictor(s) is held constant or not allowed to affect the relationship. *It is more complicated than that, but this is a good general rule for interpretation.* The Beta column of Figure 10.12 (shown in light shading) shows the first of these coefficients.

As you can see, when ethnicity is the only predictor (i.e., when it is in a block by itself), it is a significant predictor of math achievement in schools ($t = 3.192$; p < 0.002). The Beta coefficient indicates that the value of the outcome (achievement) changes by 0.418 (almost half a standard deviation unit) with changes of 1.0 in the predictor (with the other predictors held constant).

The **unstandardized coefficient** (shown in the B column) indicates the same level of impact, but it is expressed in a different metric.[8] This coefficient indicates that math achievement in schools increases by 0.361 with every 1.0 change in ethnicity. Thus schools with one additional percent of the ethnic predictor results in a 0.36 percent increase in math achievement.

When FR is added to the analysis, things change! As you can see in the second block of output (model 2 row) of Figure 10.12, the Beta(s) change drastically:

- The Beta for ethnicity drops from 0.418 (in model 1 where it is the only predictor) to 0.063 (when another predictor is added to the model). This is a very large

[8] The unstandardized coefficient expresses change in the size of the actual numbers of the outcome variable. The Beta coefficient is a standardized measure and expresses change in the outcome variable in standardized amounts. The Beta is therefore a good indicator of standardized change that can be compared when you have several different predictor variables.

change, resulting in an indication that, when FR is added to the model, ethnicity becomes almost nonexistent as an influence on achievement.

* The Beta for FR is strong and negative indicating that it is a powerful (inverse) influence on achievement regardless of the level of ethnicity at the schools.

USING MLR TO ESTABLISH CAUSALITY

In the causation chapter, we discuss spuriousness as a procedure that allows us to see if the apparent relationship between two variables disappear when a third variable is added to the analysis. In effect, that is what we have observed in this example. *When we added FR to the analysis, the effect of ethnicity on math achievement in schools became nonsignificant with a very small effect size.* Thus, if we were attempting to assess causality between ethnicity and math achievement at the school level, we would fail the test of *nonspuriousness*. To be sure, ethnicity is antecedent (*time order*) to achievement, and there is a significant *correlation*, but when FR is added to the analysis, the relationship between ethnicity and math achievement becomes nonsignificant.

We are not attempting to establish causal relationships through this MLR analysis. To be sure we could offer our results as some evidence that the relationship between ethnicity and math achievement of schools is not a causal relationship due to spuriousness. However, *we are using MLR to help us understand better the complex relationships among the study variables.*

According to other research (see Abbott 2010), ethnicity provides a small unique explanation of variance in math achievement when FR is introduced. The primary explanation, however, is due to FR, the measure of family income; the greater the percentage of low-income students at a school, the lower the schools' average math achievement. Figure 10.13 shows these relationships in brief.

As you can see, ethnicity only explains about 1 percent of the variance in school math achievement when the contribution of FR is removed. Likewise, FR uniquely

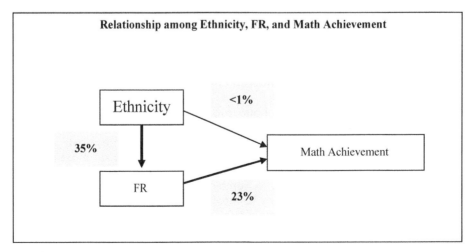

Figure 10.13. The path relationships among ethnicity, FR, and math achievement.

explains about 23 percent of the variance in school math achievement when ethnicity is removed. Thus FR is a much stronger predictor of school math achievement than ethnicity. However, also note that ethnicity explains a good deal of the variance in FR. It is fair to say that *the influence of ethnicity is contained in FR as it relates to math achievement*. Another way of stating this is that ethnicity has a (weak) direct effect on achievement, but a stronger indirect effect on achievement (through its influence on FR).

USING MLR WITH CATEGORICAL DATA

Although we do not have the space to pursue the topic in this book, MLR can be used with categorical data (predictor variables only). We show elsewhere (see Abbott 2010) how to use MLR to create **dummy variables** that are categorical predictors. Researchers use coding to assign data to different groups that can then be compared as those using analysis of variance (ANOVA) methods. Dummy, effect, and contrast code are used to specify analyses showing how categorical predictors (like experimental comparison groups, gender groups, ethnic groups, etc.) predict/explain outcome variables. This potential shows MLR's wide range of uses.

The Ecological Fallacy

Researchers using aggregate data, as we have in the previous example, must use great caution not to create a *misunderstanding* known as the *ecological fallacy*. We have defined this elsewhere as a "mistake in interpretation in which the evaluator makes conclusions about one level of data when the analyses are performed on a different level of data. For example, using school-level achievement scores, and making conclusions about individual student achievement" (Abbott 2010).

In our example, we used aggregated scores at the school level. Instead of speaking about individual students, we spoke about the *percentage of students* at a school who were qualified for free/reduced lunch, *percentages of students* with different ethnic classifications, and *percentage of students* at the school who passed the math assessment (without a previous pass). Thus our example took place at a level of abstraction above the individual, on the organizational (school) level. We were careful to interpret the findings as school-level outcomes not individual-level outcomes.

Had we made a conclusion about *students* rather than *schools*, it would have been misleading. To conclude, for example, that "low-income students do more poorly on math assessments" would not be accurate on the basis of our example findings. To be sure, *some* low-income students may do more poorly on math assessments, but not *all* of them! When we aggregate, or combine, all the math assessments by a school, we combine all the individual student performances and thereby lose the ability to speak about any one student or to conclude anything specific about whether students perform such and such a way as individuals.

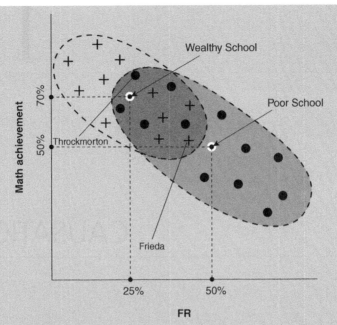

Figure 10.14. The scattergram showing two levels of data.

Figure 10.14 shows how students from two fictitious schools compare to their aggregated averages and to each other. The students that belong to each school are indicated by different symbols: Poor School students are dots "." Wealthy School students are plus signs "+". As you can see, students in both schools scatter out around their own centers (as designated by the labels). Throckmorton (from Poor School) actually performs much better on math than Frieda (from Wealthy School). Furthermore, Throckmorton actually has a higher family income than Frieda even though he is from Poor School. However, Frieda enjoys higher family income status and stronger math performance than Throckmorton *when all students are averaged together* even though Throckmorton actually has higher values on both measures as an individual student.

When we analyze entire schools as a data set by aggregating individual scores, both of our individual students get pulled back toward their school averages. Their individual achievement and income status is lost in the overall average of all students at their respective schools. All students therefore are melded into one super student score (i.e., aggregate school value) with a much narrower range of unique values. *The standard errors of both school distributions are therefore restricted* by using one school to represent an entire distribution of individual students of wide variability.

Using aggregate data does not necessarily indicate that the nature of the overall findings is inaccurate. It simply presents data with a different level of specificity and it creates the opportunity to make the ecological fallacy. When you use aggregate data, just remember to make your conclusions reflect the true nature of your data.

11

CAUSATION

The joy of social science lies in the discovery of relationships between variables. Following the Wheel of Science allows us to think about how one variable may cause changes in another variable. Using theory to propose hypotheses lets us see if variables impact each other. Up to this point we've relied on looking at simple relationships that link an independent variable with a dependent variable. In thinking about how one variable *causes* changes to another variable, however, it is seldom the case that only one variable is involved.

Let's go back to our earlier discussion regarding the relationship between education and income. Is it really that education, alone, causes changes to income? We noted previously that if we didn't believe that education impacts income, we would not invest the time and money it takes to earn a college degree. Think again about the inconsistencies (the negative cases) within the relationship between education and income. Your professors with PhDs have the highest levels of formal education. Yet those earning the most money in our society tend to be celebrities and sports figures, many of whom do not even have college educations. If those who make the most money are not the most highly educated, why do we continue to place such an emphasis on getting more education, assuming it *causes* higher levels of income?

Understanding and Applying Research Design, First Edition. Martin Lee Abbott and Jennifer McKinney.
© 2013 John Wiley & Sons, Inc. Published 2013 by John Wiley & Sons, Inc.

When we think about causes, we're looking at variables that produce a change in another variable. Thus far we've focused on single independent and dependent variables; independent variables being the variable that our theory says should cause change in the dependent variable (thus the dependent variable is affected by what happens with the independent variable first); put another way, when we know how the independent variable changes (the cause), we can predict changes in the dependent variable (the effect).

CRITERIA FOR CAUSATION

When thinking about variables that cause change in other variables we need to consider the criteria for determining if one variable (an independent variable) does in fact *cause* changes in the dependent variable. If we follow some simple procedural rules, we can have a better idea if one variable may in fact cause changes to another variable. The **criteria for causation** are composed of three rules used to establish whether there is a *causal relationship* among study variables. These three criteria include time, correlation, and nonspuriousness.

Time

The first criteria we need to determine if an independent variable causes changes in the dependent variable is **time order,** or making sure the independent variable occurs *before* the dependent variable. This seems like an obvious trait in the case of the relationship between education and income. Not many people earning higher incomes subsequently decide to go to school to earn more education. When relationships become more complex, however, which variables come first in time order can be more complicated. Just remember that in thinking through the sequence of relationships, causes must occur before their effects. In most cases, we will not know for sure which variable should precede the other (only through experiments do we see time order clearly because we put the independent variable in play and then watch how (or if) it causes changes in the dependent variable). Our theory should help us to determine which variable should logically come first in time order.

Correlation

Once we've determined the time order or sequence of variables (independent comes first, dependent second), the next task is to determine if there is a *correlation* between variables. **Correlations** illustrate that two variables naturally vary with each other, meaning they have a relationship with each other. Correlation is part of the criteria for causation because if variables are not correlated, there is no relationship between them; if there is no relationship between them, the independent variable cannot cause changes in the dependent variable. When we talked about empirical generalizations earlier in the book, we noted that there is a difference between correlation and causation. Just because two variables are correlated does not mean there is a causal relationship. However, in order for there to be a causal relationship, there must be some correlation between them (correlation is a *necessary* condition to causation, but it is not a *sufficient* condition).

A classic example of the difference between correlation and causation is the relationship between ice cream consumption and crime. There is a positive correlation between rates of ice cream consumption and crime; when one increases, so does the other. Think about why. Which variable should come first in a hypothesis? Is it that ice cream consumption leads to feelings of grandeur or a propensity for aggression, which causes people to commit crime? Or is it that good ice cream is so expensive that people commit crimes in order to support their ice cream habit? Which makes most sense? Although we could come up with several reasons why one of these variables causes the other, we need to be cautious. Just because we can predict one of these variables when we know what's going on with the other, it doesn't mean that one causes the other. Again, this example seems silly, but when we posit more complex relationships, determining correlation versus causation gets more complicated. Minimally we must recognize that if one variable causes changes in another, the variables must be correlated—the variables must vary together.

Up until this point we've been working with simple relationships, testing one variable against another. If we go back to the relationship between education and income, we should ask the question, "Do increases in education alone account for the increases in income?" Does education directly give you a paycheck? Apart from being a "funded" graduate student, where the federal and state governments give you a stipend to attend graduate school, being a student most often does not result in a paycheck. What other factors may impact (or cause changes) in levels of income? The following variables offer some possibilities and a theoretical rationale:

Occupation: The type of job a person has may have differential pay, even with the same amounts of education (e.g., having a BA in psychology versus engineering will lead to different entry-level jobs with different pay). Therefore, occupation may be what impacts income.

Gender: Research continues to show that even when taking education and experience into account, men still earn more for the same job than women. Therefore, gender may be impacting level of income (as well as educational attainment).

Age: Those who have been in the job market longer may earn more money. Therefore, we would suppose that people who are older would have higher levels of income (and possible higher levels of education).

Race: Research shows that race is an important factor in socioeconomic status, with racial/ethnic minority groups earning less (even accounting for education and experience) than whites. Therefore race may affect income (as well as education).

Multiple Causes of Income
Can you think of other variables, apart from education, that may influence levels of income and why they may influence income?

_____
_____
_____
_____
_____
_____
_____
_____

In light of thinking through how variables other than education may contribute to income, we need to combine our thinking about time order and correlation to figure out how to think about adding new variables to our equation.

First, we've hypothesized a positive relationship between education and income, stating that people first need to acquire higher levels of education before they attain higher levels of income. We note this original relationship in shorthand, illustrating that it is education that precedes (or comes before) income:

$$\text{Education} \rightarrow \text{Income}$$

Both the order of the variables and the causal arrow are important in our shorthand. The independent variable should be stated first, with the causal arrow pointing in the direction of the relationship, in effect stating:

$$\text{Independent variable [impacts] dependent variable}$$
$$\text{IV} \rightarrow \text{DV}$$
$$\text{Education} \rightarrow \text{Income}$$

Thus we've stated the time order of the variables for testing this relationship.

Next, we need to test the hypothesis to see if there is a correlation. We can use the General Social Survey (GSS) data to help us detect possible correlations between the example variables. Figure 11.1 shows how we can operationalize education and income using data from the GSS. As you can see, with lower levels of education there are lower levels of income (which still states a positive direction, since both variables are moving in the same direction).

Since we hypothesize that increases in education lead to increases in income, we can examine how the cells for categories of education, our independent variable, intersect with the categories for level of income, our dependent variable. As education increases we see the percentages for those who make the highest level of income also increasing. For example, 63.2 percent of those without high school degrees earned less than $20,000 per year compared to 16.8 percent of college graduates. Similarly, 64.8

Income categories * Education Categories Crosstabulation

	Education Categories				
Income Categories	Less Than HS	HS Grad	Some College	College grad	Total
<$20,000	79	254	33	67	433
	63.2%	43.7%	34.4%	16.8%	36.0%
$20,000 - $39,999	32	182	30	74	318
	25.6%	31.3%	31.3%	18.5%	26.5%
$40,000 or greater	14	145	33	259	451
	11.2%	25.0%	34.4%	64.8%	37.5%
Total	125	581	96	400	1202
	100.0%	100.0%	100.0%	100.0%	100.0%

Chi-Square = 218.96, p=0.000, Cramer's V = 0.302.

Figure 11.1. The relationship between education and income in the GSS data.

percent of college graduates earned at least $40,000 per year compared to only 11.2 percent of those without high school degrees. The correlation between education and income is statistically significant as indicated by the probability level ($p = 0.000$). The Cramer's V (0.304) indicates a medium effect size.

Thus far, we've inferred time order and observed a statistically significant correlation between our variables measuring income and education. We have now met two of three of the criteria for causation. Recognizing that there may be multiple causes for one effect (e.g., education along with occupation, gender, age, and/or race may cause changes to levels of income), however, social scientists need to look to one more criteria for causation: no *spuriousness*.

Nonspuriousness

The word **spurious** means that there is not a true or genuine relationship between factors; that some unobserved or unnoticed variable is correlated with the other variables, making them *appear* to have a cause-and-effect relationship. For example, earlier in the chapter we talked about the positive correlation between ice cream consumption and crime. When ice cream consumption increases, so does crime. When crime increases, so does the consumption of ice cream. These two variables are consistently correlated with each other. They do not, however, have a causal relationship. Both ice cream consumption and crime are correlated to a third factor: temperature. When temperatures rise, ice cream consumption increases (people eat more ice cream in the summer than winter). When temperatures rise, crime increases. If we take temperature into account, the relationship between ice cream consumption and crime disappears— controlling for temperature takes away the correlation between ice cream consumption and crime because the two are linked by temperature.

Research Applications: Spuriousness
Do storks cause babies? Of course not! But a time-worn example in sociology (see Bohrnstedt and Knoke 1982) is that, in Holland several years ago, communities where more storks nested had higher birthrates than in communities with fewer nesting storks. This strong association has always been used to illustrate the fact that, just because two things are strongly *related* to one another (correlation), one does not necessarily *cause* the other. There may be a third variable not included in the analysis that is related to both variables resulting in the *appearance* of a relationship.

What might that third variable be? Several possibilities exist. One might be called something like "ruralness" or "urbanness." That is, the more rural the area, the more feeding and nesting possibilities exist for storks. It also happens that there are more children outside of cities, since single individuals more than families often live in cities. Thus the apparent relationship between storks and babies is really a function of this third variable that, when introduced into the equation, reduces the original relationship (babies–storks) to zero. This is illustrated in Figure 11.2. The first panel shows the apparent relationship between babies and storks, with a two-way line connecting the variables indicating that the two are highly related to one another. The second panel shows that when the third variable (Ruralness) is introduced, the relationship between storks and babies disappears.

This situation is known as "spuriousness" in research. "Spurious" carries the definition of "false," which fits a situation in which something apparent is not accurate or true. Spuriousness is the reason "correlation does not equal causation" in research. *Just because something is related does not mean one causes*

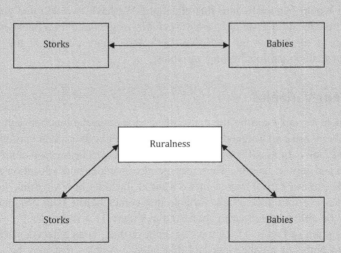

Figure 11.2. The spurious relationship between storks and babies. Identifying potentially spurious relationships is often quite difficult and comes only after extended research with a database. We discuss several research examples of spuriousness later in this chapter.

(*Continued*)

the other. You can find spurious relationships in any field of research because it is often difficult to identify whether an apparent calculated relationship between two variables is the result of another variable, or variables, not taken into account in the research. (We examine statistical procedures to help us detect spuriousness in later sections of this book.) This is one of the reasons it is important to study research and statistics procedures. We can learn the procedures that will allow us to study multiple relationships at the same time so as to identify potentially spurious situations.

If we are trying to determine when one variable causes changes to another, whenever we find a correlation between variables we need to make sure that the relationship between them is *not* spurious, that there is no unaccounted variable making the original variables appear to be correlated. Remember, if variables are not correlated, changes in one cannot cause changes in another. A variable that causes another must vary with it. In order to figure out if our correlated variables have a spurious relationship, we need to go back to the drawing board and figure out how another variable impacts the original variables, which means thinking again through time order.

In looking at the additional variables that we stated may impact the relationship between education and income, let's take the variable "gender" to see if it impacts the relationship between education and income. If we are putting together a time line of which variables come first, we have at least one hint; we know we've already determined that we believe education comes before income in our equation seen in Figure 11.3.

Where would gender fit into this equation? We have two options: gender either comes before education as an *antecedent variable*, or gender comes between education and income as an *intervening variable*. Let's look more closely at the differences between antecedent and intervening variables.

Antecedent Variables

An antecedent variable can be the cause of spurious correlations between other variables. In other words, an **antecedent variable** *precedes* both original variables. Gender is something one is born with, which means gender occurs *before* you enter the education system and earn an income. Therefore, gender precedes both education and income in time order. In our *causal argument* we want to illustrate how we think variables are related. When we introduce a new variable that comes before both original variables, we want to be able to note that in shorthand as Figure 11.4 illustrates

The diagram in Figure 11.4 shows we believe there is an antecedent variable that is related to both our independent and dependent variables that may be causing the

Education → Income

Figure 11.3. Causal diagram of one independent and dependent variable.

Figure 11.4. Causal diagram of an antecedent variable.

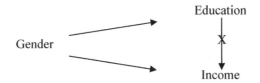

Figure 11.5. Causal diagram with gender as an antecedent variable.

independent and dependent variables to be correlated. We need to make sure, however, that we also illustrate that we believe that when we take the antecedent variable into account (or when we "control" for the new variable), we expect the relationship between the independent and dependent variables to disappear—or to be spurious. Therefore we use an "X" to "cross out" or denote that the original relationship between the independent and dependent variables is expected to be spurious (to disappear) when controlling for the antecedent variable. If the antecedent variable reveals spuriousness, there will no longer be a correlation between the independent and dependent variables.

Since we've hypothesized that gender precedes both education and income, we would diagram our causal argument like Figure 11.5.

In order to test our new causal argument, we need to go back to the data. The measure we use to control for gender in the GSS database is the variable "sex," which has two categories: male and female. In cross-tabulations, we have a two-dimensional table, which works well when you have only two variables. Now that we are adding a third variable, we need a third dimension. Unfortunately, the computer screen or the book page can only come in two dimensions. That means that we'll have to look at two additional tables, one table for each *category* of the *control variable*, when testing to see if the relationship between education and income is spurious when controlling for gender (a **control variable** is used to account for the effects of one variable on other variables, holding the control variable effects constant).

Figure 11.6 shows the original table testing the relationship between education and income, indicating the two variables are correlated.

Using chi square to include control variables is fairly straightforward. Refer to the earlier chapter on using SPSS with contingency tables to recall how to conduct a chi-square analysis (in Chapter 7 using SPSS to analyze the relationship between work

Income categories * Education Categories Crosstabulation

	Education Categories				
Income Categories	Less Than HS	HS Grad	Some College	College grad	Total
<$20,000	79	254	33	67	433
	63.2%	43.7%	34.4%	16.8%	36.0%
$20,000 - $39,999	32	182	30	74	318
	25.6%	31.3%	31.3%	18.5%	26.5%
$40,000 or greater	14	145	33	259	451
	11.2%	25.0%	34.4%	64.8%	37.5%
Total	125	581	96	400	1202
	100.0%	100.0%	100.0%	100.0%	100.0%

Chi-Square = 218.96, p=0.000, Cramer's V = 0.302.

Figure 11.6. The "original" relationship between education and income.

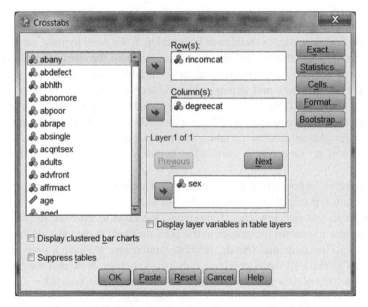

Figure 11.7. Using SPSS to introduce control variables.

autonomy and health). Utilizing the same procedure for contingency tables, we'll analyze the relationship between education and income. Introducing a control variable ("sex") to the analysis can be done by using the Layer window in the Crosstabs specification in SPSS.

When we introduce the control variable "sex" into the relationship between education (operationalizing with the variable "degreecat") and income (operationalizing with

Income categories * Education Categories * RESPONDENTS SEX Crosstabulation

RESPONDENTS SEX			Education Categories				Total
			Less Than HS	HS Grad	Some College	College grad	
MALE	Income categories	<$20,000	39	88	7	25	159
			52.0%	32.7%	20.0%	13.6%	28.2%
		$20,000 - $39,999	23	82	9	27	141
			30.7%	30.5%	25.7%	14.7%	25.0%
		$40,000 or greater	13	99	19	132	263
			17.3%	36.8%	54.3%	71.7%	46.7%
	Total		75	269	35	184	563
			100.0%	100.0%	100.0%	100.0%	100.0%
FEMALE	Income categories	<$20,000	40	166	26	42	274
			80.0%	53.2%	42.6%	19.4%	42.9%
		$20,000 - $39,999	9	100	21	47	177
			18.0%	32.1%	34.4%	21.8%	27.7%
		$40,000 or greater	1	46	14	127	188
			2.0%	14.7%	23.0%	58.8%	29.4%
	Total		50	312	61	216	639
			100.0%	100.0%	100.0%	100.0%	100.0%
Total	Income categories	<$20,000	79	254	33	67	433
			63.2%	43.7%	34.4%	16.8%	36.0%
		$20,000 - $39,999	32	182	30	74	318
			25.6%	31.3%	31.3%	18.5%	26.5%
		$40,000 or greater	14	145	33	259	451
			11.2%	25.0%	34.4%	64.8%	37.5%
	Total		125	581	96	400	1202
			100.0%	100.0%	100.0%	100.0%	100.0%

Figure 11.8. The stacked results table for chi square with a control variable.

the variable "rincomcat"), SPSS produces the output table shown in Figure 11.8. As you can see, there are three tables stacked together.

- Top table (light shading): The relationship between education and income with MALE respondents.
- Middle table (darker shading): The relationship between education and income with FEMALE respondents.

- Bottom table (darkest shading): The *original* relationship between education and income.

As you can see, the table in Figure 11.8 presents the three tables together so that you can compare the results of the two control categories (MALE and FEMALE) with the original table on the bottom of the stack. If you look at each of the control tables, you will find the same results as you observed for the original table. Thus, for example,

- For males, 52 percent of those with less than high school earned less than $20,000 compared to only 13.6 percent of college graduates. Further, 71.7 percent of college graduates earned at least $40,000 compared to only 17.3 percent of those with less than high school.
- For females, 80.0 percent of those with less than high school earned less than $20,000 compared to only 19.4 percent of college graduates. Further, 58.8 percent of college graduates earned at least $40,000 compared to only 2.0 percent of those with less than high school.
- For the original table (combining male and female), you will recall that 63.2 percent of those with high school earned less than $20,000 compared with only 16.8 percent of college graduates. Further, 64.8 percent of college graduates earned at least $40,000 compared to only 11.2 percent of those with less than high school.

Using this procedure in SPSS also produces the significance and effect size results in stacked tables. Figures 11.9 and 11.10 show these tables separately using corresponding shading to that in Figure 11.8. As you can see, both control tables indicate significant chi-square tests (with probability = 0.000 in each case) and near to medium effect sizes (Cramer's V values of 0.280 or larger).

Using Control Variables to Detect Spuriousness

You can see from the preceding chi-square analyses that we added sex as the control variable in the relationship between education and income to see whether the original relationship became spurious (in other words, we tested to see if the originally significant relationship was no longer significant when we added sex as a control or antecedent variable). When looking at two-dimensional displays, keep in mind that we are only controlling for one variable, even though we now have multiple (stacked) tables (one table for each category of the control variable). Because we have more than one table, we have to take all of them into account when determining if the relationship between education and income is spurious when controlling for the variable sex.

If our antecedent variable (sex) does impact the original relationship, we expect the original significant relationship between education and income to disappear. That would mean that *neither* of the two new tables (i.e., for male and female) would show a significant relationship between education and income. Since the probability statistic tells us whether or not a relationship is statistically significant, what do you take away

Chi-Square Tests

RESPONDENTS SEX		Value	df	Asymp. Sig. (2-sided)
MALE	Pearson Chi-Square	88.373[a]	6	.000
	Likelihood Ratio	91.196	6	.000
	Linear-by-Linear Association	78.056	1	.000
	N of Valid Cases	563		
FEMALE	Pearson Chi-Square	158.915[b]	6	.000
	Likelihood Ratio	162.617	6	.000
	Linear-by-Linear Association	139.664	1	.000
	N of Valid Cases	639		
Total	Pearson Chi-Square	218.960[c]	6	.000
	Likelihood Ratio	222.664	6	.000
	Linear-by-Linear Association	196.063	1	.000
	N of Valid Cases	1202		

Figure 11.9. Chi-square test results including control variable.

Symmetric Measures

RESPONDENTS SEX			Value	Approx. Sig.
MALE	Nominal by Nominal	Phi	.396	.000
		Cramer's V	.280	.000
		Contingency Coefficient	.368	.000
	N of Valid Cases		563	
FEMALE	Nominal by Nominal	Phi	.499	.000
		Cramer's V	.353	.000
		Contingency Coefficient	.446	.000
	N of Valid Cases		639	
Total	Nominal by Nominal	Phi	.427	.000
		Cramer's V	.302	.000
		Contingency Coefficient	.393	.000
	N of Valid Cases		1202	

Figure 11.10. Chi-square results showing Cramer's V values for control variable.

from the results of the tables in Figures 11.9 and 11.10? Both the tables for male and female have probability levels that show statistical significance between education and income ($p = 0.000$). Since both tables show significance, can our causal argument that gender is an antecedent variable be true? No. Even when controlling for the effects of gender, the relationship between education and income is still significant. If gender were indeed an antecedent variable, both tables for the variable sex (for categories male and female) would show that there was no statistically significant relationship between education and income. Thus our causal argument (and our causal diagram) is not supported by the data.

Intervening Variables

There is a second possibility for time order when adding a third variable to an equation. There could be an *intervening variable* impacting the relationship between education and income. Let's look at the variable occupation. In thinking about time order, where do you see occupation and its relationship to education and income? If it was an antecedent variable, we would expect that a person obtains an occupation, then gets an education, and then earns more money. In our culture, however, it seems that getting an education comes before getting an occupation, and it's the occupation that then allows people to earn more money. Therefore, occupation would be an intervening variable. *An **intervening variable** is a variable that comes between the independent and dependent variables.* In effect, an intervening variable is a link between independent and dependent variables. Figure 11.11 illustrates the causal diagram for an intervening relationship.

If we believed that occupation was an intervening variable, coming *between* the original independent variable (education) and dependent variable (income), then our causal diagram would look like Figure 11.12.

Figure 11.12 shows how we expect the relationship between education, occupation, and income to work. The causal arrows drawn to show the direction of the causation, allow us to follow the logic of the causal argument (that education impacts occupation and that occupation subsequently impacts income). In this analysis we expect that when

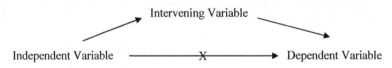

Figure 11.11. Causal diagram of an intervening variable.

Figure 11.12. Causal diagram with occupation as an intervening variable.

Income categories * Education Categories Crosstabulation

	Education Categories				
Income Categories	Less Than HS	HS Grad	Some College	College grad	Total
<$20,000	79	254	33	67	433
	63.2%	43.7%	34.4%	16.8%	36.0%
$20,000 - $39,999	32	182	30	74	318
	25.6%	31.3%	31.3%	18.5%	26.5%
$40,000 or greater	14	145	33	259	451
	11.2%	25.0%	34.4%	64.8%	37.5%
Total	125	581	96	400	1202
	100.0%	100.0%	100.0%	100.0%	100.0%

Chi-Square = 218.96, p=0.000, Cramer's V = 0.302.

Figure 11.13. The relationship between education and income without prestige.

controlling for occupation, the original relationship between education and income will disappear (thus the added "X" to the casual diagram in Figure 11.12).

In order to test to see if occupation is in fact an intervening variable linking education and income, we need to go to the data to find a measure operationalizing occupation. The variable "prestige" measures the prestige of a person's occupation on a scale from 1 to 100. The categories for the variable are respondent's occupational prestige less than 43 or 43 and higher.[1] Since there are two categories of the control variable, we will need to look at two new tables to see if our causal argument is correct.

Figure 11.13 shows the original table testing the relationship between education and income.

Figure 11.14 shows the stacked table including the original relationship within categories of the control variable prestige ("PrestigeHiLo"). As you can see, we used the same shading scheme to highlight each of the categories of the control variable "prestige."

The two top tables (the lightest top table and the medium shaded middle table) each report the relationship between education and income when controlling for the effects of occupational prestige. The first table shows the relationship accounting for the category of prestige less than 43, the second for the category of prestige 43 or higher.

If occupational prestige is in fact a link between education and income, we would expect to see that the two tables measuring occupational prestige would show no statistical significance between education and income. Yet both tables (prestige lo and prestige high) in Figure 11.15 indicate that even when controlling for occupational prestige, there remains a statistically significant relationship between education and income. Examining Figure 11.15 shows that, despite either high or low prestige

---

[1] We created these categories using SPSS to identify values that split the overall set of respondents into equal groups.

Income categories * Education Categories * PrestigeHiLo Crosstabulation

PrestigeHiLo	Income	Less Than HS	HS Grad	Some College	College grad	Total
Low	<$20,000	69	191	19	29	308
		65.7%	51.3%	46.3%	33.7%	51.0%
	$20,000 - $39,999	27	116	15	28	186
		25.7%	31.2%	36.6%	32.6%	30.8%
	$40,000 or greater	9	65	7	29	110
		8.6%	17.5%	17.1%	33.7%	18.2%
Total		105	372	41	86	604
		100.0%	100.0%	100.0%	100.0%	100.0%
High	<$20,000	10	63	14	38	125
		50.0%	30.1%	25.5%	12.1%	20.9%
	$20,000 - $39,999	5	66	15	46	132
		25.0%	31.6%	27.3%	14.6%	22.1%
	$40,000 or greater	5	80	26	230	341
		25.0%	38.3%	47.3%	73.2%	57.0%
Total		20	209	55	314	598
		100.0%	100.0%	100.0%	100.0%	100.0%
Total	<$20,000	79	254	33	67	433
		63.2%	43.7%	34.4%	16.8%	36.0%
	$20,000 - $39,999	32	182	30	74	318
		25.6%	31.3%	31.3%	18.5%	26.5%
	$40,000 or greater	14	145	33	259	451
		11.2%	25.0%	34.4%	64.8%	37.5%
Total		125	581	96	400	1202
		100.0%	100.0%	100.0%	100.0%	100.0%

Figure 11.14. The chi-square table including the control variable "prestige."

rankings, higher education categories show higher incomes than lower education rankings. Figures 11.15 and 11.16 show that both the control categories indicate significant
relationships and small to medium effect sizes (Cramer's V).

Thus our causal argument and diagram (Figure 11.12) are incorrect: the effects of
occupation do not intervene between education and income.

We've now controlled for two variables, gender and occupation, that do not make
the original relationship between education and income disappear. This is evidence that
education is a causal factor for income. There is a third possibility: it could be that

Chi-Square Tests

PrestigeHiLo		Value	df	Asymp. Sig. (2-sided)
Lo	Pearson Chi-Square	27.965[a]	6	.000
	Likelihood Ratio	27.330	6	.000
	Linear-by-Linear Association	25.239	1	.000
	N of Valid Cases	604		
High	Pearson Chi-Square	77.854[b]	6	.000
	Likelihood Ratio	78.100	6	.000
	Linear-by-Linear Association	67.364	1	.000
	N of Valid Cases	598		
Total	Pearson Chi-Square	218.960[c]	6	.000
	Likelihood Ratio	222.664	6	.000
	Linear-by-Linear Association	196.063	1	.000
	N of Valid Cases	1202		

Figure 11.15. Significance tests for prestige categories.

Symmetric Measures

PrestigeHiLo			Value	Approx. Sig.
Lo	Nominal by Nominal	Phi	.215	.000
		Cramer's V	.152	.000
		Contingency Coefficient	.210	.000
	N of Valid Cases		604	
High	Nominal by Nominal	Phi	.361	.000
		Cramer's V	.255	.000
		Contingency Coefficient	.339	.000
	N of Valid Cases		598	
Total	Nominal by Nominal	Phi	.427	.000
		Cramer's V	.302	.000
		Contingency Coefficient	.393	.000
	N of Valid Cases		1202	

Figure 11.16. The effect size results for the control categories of prestige.

PolParty * Income categories Crosstabulation

		Income categories			
		<$20,000	$20,000 - $39,999	$40,000 or greater	Total
PolParty	Democrat	146	117	128	391
		34.7%	38.0%	29.0%	33.4%
	Independent	190	127	171	488
		45.1%	41.2%	38.7%	41.7%
	Republican	85	64	143	292
		20.2%	20.8%	32.4%	24.9%
Total		421	308	442	1171
		100.0%	100.0%	100.0%	100.0%

Chi Square = 22.613, p=0.000, Cramer's V = 0.100.

Figure 11.17. Chi-square results between income and political party.

income has more than one cause. Perhaps gender and/or occupation, along with education, impact income. Unfortunately, when using a two-dimensional table like a crosstab, it is difficult to test for this possibility. Although we could add more than one control variable to a cross-tabular equation, it becomes increasingly difficult to read the multiple tables. Regression analysis is a better statistical technique to test for the possibility of multiple causes, which we describe later in the chapter. Before we do, let's revisit antecedent and intervening arguments using a new example.

The Effect of Income on Politics

Let us look at the relationship between socioeconomic status and politics. Previous research has found links between these two variables; those with higher socioeconomic status tend to identify with the Republican Party; while those with lower socioeconomic status tend to identify with the Democratic Party. Using income as a measure of socioeconomic status and the variable "PolParty" for political party, we hypothesize that those with higher income will be more likely to report being a member of the Republican Party. Testing this hypothesis we find the results shown in Figure 11.17.

First, look at the percentages in the table. Does it appear that as income increases, more people identify as Republican? Remember, begin with the independent variable and compare across columns (what happens to the percentage of people who report being Republican as the categories of income change/increase). The direction of our hypothesis appears to be borne out in the table data. As income goes from less than $20,000 (20.2 percent), to those making $20,000 to $39,999 (20.8 percent), to those making $40,000 or higher (32.4 percent), more people identify as Republican. Is there a statistical relationship, however? The significance statistic is the probability, which is less than 0.05 ($p = 0.000$), and it means that the relationship is statistically significant. Is the strength of the relationship relatively weak, moderate, or strong? The Cramer's

V indicates a small but meaningful effect size. What do you conclude? There is a significant relationship between income and political party.

Two of the steps for determining causation have been put in place. By choosing income as our independent variable, we are inferring that income has an effect on political party preference (thus, income/SES comes first in time). Next, we've determined that there is a significant relationship (correlation) between income and political party preference. Finding a significant correlation, however, is not enough to help us determine causation. We need to test for spuriousness.

What variable may impact the relationship between income and political party? Some research suggests that race may be a factor for political party preference, that racial/ethnic minority groups are more likely to identify as Democrat while whites are more likely to identify as Republican. How might race impact the original relationship between income and political party preference? Would race be an antecedent variable (a variable that precedes both income and political party preference—something about race causes changes in both income and party preference) or an intervening variable (a variable that comes between income and party preference—something about income causes changes to race and race then causes changes to party preference)?

In thinking about the causal logic, race is a characteristic that is ascribed at birth; thus it has to come before both of the other variables (level of income cannot change a person's racial/ethnic category). Therefore, race must be an antecedent variable (Figure 11.18).

Let us test the causal relationship between income and political party preference using the variable RaceCat as an antecedent variable, following the process we used earlier. Figure 11.19 shows the stacked table including the categories for race (white and black) using the shading process we have used with earlier tables.[2]

Figures 11.20 and 11.21 show the significance and effect size results, respectively. As you can see in Figure 11.20, the table showing the control condition for "White" is significant ($p = 0.012$); the control condition for "Black" is not significant ($p = 0.922$). The effect sizes shown in Figure 11.21 also show a divergence between the control categories. The effect size for the first category (White) is $V = 0.085$; the effect size for the second category (Black) is $V = 0.052$. Both of these effect sizes would be considered small, with the second ($V = 0.052$ for "Black") much less significant and meaningful.

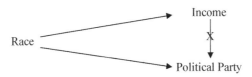

Figure 11.18. Race as an antecedent variable.

---

[2] We used these GSS categories to indicate race in this analysis, although other analyses could introduce additional race categories.

PolParty * Income categories * RaceCat Crosstabulation

RaceCat		<$20,000	Income categories		Total
			$20,000 - $39,999	$40,000 or greater	
White	Democrat	78	70	96	244
		26.8%	30.4%	25.7%	27.3%
	Independent	141	100	144	385
		48.5%	43.5%	38.6%	43.1%
	Republican	72	60	133	265
		24.7%	26.1%	35.7%	29.6%
	Total	291	230	373	894
		100.0%	100.0%	100.0%	100.0%
Black	Democrat	46	41	23	110
		64.8%	65.1%	62.2%	64.3%
	Independent	19	19	11	49
		26.8%	30.2%	29.7%	28.7%
	Republican	6	3	3	12
		8.5%	4.8%	8.1%	7.0%
	Total	71	63	37	171
		100.0%	100.0%	100.0%	100.0%
Total	Democrat	124	111	119	354
		34.3%	37.9%	29.0%	33.2%
	Independent	160	119	155	434
		44.2%	40.6%	37.8%	40.8%
	Republican	78	63	136	277
		21.5%	21.5%	33.2%	26.0%
	Total	362	293	410	1065
		100.0%	100.0%	100.0%	100.0%

Figure 11.19. The chi-square analysis showing the categories of race.

In looking at the tables in Figures 11.20 and 11.21, there are mixed results. How do we interpret these findings? In using a third factor as a control variable, we are looking for the effect on the original significant relationship when adding a third variable. In terms of causation, we are looking to see if the original relationship maintains statistical significance, that even when controlling for a third factor the relationship is not spurious. One table for one category of race shows significance; the other table does not. What do we conclude? *If any table of the control variable shows that the original relationship is still significant, the original relationship cannot be spurious.* A

Chi-Square Tests

RaceCat		Value	df	Asymp. Sig. (2-sided)
White	Pearson Chi-Square	12.771[a]	4	.012
	Likelihood Ratio	12.637	4	.013
	Linear-by-Linear Association	4.513	1	.034
	N of Valid Cases	894		
Black	Pearson Chi-Square	.916[b]	4	.922
	Likelihood Ratio	.960	4	.916
	Linear-by-Linear Association	.007	1	.934
	N of Valid Cases	171		
Total	Pearson Chi-Square	19.281[c]	4	.001
	Likelihood Ratio	18.953	4	.001
	Linear-by-Linear Association	9.764	1	.002
	N of Valid Cases	1065		

Figure 11.20. The significance tables for the chi-square analysis.

Symmetric Measures

RaceCat			Value	Approx. Sig.
White	Nominal by Nominal	Phi	.120	.012
		Cramer's V	.085	.012
		Contingency Coefficient	.119	.012
	N of Valid Cases		894	
Black	Nominal by Nominal	Phi	.073	.922
		Cramer's V	.052	.922
		Contingency Coefficient	.073	.922
	N of Valid Cases		171	
Total	Nominal by Nominal	Phi	.135	.001
		Cramer's V	.095	.001
		Contingency Coefficient	.133	.001
	N of Valid Cases		1065	

Figure 11.21. The effect size tables for the chi-square analysis.

spurious relationship is one that is no longer statistically significant. All tables (in this case both tables) would have to show *no* significant relationship in order for us to conclude that the relationship between income and political party is spurious when controlling for race. This is not the case. We conclude that the relationship between income and political party is not spurious, even when controlling for race. Since one table of the control variable maintains statistical significance, the relationship between income and political party is not spurious when controlling for race.

What does this tell you about the causal diagram? Do the table results support the hypothesis that race is an antecedent variable? If a variable is an antecedent variable, the original relationship would no longer be significant. Since one table of the control variable shows significance, race cannot be an antecedent variable. The diagram is not supported by the data.

Income and Voting Example

Let's explore the relationship between social class and political participation. Using income as a measure of social class and whether or not a person voted as a measure of political participation, we hypothesize that as income increases, people will be more likely to have voted. Figure 11.22 shows the relationship between income and voting.

First, are the data arrayed in the direction we hypothesized? As levels of income increase, what happens to the percentage of people who say they voted? The direction of the data support the hypothesis. The percentage of the those who voted increases across the income categories from 63 percent (less than \$35,000) to 71.5 percent (\$35,000 to \$64,999) to 78.9 percent (\$65,000 and higher). Clearly, as level of income increases, more people report voting. Is there a statistically significant relationship

		SUMMARY: HOUSEHOLD INCOME			
		<35,000	\$35K-64.9K	65,000 +	Total
VOTED		250	291	370	911
		63.0%	71.5%	78.9%	71.6%
DID NOT		147	116	99	362
		37.0%	28.5%	21.1%	28.4%
Total		397	407	469	1273
		100.0%	100.0%	100.0%	100.0%

Chi Square = 26.775, p = 0.000, Cramer's V = 0.145.

Figure 11.22. The chi-square results between income and voting.

Figure 11.23. Education as an antecedent variable.

between income and voting? Yes, the probability level is less than 0.05 ($p = 0.000$), illustrating a significant relationship between income and voting. The effect size, however, is somewhat small (Cramer's V = 0.146). The data support our hypothesis: as the level of income increases, people are more likely to report having voted.

Again, two of the three criteria for causation have been met: time order and correlation. Now we need to test for spuriousness. Using education as the control variable, how would you diagram a causal argument that though income impacts voting, education impacts both income and voting? As a variable that impacts both original variables, education would be an antecedent variable.

When controlling for education, we find the results shown in Figure 11.24. As before, the stacked tables are shown with different shading to allow you to identify the control conditions. In looking at the two tables, there are consistent results.

The tables in Figure 11.25 show statistically insignificant relationships for all categories of the control variable ($p = 0.105$ for less than a high school education, $p = 0.336$ for a high school education, $p = 0.178$ for a college education, and $p = 0.124$ for having an advanced degree). How do we interpret these findings? When controlling for a third factor the question we ask is, "Does the original relationship become spurious?" All of the tables of the control variable show that there is no significant relationship between income and voting when controlling for each category of education. A spurious relationship exists between income and voting when there is no longer any statistical significance. Since none of the tables show statistical significance, we conclude that when controlling for education there is no longer a statistically significant relationship between income and voting. The relationship between income and voting is spurious when controlling for education.

What do these results tell us about the causal diagram? Do the table results support the hypothesis that education is an antecedent variable? If a variable is an antecedent variable, the original relationship would no longer be significant. Since the tables show that the relationship is no longer significant, we determine that education is, in fact, an antecedent variable. The causal diagram is supported by the data.

Keep in mind that there is nothing about the data specifically that tells us if a third variable is an antecedent or intervening variable. For either case we look to the same indicators to determine if an original relationship exists or not when controlling for another variable. In thinking through causal logic, we have to determine theoretically if a new variable comes before (antecedent) or between (intervening) the original variables.

SUMMARY: RESPONDENT EDUCATION		SUMMARY: HOUSEHOLD INCOME			
		<35,000	$35K-64.9K	65,000 +	Total
NO H.S. DEGREE	VOTED	28	9	4	41
		45.9%	45.0%	100.0%	48.2%
	DID NOT	33	11	0	44
		54.1%	55.0%	.0%	51.8%
	Total	61	20	4	85
		100.0%	100.0%	100.0%	100.0%
H.S. DEGREE	VOTED	150	130	112	392
		62.0%	65.0%	69.1%	64.9%
	DID NOT	92	70	50	212
		38.0%	35.0%	30.9%	35.1%
	Total	242	200	162	604
		100.0%	100.0%	100.0%	100.0%
COLLEGE	VOTED	63	112	166	341
		76.8%	77.2%	84.3%	80.4%
	DID NOT	19	33	31	83
		23.2%	22.8%	15.7%	19.6%
	Total	82	145	197	424
		100.0%	100.0%	100.0%	100.0%
ADV DEGREE	VOTED	9	37	82	128
		75.0%	94.9%	83.7%	85.9%
	DID NOT	3	2	16	21
		25.0%	5.1%	16.3%	14.1%
	Total	12	39	98	149
		100.0%	100.0%	100.0%	100.0%
Total	VOTED	250	288	364	902
		63.0%	71.3%	79.0%	71.5%
	DID NOT	147	116	97	360
		37.0%	28.7%	21.0%	28.5%
	Total	397	404	461	1262
		100.0%	100.0%	100.0%	100.0%

Figure 11.24. Chi-square test results including control variable education.

Chi-Square Tests

SUMMARY: RESPONDENT EDUCATION		Value	df	Asymp. Sig. (2-sided)
NO H.S. DEGREE	Pearson Chi-Square	4.510^a	2	.105
	Likelihood Ratio	6.050	2	.049
	Linear-by-Linear Association	1.805	1	.179
	N of Valid Cases	85		
H.S. DEGREE	Pearson Chi-Square	2.180^b	2	.336
	Likelihood Ratio	2.196	2	.334
	Linear-by-Linear Association	2.159	1	.142
	N of Valid Cases	604		
COLLEGE	Pearson Chi-Square	3.451^c	2	.178
	Likelihood Ratio	3.490	2	.175
	Linear-by-Linear Association	2.826	1	.093
	N of Valid Cases	424		
ADV DEGREE	Pearson Chi-Square	4.172^d	2	.124
	Likelihood Ratio	4.684	2	.096
	Linear-by-Linear Association	.105	1	.746
	N of Valid Cases	149		
Total	Pearson Chi-Square	26.748^e	2	.000
	Likelihood Ratio	26.840	2	.000
	Linear-by-Linear Association	26.713	1	.000
	N of Valid Cases	1262		

Figure 11.25. Significance tests for education categories.

REGRESSION ANALYSIS AND TESTING FOR SPURIOUSNESS

In Chapter 10 we discussed regression analysis. Regression analysis allows us to test the effects of multiple independent variables on the dependent variable. While our earlier examples of testing for spuriousness in cross-tabs limited the number of

Symmetric Measures

SUMMARY: RESPONDENT EDUCATION			Value	Approx. Sig.
NO H.S. DEGREE	Nominal by Nominal	Phi	.230	.105
		Cramer's V	.230	.105
		Contingency Coefficient	.224	.105
	N of Valid Cases		85	
H.S. DEGREE	Nominal by Nominal	Phi	.060	.336
		Cramer's V	.060	.336
		Contingency Coefficient	.060	.336
	N of Valid Cases		604	
COLLEGE	Nominal by Nominal	Phi	.090	.178
		Cramer's V	.090	.178
		Contingency Coefficient	.090	.178
	N of Valid Cases		424	
ADV DEGREE	Nominal by Nominal	Phi	.167	.124
		Cramer's V	.167	.124
		Contingency Coefficient	.165	.124
	N of Valid Cases		149	
Total	Nominal by Nominal	Phi	.146	.000
		Cramer's V	.146	.000
		Contingency Coefficient	.144	.000
	N of Valid Cases		1262	

Figure 11.26. The effect size results for the control categories of education.

variables we can use, regression gives us the opportunity to see the effects of several independent variables. Using census data (aggregate-level data) for the states, let's look at how to use regression analysis to test the relationship we used as an example earlier in the chapter—between education and income. Since we are using aggregate level data, we operationalize education using a measure for "percent of the population with a bachelor's degree or higher" and we operationalize income using "median family income." Following our previous example, we hypothesize that as percentage

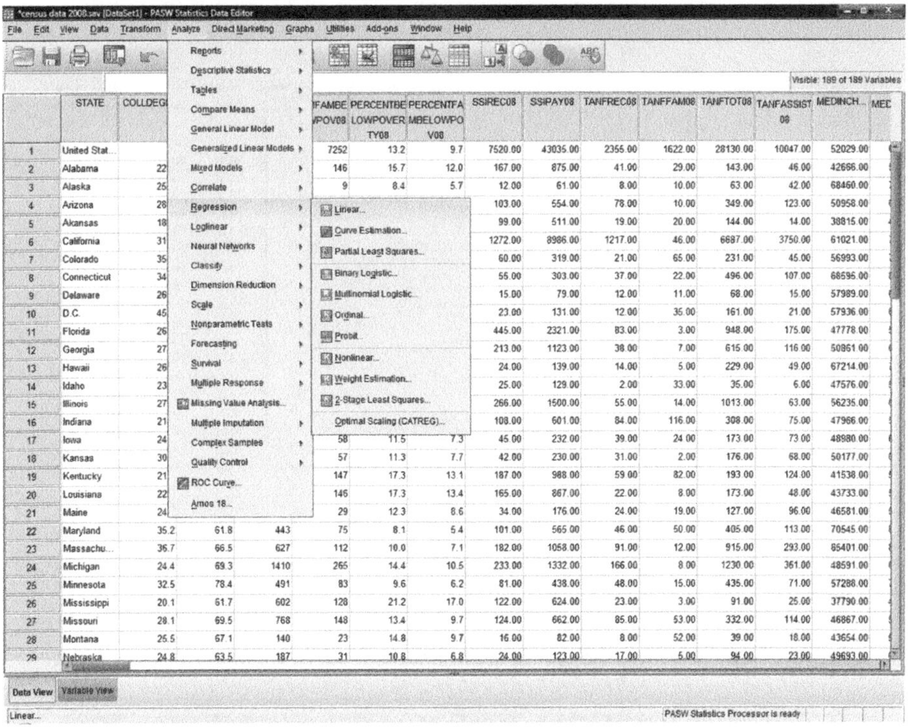

Figure 11.27. The Analyze task in SPSS using linear regression.

of the population with a bachelor's degree or higher increases, the median family income in states will also increase, a positive directionality. Let's go to SPSS to analyze the data.

Figure 11.27 shows you how to use the Analyze menu in SPSS to select regression analysis for linear regression. Once you've selected your independent variable (percentage college degrees) and dependent (median family income) variable (Figure 11.28), click the statistics button in order to select "Model fit" and "R squared change" (Figure 11.29).

Once these statistics have been specified, SPSS will take you to the output screen where three tables (Figure 11.30) show the results of the regression analysis of percentage of the population with college degrees in a state and median family income.

We focus on just two of these output tables, the table for "Model Summary" and for "Coefficients." Figure 11.31 shows the model summary for the regression analysis. We have used shading to highlight the important pieces of the analysis and to focus your attention. The "R Square" figure (0.505) is very important because it indicates *how much variance in the dependent variable (family income) is explained (or contributed) by the independent variable (education).* These results show that over 50 percent of the variance in family income is due to education. (Translating the R square figure

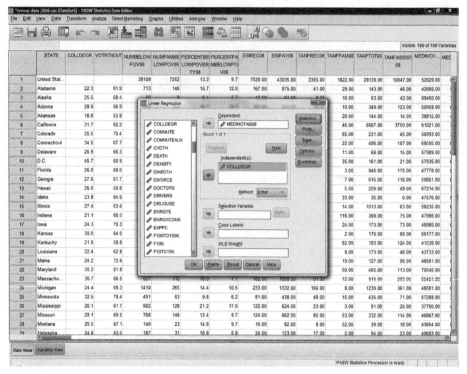

Figure 11.28. Selecting the independent and dependent variables for regression.

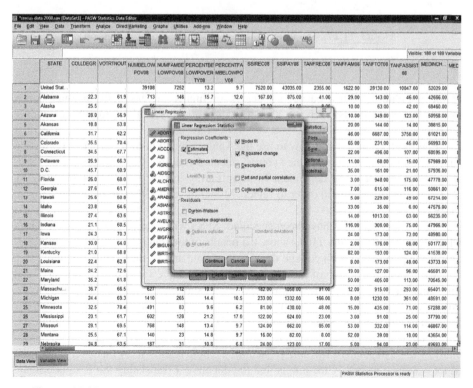

Figure 11.29. Selecting "Model fit" and "R squared change" in the Statistics task.

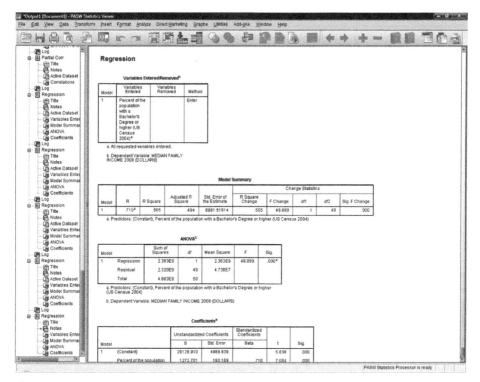

Figure 11.30. SPSS regression output for percentage college degree and median family income.

Model Summary

Model	R	R Square	Adjusted R Square	Std. Error of the Estimate	Change Statistics				
					R Square Change	F Change	df1	df2	Sig. F Change
1	.710[a]	.505	.494	6881.51914	.505	49.899	1	49	.000

[a] Predictors: (Constant), Percent of the population with a Bachelor's Degree or higher (US Census 2004).

Figure 11.31. Model summary of percentage college degrees on median family income.

of 0.505 to the amount of explained variance is simply a matter of moving the decimal place to the right two places so that it becomes 50.5 percent). This is a very large effect and indicates a *strong* relationship between the variables.

Figure 11.32 shows the findings that help us to see whether the predictor (percentage of the population with bachelor's degrees or higher) is significant in its relationship to median family income. The significance statistic shown in the table for the independent variable is $p = 0.000$ (see shaded portion). Note, however, that there is also a

Coefficients[a]

Model	Unstandardized Coefficients		Standardized Coefficients			Correlations		
	B	Std. Error	Beta	t	Sig.	Zero-order	Partial	Part
1 (Constant)	29129.910	4989.939		5.838	.000			
Percent of the population with a Bachelor's Degree or higher (US Census 2004)	1272.701	180.169	.710	7.064	.000	.710	.710	.710

[a] Dependent variable: MEDIAN FAMILY INCOME 2008 (DOLLARS).

Figure 11.32. Coefficients for percentage college degrees on median family income.

significance statistic for the constant. When evaluating the relationship between the independent and dependent variable, make sure to use the correct significance statistic (for the independent variable, not the constant). Based on significance, we can see that there is a statistically significant relationship between the percentage of the population with college degrees and median family income.

What about the effect size? We described effect size in the chapter on regression as the squared correlation (R Squared, as just described). We can also look at another indicator of effect size, the "standardized Beta." The standardized Beta is a correlation coefficient measured like a Pearson's r to indicate the importance (i.e., effect size) of the predictor. The standardized Beta for percentage of the population with college degrees is 0.710 (see shaded portion), which is considered relatively strong. The standardized Beta is also a *positive* number, indicating that the relationship between percentage of college degrees and median family income in states is positive: as the percentage of the population with a college degree increases, so does the median family income in states, which is what we hypothesized.

Detecting Spuriousness

Now that we've determined that there is a significant correlation between our two variables, we need to think about the possibility that we have a spurious relationship. When we tested this relationship using cross-tabs, we controlled for the effects of gender, hypothesizing that gender served as an antecedent variable. Using the same causal logic, that gender serves as an antecedent variable (since it comes before percentage of college degrees and median family income), let's use the variable "percentage of the population that is female" to test the effect of gender and be consistent between analyses.

In regression analysis, we treat our new variable as a second *independent variable*. When we go to the Analyze task and select linear regression, we simply add our gender variable to the independent variable section as shown in Figure 11.33.

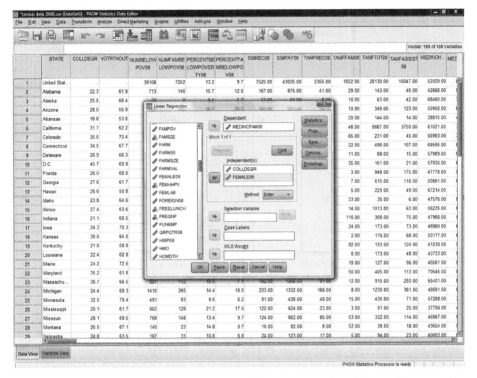

Figure 11.33. Adding the control variable of "percentage female" to the regression analysis.

Model Summary

Model	R	R Square	Adjusted R Square	Std. Error of the Estimate	Change Statistics				
					R Square Change	F Change	df1	df2	Sig. F Change
1	.760ᵃ	.578	.561	6415.65326	.578	32.892	2	48	.000

ᵃ Predictors: (Constant), percent of the population who are female 2009 (SA 2011), Percent of the population with a Bachelor's Degree or higher (US Census 2004).

Figure 11.34. Model summary with gender included in the analysis.

The appropriate statistics should still be selected (Model Summary and R squared change). Again, we focus on just two tables from the SPSS output. The SPSS Model Summary is shown in Figure 11.34.

As you can see, the R Square figure *when both of the predictors are acting together to correlate with the dependent variable* (family income), is 0.578. Therefore, adding a second predictor to the regression analysis adds 0.073 to the original R Square figure (0.505) where education was the only predictor of family income. (We arrived at 0.073

Coefficients[a]								
	Unstandardized Coefficients		Standardized Coefficients			Correlations		
Model	B	Std. Error	Beta	t	Sig.	Zero-order	Partial	Part
1 (Constant)	195314.760	57614.466		3.390	.001			
Percent of the population with a Bachelor's Degree or higher (US Census 2004)	1362.552	170.818	.760	7.977	.000	.710	.755	.748
Percent of the population who are female 2009 (SA 2011)	-3330.707	1150.949	-.276	-2.894	.006	-.138	-.385	-.271

[a] Dependent variable: MEDIAN FAMILY INCOME 2008 (DOLLARS).

Figure 11.35. Coefficients for percentage of the population with a college degree and percentage of the population that is female.

simply by subtracting -0.505 from 0.578, yielding 0.073). Thus adding the second predictor (Female percentage of the population) increases how much additional variance in the dependent variable is explained by virtue of adding one additional predictor. In this case, adding female percentage of the population to the analysis increases how much we know of the variance in income by 7.3 percent. This may seem like a small amount, but it is meaningful. The next question is whether this additional amount is a *significant* (nonchance) amount.

Since we are using percentage of the population that is female as a control variable, the first statistic we need to check is the significance statistic for percentage of the population with a college degree. This is shown in Figure 11.35. The probability shown in the top of the panel, with lighter shading ($p = 0.000$), indicates that even when controlling for percentage female, the relationship between percentage of the population with a college degree and median family income is still significant, and therefore not spurious. (Thus percentage female is not an antecedent variable.) The Standardized Beta (0.760) is also strong and positive. If the significance of education disappeared (i.e., showing a significance value greater than 0.05) when we added percentage female, then education would be considered spurious.

One benefit of using regression analysis is that you notice in Figure 11.35 that more information is given about the effect of percentage female on median family income. Whereas in cross-tabs, it is difficult to know the precise relationship of the control variable to the dependent variable, in regression we can see that percentage of the population that is female is also a significant factor in median family income ($p = 0.006$). The standardized Beta for percentage female is equal to −0.276; it is a negative and small effect illustrating that as percentage female in states increases, median family income *decreases*.

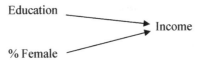

Figure 11.36. The regression results of predicting family income from Education and % Female.

Figure 11.37. The regression results indicating a potential spurious relationship.

Using our earlier models, we can show these results. The model in Figure 11.36 shows the results of our regression analysis. Both % Female and Education have a significant relationship to Income. Adding % Female did not reduce or eliminate the original relationship between Education and Income.

The next model, shown in Figure 11.37, indicates that had the relationship between Education and Income disappeared when % Female was added (indicated by the disappearance of the arrow between Education and Income), then the Education–Income relationship would be spurious.

Regression Example

Let's use another example from the cross-tabs section to test a second regression analysis. Earlier in the chapter we hypothesized that those with higher levels of income would be more likely to participate in politics. If this is true at the individual level, then we would expect to see a similar relationship at the aggregate level. Operationalizing social class with median family income, and operationalizing political participation with "percentage of the population who voted in the 2008 presidential election," we hypothesize that as median family income increases, so will the voter turnout in states.

Having selected the appropriate variables and statistics in SPSS, we look at the output for the relationship between median family income and voter turnout in states. Figure 11.38 gives the model summary for the analysis.

As you can see from Figure 11.38, the R Square is 0.100, indicating that 10 percent of the variance in voter turnout is explained by family income. This is a small, but meaningful, effect size.

Next, we turn to Figure 11.39 to look at the significance statistics and correlation coefficients to see additional indicators of effect size and directionality. In the Coefficients table we see that median family income has a significant impact on voter turnout ($p = 0.024$). The standardized Beta of 0.317 shows a positive relationship with a moderate effect size. Therefore we can conclude that the data support our hypothesis; states with higher median family income have higher voter turnout.

Now that we have determined that the relationship is significant, we need to think about testing for spuriousness. In our cross-tab example addressing this relationship,

Model Summary

Model	R	R Square	Adjusted R Square	Std. Error of the Estimate	Change Statistics				
					R Square Change	F Change	df1	df2	Sig. F Change
1	.317[a]	.100	.082	5.4113	.100	5.458	1	49	.024

[a] Predictors: (Constant), MEDIAN FAMILY INCOME 2008 (DOLLARS).

Figure 11.38. Model summary for median family income on voter turnout.

Coefficients[a]

Model		Unstandardized Coefficients		Standardized Coefficients	t	Sig.
		B	Std. Error	Beta		
1	(Constant)	52.148	5.095		10.236	.000
	MEDIAN FAMILY INCOME 2008 (DOLLARS)	.000	.000	.317	2.336	.024

[a] Dependent variable: Percent of population turning out for the 2008 presidential election.

Figure 11.39. Coefficients for median family income on voter turnout.

Model Summary

Model	R	R Square	Adjusted R Square	Std. Error of the Estimate	Change Statistics				
					R Square Change	F Change	df1	df2	Sig. F Change
1	0.345[a]	0.119	0.082	5.4095	0.119	3.247	2	48	0.048

[a] Predictors: (Constant), Percent of the population with a Bachelor's Degree or higher (US Census 2004), MEDIAN FAMILY INCOME 2008 (DOLLARS).

Figure 11.40. Model summary with median family income and percentage college degrees.

we used a variable for education to test for spuriousness, hypothesizing that education served as an antecedent variable (coming before the other two variables). For the sake of consistency, let's rely on our previous causal logic to use education as a control variable, operationalizing it as "percentage of the population with a bachelor's degree or higher."

Having added the variable for percentage college degree to our original analysis, Figure 11.40 shows the model summary for the analysis. You can see that adding another predictor to the model does not increase the R Square very much. Only 0.019 (or 1.9%) is added to the explanation of variance in voter turnout by adding education (0.019 = 0.119 – 0.100 from the R Square figures).

Figure 11.41 shows the significance and correlation coefficients for the analysis. Since we are testing for spuriousness, using percentage of college degrees as a control

Coefficients[a]

Model		Unstandardized Coefficients		Standardized Coefficients		
		B	Std. Error	Beta	t	Sig.
1	(Constant)	51.756	5.108		10.133	.000
	MEDIAN FAMILY INCOME 2008 (DOLLARS)	.000	.000		.923	
	Percent of the population with a Bachelor's Degree or higher (US Census 2004)	.205	.201	.196	1.016	.315

[a] Dependent variable: Percent of population turning out for the 2008 presidential election.

Figure 11.41. Coefficients for median family income and percentage college degrees on voter turnout.

variable, we first want to evaluate whether or not the original independent variable maintains statistical significance. The significance statistic for median family income, when controlling for percentage college degrees, is now equal to 0.361; median family income is no longer significant when controlling for percentage college degrees. Therefore we conclude that the relationship between median family income and voter turnout in states is spurious when controlling for percentage of college degrees.

Again, because we are using regression analysis, we can also evaluate the effect of the control variable on the dependent variable. The significance statistic ($p = 0.315$) illustrates that percentage college degrees does not have an effect on voter turnout. The standardized Beta shows a positive but small effect size (0.196).[3]

Necessary and Sufficient Conditions
When researchers think about causal relationships, they must consider structured ways of thinking about the relationships among the study concepts. In sections of this book, we illustrated such concepts as spuriousness, indicating that the causal relationship between two variables (e.g., ice cream sales and crime rates) cannot be established if there is another variable confounding this relationship (season of the year). Considering necessary and sufficient conditions helps us to think more clearly and logically about the causal relationships among study variables.

(Continued)

---

[3] You will note that neither variable, median family income or percentage college degree, is significant in the analysis when they are both present as predictors. This may be due to an interaction effect that impacts the analysis. For our purposes in this analysis, we will only highlight that when both variables are present; the original relationship between median family income and voter turnout is spurious.

Necessary conditions are those that must be present for another condition to follow. For example, measuring employee satisfaction in a certain company requires that the study subjects be those who are hired by the company. Or, in order to be diagnosed with cholera, a patient must be infected by *Vibrio cholerae.*

Sufficient conditions are those that, if they are present, will result in a certain effect. Drinking a triple espresso coffee drink will ensure that my blood will have a higher level of caffeine. However, there are other ways to get caffeinated, like drinking energy drinks, eating chocolate, drinking colas, and so on. The espresso drink is sufficient but not necessary for caffein.

In social research, it is difficult to think of situations in which both necessary and sufficient conditions are present. If so, it would help to ensure confidence in causal relationships. If you are interested in the relationship between poverty and crime, for example, you could speak causally if you could establish that crime could only occur when poverty is present (which cannot be established) and that, if poverty is present, it would inevitably lead to crime (which it would not). In real-world social research, there are typically constellations of influencing factors on a study problem that detract from a researcher's ability to make causal conclusions.

Exercise: Testing Spuriousness in Crosstabs
Our public discourse often addresses the relationship between faith and science. Many think faith and science are incompatible meaning systems, yet others see the two systems as complementary (faith picks up where science leaves off). The general understanding is that the more conservative a person's religion, the more distrustful of science he or she is. Using data from the 2010 GSS, we can test this relationship. We operationalize attitudes toward science with a question that asks if respondents agree or disagree with this statement: "We believe too often in science, and not enough in feelings and faith." We operationalize conservative religion with a measure of respondents' fundamentalism/liberalism of religion (fundamentalist being the most conservative). We hypothesize that the more fundamentalist a respondent's religion, the more likely the respondent will be to agree that we believe too often in science, and not enough in faith.

Based on this hypothesis, answer the following questions.

1. What are the units of analysis for this hypothesis? _____
2. What is the independent variable for this hypothesis? _____
3. What is the hypothesized direction of the relationship? _____

Turning to SPSS we use the crosstab analysis to select the variables for analysis. In analyzing the relationship, we find the following results:

Faith in science * HOW FUNDAMENTALIST IS R CURRENTLY Crosstabulation					
		HOW FUNDAMENTALIST IS R CURRENTLY			
		FUNDAMENT ALIST	MODERATE	LIBERAL	Total
Faith in science	Agree	213	240	117	570
		60.7%	45.3%	27.7%	43.7%
	Neither	76	144	117	337
		21.7%	27.2%	27.7%	25.9%
	Disagree	62	146	188	396
		17.7%	27.5%	44.5%	30.4%
Total		351	530	422	1303
		100.0%	100.0%	100.0%	100.0%

Chi-Square = 99.342, p=0.000, Cramer's V = 0.195.

Based on the data presented, answer the following questions:

1. What is the significance statistic for the independent variable? _____
2. What is the value of the correlational coefficient? _____
3. What is the effect size for the coefficient? _____
4. What is the direction of the relationship (based on table percentages)? _____
5. Do these results support the hypothesis? _____
6. What do you conclude about the relationship between religious funda- mentalism and attitudes toward science (that is, what do these findings tell us)? _____

Based on these findings, we may need to test for spuriousness. Perhaps level of education is what impacts attitudes toward science; that the fundamentalism/liberalism of a respondent's religion impacts the level of edu- cation obtained (fundamentalists obtaining lower levels of education; liberals obtaining higher levels of education), and then it's the level of education that impacts a person's attitude toward science. Therefore, education is a link between respondent's religion and attitude toward science. Diagram this causal argument.

(Continued)

When controlling for level of education, we find the following results:

Faith in science * HOW FUNDAMENTALIST IS R CURRENTLY * EducationCat Crosstabulation

EducationCat			HOW FUNDAMENTALIST IS R CURRENTLY			
			FUNDAMENTALIST	MODERATE	LIBERAL	Total
<HS	Faith in science	Agree	12	24	5	41
			63.2%	60.0%	38.5%	56.9%
		Neither	2	6	4	12
			10.5%	15.0%	30.8%	16.7%
		Disagree	5	10	4	19
			26.3%	25.0%	30.8%	26.4%
	Total		19	40	13	72
			100.0%	100.0%	100.0%	100.0%
HS	Faith in science	Agree	116	100	52	268
			68.6%	50.5%	40.9%	54.3%
		Neither	32	57	41	130
			18.9%	28.8%	32.3%	26.3%
		Disagree	21	41	34	96
			12.4%	20.7%	26.8%	19.4%
	Total		169	198	127	494
			100.0%	100.0%	100.0%	100.0%
Coll grad	Faith in science	Agree	77	103	42	222
			55.4%	46.2%	20.7%	39.3%
		Neither	37	54	60	151
			26.6%	24.2%	29.6%	26.7%
		Disagree	25	66	101	192
			18.0%	29.6%	49.8%	34.0%
	Total		139	223	203	565
			100.0%	100.0%	100.0%	100.0%
Adv degree	Faith in science	Agree	8	11	18	37
			33.3%	16.4%	23.1%	21.9%
		Neither	5	27	11	43
			20.8%	40.3%	14.1%	25.4%
		Disagree	11	29	49	89
			45.8%	43.3%	62.8%	52.7%
	Total		24	67	78	169
			100%	100%	100%	100%

Chi-Square Tests

EducationCat		Value	df	Asymp. Sig. (2-sided)
<HS	Pearson Chi-Square	3.146[a]	4	.534
	Likelihood Ratio	2.992	4	.559
	Linear-by-Linear Association	.734	1	.391
	N of Valid Cases	72		
HS	Pearson Chi-Square	24.738[b]	4	.000
	Likelihood Ratio	25.139	4	.000
	Linear-by-Linear Association	21.569	1	.000
	N of Valid Cases	494		
Coll grad	Pearson Chi-Square	57.475[c]	4	.000
	Likelihood Ratio	60.420	4	.000
	Linear-by-Linear Association	53.465	1	.000
	N of Valid Cases	565		
Adv degree	Pearson Chi-Square	15.216[d]	4	.004
	Likelihood Ratio	14.992	4	.005
	Linear-by-Linear Association	2.336	1	.126
	N of Valid Cases	169		

Symmetric Measures

EducationCat			Value	Approx. Sig.
<HS	Nominal by Nominal	Phi	.209	.534
		Cramer's V	.148	.534
	N of Valid Cases		72	
HS	Nominal by Nominal	Phi	.224	.000
		Cramer's V	.158	.000
	N of Valid Cases		494	
Coll grad	Nominal by Nominal	Phi	.319	.000
		Cramer's V	.226	.000
	N of Valid Cases		565	
Adv degree	Nominal by Nominal	Phi	.300	.004
		Cramer's V	.212	.004
	N of Valid Cases		169	

(Continued)

Based on the results presented in these three tables, answer the following questions:

1. What are the significance statistics for each category of the control variable?
   a. Less than high school _____
   b. High school _____
   c. College _____
   d. Advanced degree _____
2. What are the correlational coefficients for each category of the control variable?
   a. Less than high school _____
   b. High school _____
   c. College _____
   d. Advanced degree _____
3. When controlling for level of education, is there a relationship between respondent fundamentalism/liberalism of religion and attitude toward science? Yes   No
4. Do these results support the causal diagram that you drew? Yes   No
5. Why or why not? _____
   _____
   _____
   _____
   _____

Exercise: Testing for Spuriousness in Regression

Social scientists are often concerned about the impact of poverty on individuals and societies. Research had linked poverty levels with higher teen birthrates. Using census data (aggregate data), we test the relationship between poverty and teen birthrates. Operationalizing poverty with a measure of "percentage of individuals living below the poverty line" and operationalizing teen births with "births to teenage mothers as a percentage of the total birth rate," we hypothesize that states with a higher percentage of people living in poverty will have higher teen birthrates.

Based on this hypothesis, answer the following questions.

1. What are the units of analysis for this hypothesis? _____
2. What is the independent variable for this hypothesis? _____
3. What is the hypothesized direction of the relationship? _____

Turning to SPSS, we use the linear regression analysis in the program to select the variables for analysis. In analyzing the variables, we find these results:

Model Summary

Model	R	R Square	Adjusted R Square	Std. Error of the Estimate	Change Statistics				
					R Square Change	F Change	df1	df2	Sig. F Change
1	.831[a]	.691	.685	1.40599	.691	111.727	1	50	.000

[a] Predictors: (Constant), percent of individuals living below the poverty line in 2008.

Coefficients[a]

Model		Unstandardized Coefficients		Standardized Coefficients	t	Sig.
		B	Std. Error	Beta		
1	(Constant)	1.304	.879		1.484	.144
	percent of individuals living below the poverty line in 2008	.706	.067	.831	10.570	.000

[a] Dependent variable: births to teenage mothers as a percent of the total 2007 (SA 2011).

Based on the data presented, answer the following questions:

4. What is the significance statistic for the independent variable? _____
5. What is the value of the appropriate correlational coefficient? _____
6. What is the direction of the coefficient? _____
7. What is the effect size for the coefficient? _____
8. How much of the variance in the dependent variable is explained by the independent variable? _____
9. Do these results support the hypothesis? _____
10. What do you conclude about the relationship between poverty and teen birthrates (that is, what happens to one when we know what's happening to the other)? _____
_____
_____
_____

Based on these data, we may need to test for spuriousness. Let's hypothesize that it's not actually poverty that impacts teen birthrates, but level of education that impacts both levels of poverty and teen birthrates. Diagram this causal argument.

We operationalize education using the measure for "percentage of population with a bachelor's degree or higher" to test for spuriousness in the relationship between poverty and teen birthrates and see the following results:

(Continued)

				Std. Error	Change Statistics				
Model	R	R Square	Adjusted R Square	of the Estimate	R Square Change	F Change	df1	df2	Sig. F Change
1	.851ª	.724	.712	1.35689	.724	62.816	2	48	.000

Model Summary

ª Predictors: (Constant), Percent of the population with a Bachelor's Degree or higher (US Census 2004), percent of individuals living below the poverty line in 2008.

Coefficientsª

Model		Unstandardized Coefficients		Standardized Coefficients		
		B	Std. Error	Beta	t	Sig.
1	(Constant)	4.971	1.757		2.829	.007
	percent of individuals living below the poverty line in 2008	.624	.073	.734	8.526	.000
	Percent of the population with a Bachelor's Degree or higher (US Census 2004)	-.096	.040	-.205	-2.382	.021

ª Dependent variable: births to teenage mothers as a percent of the total 2007 (SA 2011).

Based on the results presented in these last two tables, answer the following questions:

6. What is the significance statistic for the original independent variable?

7. What is the value of the appropriate correlational coefficient for the original independent variable? _____

8. What is the direction of the coefficient? _____

9. What is the effect size for the coefficient? _____

10. What is the significance statistic for the second independent variable?

11. What is the value of the appropriate correlational coefficient for the second independent variable? _____

12. What is the direction of the coefficient? _____

13. What is the effect size for the coefficient? _____

14. How much of the variance in the dependent variable is explained by both independent variables? _____

15. Do these results support the causal diagram that you drew? Why or why not? _____

PART III
WHEEL OF SCIENCE: DESIGNS OF RESEARCH

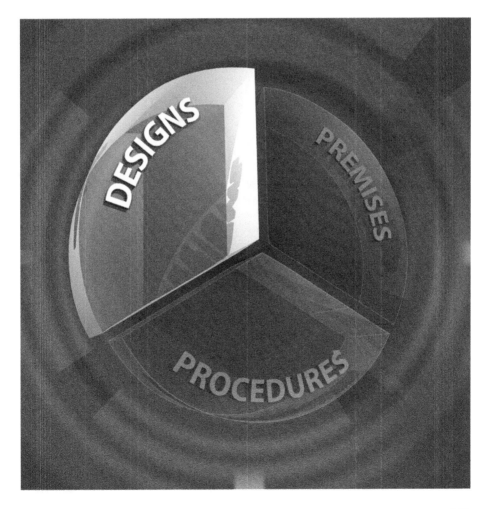

12

SURVEY RESEARCH

Survey research is the bread and butter of mainstream social science. Surveys allow us to draw on representative samples to learn about people's beliefs, behaviors, and experiences—at least this is the goal for scientific research. Unfortunately, as we have discussed previously, many people do not understand the scientific logic and process that underlies surveys as a research design. Since virtually everyone has taken a survey in some form or another, there's a general understanding that any data gathered from a survey is useful. That is not always the case, however. Surveys cannot be separated from good instrument design (the survey itself is referred to as the "survey instrument"), nor can they be separated from our earlier discussion of censuses and samples. Just because you get information from a survey does not mean that information is useful—if the survey is poorly designed and/or if the survey does not rely on representative samples (or a significant response rate from a population census), then nothing "learned" from the survey can be generalized to a population.

Good survey design is complex and linked to a theoretical literature. When should surveys be the research design chosen? How do you write a good survey? How do you make sure you get the survey to a representative sample (or a census)? We spend this chapter looking into the complexities of what it takes to do good survey research.

Understanding and Applying Research Design, First Edition. Martin Lee Abbott and Jennifer McKinney.
© 2013 John Wiley & Sons, Inc. Published 2013 by John Wiley & Sons, Inc.

NATURE OF THE SURVEY

The **survey** is a nonexperimental design that uses a series of written and verbal prompts/items to quantify the personal opinions, beliefs, and ideas from a group of respondents. The **survey instrument** (typically a questionnaire or interview schedule) translates unobservable content (e.g., beliefs) into numerical or other empirical referents into order to observe patterns across a group of respondents.

THREE TYPES OF SURVEYS

The word *survey* seems to cover a lot of ground, and yet is still nebulous enough that it gets a lot of play without a lot of understanding. Generally people think of a survey as a questionnaire, a piece of paper or a Web page with a lot of questions. There are, however, three distinct types of surveys: face-to-face interviews, telephone interviews, and self-administered questionnaires (that include online surveys). Each survey type offers researchers a tool to measure more adequately what they are trying to measure. There are good reasons for selecting one type of survey over another, depending on the research question(s) you are attempting to answer with this particular research design.

Face-to-Face Interviews

Face-to-face interviews are the most intensive type of survey, in terms of time and cost. Creating the survey instrument, training interviewers, and interviewing respondents individually can be costly in both time and money. Researchers choose to do face-to-face interviews for a variety of reasons. Often interviews are chosen because they allow researchers a chance to help clarify survey questions, or because in person interviewers can observe the respondent, or because the research question is not as salient to a population, therefore a face-to-face survey may help to prod a person's memory or get to more accurate responses.

Several times in this book we have drawn on data from the 2010 General Social Survey (GSS). Although the data may seem like they are collected using a questionnaire, GSS actually employs face-to-face interviews to collect data. The **interview schedule** used to collect data in a face-to-face interview looks remarkably like a self-administered questionnaire. Yet for an interview, the interviewer reads the questions on the interview schedule (along with the answer categories) and records the respondent's answers. This is part of the costliness of doing a face-to-face survey—not only must an interview schedule be well designed; the interviewers who administer the survey must be well trained.

Part of being a well-trained interviewer is in being extremely familiar with the interview schedule. Interviewers must read each question to each respondent with the same tone or inflection, as well as making sure to neither add nor subtract anything from the language of each question on the interview schedule. If an interviewer deviates from the exact wording of a question, it could change the meaning of the question, leading one respondent to answer differently from another (causing reliability issues) and subsequently misclassifying a respondent. For example, in a project designed to measure how people responded to the debates between science and religion, an open-ended question included in the interview schedule asked something like this: "Did you know that some people believe in creationism while others believe in evolution as a way of explaining the origins of people?" During interviewer training, one of the interviewers asked this question: "Did you know that some people *actually* believe in creationism instead of evolution?" Look at the two questions. Do you see any problems with the way the practice interviewer asked the question? Can you predict what he believed in regard to the debate between creationism and evolution? Many of the people being interviewed for this project included religious fundamentalists, who often favor a creationist explanation of the origins of people over an evolutionary explanation. Keep in mind that science is not about finding evidence to support our own views. The process of science allows us to study the world as it is, not as we may want it to be. Not only did the interviewer in this case entirely change the meaning of the question, what is the likelihood that anyone believing in creationism is going to give honest feedback on this or any additional survey question? Changing even minor words, inflection, and so on, changes the nature of the question itself, hurting the reliability and validity of the survey results.

Another point this example highlights is that good interviewers subordinate their own opinions or feelings about topics in order to accurately select a respondent's choices. When reviewing the practice interview session with this neophyte interviewer— even after it was pointed out that there was a problem with how he asked this question— he could not see the problem. The interviewer's biases against religious fundamentalism ran so deeply that even when it was pointed out, he could not tell that he had biased the question; he saw nothing wrong with the question as he asked it. This is not unusual. We've noted the importance of social location in constraining our ability to see the world randomly. We've also noted how the process of science should guard against our personal biases. Not everyone makes a good scientist or a good interviewer. Interviewing is an art that requires great skill. The problematic interviewer, who took for granted that educated people disdained religious fundamentalist ideologies, was not, by the way, sent out to conduct any interviews with religious fundamentalists.

It's generally a good idea to hire and train interviewers who can master the art of appearing neutral on any topic covered by the interview schedule (e.g., politics, religion, etc.). This does not mean that interviewers do not personally have strong opinions about the topics addressed on a survey. They do, however, recognize their biases and are able to set them aside to mask any personal preferences in order to be conscientious data collectors who are also friendly good listeners. This allows interviewers to be prepared to help respondents understand questions as they are asked. Interviewers can also probe for the appropriate information. If you've ever taken a survey where the responses were "If No, then skip to question 104; if Yes, skip to questions 111," then you've probably experienced some frustration. We can become easily irritated when taking a survey with several "skip" (or contingency) questions because we wonder what it is we might be missing in the skipped questions. These questions can be elaborate and begin to annoy respondents, meaning there is less chance of them completing the survey. This problem is ameliorated with face-to-face interviews because good interviewers know the survey instrument so well, they are able to automatically skip to the next appropriate questions, with the respondent having no idea that is what is happening. So when an interviewer asks, "Do you have any children? Yes or No" depending on the respondent's answer, the interviewer can go to the next appropriate question (for example, if the respondent answers "Yes," the interviewer can ask, "How many children do you have?", whereas the respondent who replies "No" moves on to the next relevant question).

Face-to-face interviews have one more advantage; interviewers can observe their surroundings to serve as a check on the honesty of the respondent. If you are conducting an interview in a public place and the respondent arrives in a Tesla (an all-electric sports car worth $150,000) but when asked about household income selects the category for $25,000 to $49,999, an interviewer may note at the end of the schedule that the respondent may not have been honest. There are also some mainstay items for interviewers to observe, including the hostility of the respondent. If you look at the GSS for the variable COOP, you will see the response categories of "Friendly and interested," "Cooperative but not particularly interested," "Impatient and restless," and "Hostile." This variable allows interviewers to report the overall cooperation of the respondent, which then allows researchers to take into account the possible invalidity/unreliability of an individual survey.

Another set of interview schedule questions specifically geared toward interviewers are the sex and race/ethnicity of the interviewers themselves. Interestingly, even for good interviewers, there is an unconscious respondent bias that can arise based on an interviewer's sex and/or race/ethnicity. When a female interviewer asks respondents questions related to gender, respondents often give answers that are slightly more positive than average. Likewise, for interviewers who belong to racial or ethnic minority groups, interview questions related to civil rights or race/ethnic issues also tend to receive more positive than average scores.

We've already mentioned that poor interviewers can bias interviews by the way they misstate questions or by the way their own biases seep into the interview. Interviewers can also cheat. For many market-driven surveys, interviewers are paid by the interview. If a respondent fails to show up, the interviewer does not get paid. This is strong incentive for an interviewer to simply make up an entire interview, seemingly marking random answers in order to complete the survey and get paid. There could also be other biases at work. McKinney was once asked to participate in a marketing survey where questions centered on household cleaning products. The interviewer asked which products her household used for cleaning and then read a list of possible cleaning products she might use (the answer categories). She selected those products used in her household. The interviewer paused and asked, "Are you sure you have not used Product X?" "Yes," she responded, sure that she had never purchased Product X. The interviewer paused and then said, "Well, this survey is being sponsored by the makers of Product X, so I'm going to go ahead and put that down." While this interviewer did not make up a whole interview, he disregarded the respondent's answer to select his own. Even if the interviewer thought it was in his best interest or the best interest of the survey sponsor, the result is the same: the data were misclassified and therefore did not measure the respondent's actual behavior.

Bias in face-to-face interviews, however, is a two-way street. Respondents can be just as likely as interviewers to distort their answers—unconsciously as mentioned earlier when they give answers that are more in line with the sex or race/ethnicity of an interviewer, or consciously to make sure they follow more politically correct or socially acceptable norms. Luckily, there are patterns to how respondents distort answers (and there are ways to minimize these issues by good survey instrument design, which we look at later in the chapter). For example, male respondents are more likely to overstate their age, income, and number of sex partners; female respondents are more likely to understate their age, income, and number of sex partners. These patterns follow basic cultural expectations and can easily be corrected through survey design or statistical normalization.

One of the critiques of surveys is that they are "self-report." That is, why should we believe anything people say about themselves, especially in light of the discussion on how both interviewers and respondents can bias the data collected? Because we use large representative samples of individuals to collect survey data, we expect the data to be robust—that if social patterns exist, the data will illustrate the appropriate pattern. When you take a survey and are asked what sex you are (male or female), do you think you put down the appropriate answer? Overall, we expect that most people completing surveys are honest in their responses. Even if many are not honest, the nature of taking data from large samples of individuals allows researchers to discern social patterns in the data.

Telephone Interviews

Telephone interviews are similar to face-to-face interviews and a good alternative when the research question is more salient to respondents. Oftentimes using a random-digit dialing sampling procedure, the telephone interview enjoys many of the advantages of the face-to-face interview. By knowing their interview schedules well, telephone interviewers are able to help respondents understand the questions, probe for the appropriate information, as well as being able to observe (admittedly, telephone interviewers observe less than face-to-face interviewers; however, basic observations like the hostility of the respondent are still valid). The biggest advantage of the telephone interview is that it's cheaper than the face-to-face interview. This is only an advantage, however, if your research question is appropriate to telephone surveys.

There are some unique disadvantages to telephone surveys. Like random-digit dialing sampling, it can be hard to get past the voice mail and/or have people pick up the phone with an unfamiliar number. Thus there tends to be a higher rate of nonresponse. During face-to-face interviews, respondents develop a rapport with the interviewer. Since there is no actual person sitting across from a respondent during a telephone interview, this rapport is significantly less likely to develop. If a respondent does not like a question or gets distracted while in the interview, the likelihood of the respondent simply disconnecting is higher. This means that telephone interviews must be shorter than other surveys because respondents are less willing to spend long periods of time on the telephone. Another disadvantage is the inability to get sensitive information during a telephone interview. For both face-to-face interviews and questionnaires, people develop a rapport with either the interviewer or the survey instrument itself (in the case of a self-administered questionnaire) and will respond by giving very detailed information on sensitive subjects. This does not happen in telephone interviews because the telephone itself is a mediating factor inhibiting rapport, resulting in less access to sensitive data.

Questionnaires

Not everyone has participated in a face-to-face or telephone interview, whereas most people have filled out a **questionnaire,** written questions to be completed by respondents. Often questionnaires are conducted online or through the mail. Questionnaires look just like their interview schedule counterparts, with written-out questions followed by answer categories (in the case of closed-ended questions) with a few questions that are open-ended (we look at the difference between closed- and open-ended questions when we turn to survey construction).

Because reading comprehension is higher than the aural comprehension needed in an interview, questionnaires can be much longer and more complex than either type of interview. Interestingly, questionnaires can also ask more sensitive information than either type of interview, due to the rapport just described that develops between the respondent and the survey instrument. Another advantage of the questionnaire is that is much less costly than interviews.

There can be a downside to questionnaires: the elaborate and complex contingency (or "skip") questions. Respondents can lose interest in completing the questionnaire if they are constantly navigating sets of contingency questions. Online questionnaires can alleviate this problem by programming moves to the next appropriate question, so respondents are not distracted and frustrated by the complex skipping. Questionnaires offer virtually no form of observation (except wondering about the validity of a returned mail questionnaire completed in crayon or reading the rants of a respondent who uses the questionnaire as a "bully pulpit") and are limited to literate respondents. Although it would seem like the percentage of literate adults would be quite low, questionnaires underestimate the beliefs, behavior, and experiences of those cannot read at all or those who cannot read the language of the questionnaire. One last critique of questionnaires is that they can have a lower response rate than interviews. This, however, should not longer be a criticism. Survey design has improved to show that using the proper techniques—aggressive preparation and follow-up—will result in similar response rates to interviews (see Dillman 1999).

ONLINE SURVEY METHODS

Thus far, we have discussed survey methods that involved the use of questionnaires as part of face-to-face, telephone, and self-administered questionnaires. To these we add online survey methods. Online surveys represent nothing substantially new structurally; they still involve a self-administered questionnaire instrument. Online surveys, however, are unique because of the way in which they are delivered to respondents. Due to widespread access to electronic communication, online methods afford the researcher additional access to collecting data relevant to a research problem.

Online Surveys

Online surveys typically involve a questionnaire that is identical to the questionnaire used by a face-to-face interviewer or in a self-administered survey. It could be argued that it offers significant improvement over the hard copy survey instrument, however, in the complexity that can be assessed, given the user friendliness of online branching, piping, and so on. The same issues of instrument reliability and validity apply when constructing an online survey, and the overall research process relies on the same process of sampling (for representativeness) and is subject to the same constraints for response rates (for inferential conclusions and generalizability).

What is distinctive about online surveys is the way they are delivered, with both advantages and disadvantages over traditional methods. Typically, questionnaires are posted online in a Web format so that the respondent can access the questionnaire through his or her own computer at any time of day or night, in any location worldwide. This would be difficult for most researchers to create on their own, so several online companies have developed the architecture to accommodate such efforts. Some of the most popular online companies are SurveyMonkey (http://www.surveymonkey.com/) and Zoomerang (http://www.zoomerang.com/online-surveys/).

The advantages of online surveys include greater geographic access, greater adaptability to respondent subgroups, and reduced costs. With widespread accessibility to computer technology, many people rely on the use of computers in their everyday work and private lives. Accessing a survey through this avenue is therefore a natural way to introduce a questionnaire. The request to participate in such a study is therefore no longer an atypical request, with some exceptions for those living remotely without computer access or those who choose not to use online processes.

The online questionnaire is also adaptable to respondent subgroups. For instance, if the respondent group includes veterans and nonveterans, questionnaire items that relate to these subgroups can be fashioned so that only the appropriate items appear to the appropriate group. Thus, for example, if a purpose of the survey was to assess discrimination over military service, only the veterans would be given items about military discrimination. This process is conceptually no different from the contingency questions in self-administered surveys, but on an experiential level the respondent not affected by the inquiry need never see some items that decrease response time, thus maintaining the appropriate face validity of the questionnaire. The branching is constructed in HTML and transparent to the user.

Overall costs are reduced because the online questionnaire is not delivered on paper through the mail and requires less handling. We say "reduced" because there are still costs associated with the online version. The costs primarily involve the price of the software necessary to host the questionnaire and the necessity to mail hardcopy questionnaires to those randomly selected respondents who have no e-mail address. Despite these costs, however, the overall process of the online survey is generally significantly more cost effective, especially in large-scale research projects.

There are also disadvantages to online surveys beyond those already noted. There is a bias toward those with e-mail addresses as there is among respondents of telephone surveys with nonlisted numbers. The jury is still out about whether online questionnaires are more likely to be completed than the non-online variety, but both share the problem of capturing an appropriate response rate.

For all the reasons we have discussed, it is important to use the same method for contacting mail and telephone respondents with both online and hardcopy questionnaires. We prefer the general method of (1) prenotification of selected respondents for questionnaires, creating both online and mailed notification for those with and without e-mail addresses, (2) e-mailing and hardcopy mailing the questionnaire to those with and without e-mail addresses, respectively, (3) sending reminder notes to those who do not submit their completed questionnaires (either electronically or through the mail), and (4) creating a second wave of the questionnaire to both groups.

ONLINE FORUMS

Online forums are essentially electronically hosted focus groups. They do not necessarily (but may) involve questionnaire items, but we note these here because they share the same electronic delivery method as online surveys. The advantages include cost savings, especially with large diverse respondent groups that would require special arrangements for a common meeting time if electronic means were not available. With electronic forums, respondents do not have to travel to participate and they have similar assurances of confidentiality. These forums can be asynchronous, enabling participants to have additional time to reflect on prior conversation. Like online surveys, several electronic hosts exist to conduct online forums. Survey Gizmo (http://www.surveygizmo.com/), HipChat (https://www.hipchat.com/), and artafact (http://www.artafact.com/) are some of the most popular host sites for online forums.

SURVEY ITEM CONSTRUCTION

In looking at the types of surveys to choose from when collecting data, we seem to have taken for granted that surveys are made of questions with various response options. These questions (also referred to as survey items) cannot be taken for granted. We've noted that surveys are only as good as the instrument constructed. We've both seen enough bad survey questions to last us a lifetime; just because someone has taken a survey does not mean they can write one well. Survey construction is a scientific art—you must be able to write questions that measure the concepts you intend to measure, questions that take into account the population you are studying, and questions that contain no biased language, no anchors, and no double-barreled questions.

As noted, surveys can be very cost effective. One of the issues survey methodologists evaluate when creating a new survey is the difference between standard items versus original items. Each survey will have a unique balance of standard and original items. **Standard items** are survey items that have been used in previous surveys and have been found to be reliable and valid. Because they are items that have been tested, standard items are time and cost effective. Standard items also allow researchers to compare within and between populations who have answered those survey questions in other surveys, providing meaningful comparisons across studies.

If you are undertaking survey research and creating a new survey, you do so because you have **original items** to test. When you need to measure concepts in new ways, new items need to be created. Since new items are untested, we do not know if they are reliable or valid. Any time original items are used, they are incorporated into the survey and the entire survey is **pretested** using a small sample of the target population. This, of course, takes more time and money. Pretesting original items, however, is vital to doing the research. If you find that your items are not reliable or valid after the fact, then you've wasted time and money and will have two choices: either redo the survey or do not use the new items. Neither of these choices is ideal. When doing scientific survey research, you must make sure your survey instrument (be it interview schedule or questionnaire) is pretested using an independent sample of the target population before the actual survey (test) is administered.

As you evaluate the type of items included in your survey, you also need to balance closed-ended and open-ended questions. Most surveys use **closed-ended questions,** which have clear response categories, such as the variable SOCBAR:

"How often do you visit a bar or tavern?"
1. Almost every day
2. Once or twice a week
3. Several times a month
4. About once a month
5. Several times a year
6. Once a year

Relying on closed-ended questions allows researchers to maximize their ability to compare answers across cases to see patterns emerge from the data. Closed-ended questions are also more cost effective and efficient because the coding is built into the answer categories themselves. When cleaning or entering the data, the category number is already selected and ready for input. There are a few drawbacks to having only closed-ended questions, for example, suppressing variation. We talked about variation in Chapter 5 on measurement. We noted that when asking questions it is best to *maximize* the variation—at least when asking a question (for example, "How old are you?"). Surveys, however, seek to maximize comparability by limiting the answer categories in most cases (thus the closed-ended questions).

Surveys are about collecting the maximum amount of data from the selected sample, so using closed-ended questions is the best way to do that. Even so, surveys should include some **open-ended questions to** maximize variation and give respondents a chance to fully express themselves,. Open-ended questions are better suited to sensitive subjects and allow the respondents to have a voice. Open-ended questions are particularly useful in exploratory studies, allowing researchers to see the full variation and nuance of how people answer questions, giving them a tool to create valid and reliable closed-ended questions on future surveys.

One unexpected function of having an open-ended question section is the textual data that you get, the so-called juicy quotes. Minimally a survey should contain a section at the end that gives respondents a chance to respond to the survey subject or the survey instrument itself. This can be quite harrowing for survey researchers. The data contained in this open-ended section allow respondents to express themselves and the data they give you may help elucidate some of the quantitative findings. For example, in McKinney's (2001) survey, "Clergy Connections," measuring clergy network ties to Evangelical Renewal Movements (ERMs), the data suggested that many clergy felt disconnected from the larger denomination and favored joining ERMs. The back page of her survey gave ample space for people to respond to the subject of their network ties, as well as the survey instrument itself. Although McKinney had at least one respondent say, "This is the stupidest survey I've ever taken," many more written comments gave insight into how clergy were feeling about their connections, or rather disconnections, to their denomination. Here are some comments from the open-ended section:

> Sometimes I think the [denomination] would serve the kingdom of God if it dissolved.

> I am firmly committed to ministry within the [denomination]. Our discipline, doctrine and history are outstanding and excellent. However, with the present controversies over non-biblical and sinful lifestyles being encouraged to be accepted as normal and o.k. I feel there is potential to fragment the denomination. If this happens . . . I affiliate with the [ERM] . . . and so does the church I presently serve.

> I see the signs of the church affirming Biblical authority as a positive sign. I remain loosely committed to the denomination because the authority of Scripture is higher in my life than the authority of the Church. Without a foundation, the Church will not survive and without a common purpose we cannot truly be a united Church. When I can no longer agree with the position of the church—I will leave. Until then, I thank God for the opportunity to make a difference for eternity in the life of others through the instrument of the [denomination].

> I love the . . . denomination, however, am *deeply* concerned about the direction I see *many* of our leaders taking in regard to lack of Biblical convictions. It concerns me and I keep my eyes open in case I feel I can no longer identify with an apostate church.

These comments are fascinating, but they are not generalizable. Since not everyone commented in this section, data from the open-ended question could not be generalized to the population. These comments, however, would be a good starting point for future research; they help give direction to learning about how some clergy from the sample felt about their connections and disconnections to their denomination.

Employing too many open-ended questions can become quite expensive and time consuming because someone will have to read all of the responses and try to code them as they are. Responses to open-ended questions are much less comparable than the precoded closed-ended questions and are not as useful in survey research. There is a place to ask long open-ended questions within the context of research (qualitative research designs), which we discuss later in this book.

Evaluating Closed-Ended Questions

Since closed-ended questions are what survey researchers specialize in creating, there are a few guidelines to use when evaluating the items you select from previous surveys or when creating original items. Let's look at some examples of survey items.

In 2010 the political blogosphere addressed a controversial document titled, "2010 Congressional District Census" (Figure 12.1). The document that was pictured online (Knutson 2010)[1] began with a "Special Notice" at the top, specifying that the respondent had been chosen to represent Republican voters from congressional districts in one state. The survey looked very official, with what appeared to be codes for congressional district and census tracking. Very quickly, however, it was clear this was no ordinary congressional district census. The instructions stated,

Strengthening our Party for the 2010 Elections will take a massive grassroots effort. As a key facet of our overall campaign strategy, the Republican Party is conducting a Census of Congressional Districts all across America.

What do you take away from the instructions? Is this a scientific survey? Well, it could be as long as the survey was sent to all registered Republicans (a census) or a representative sample of registered Republicans. It could even have been sent to constituents of both parties; we just do not know. Unfortunately, there is nothing in the four-page survey to tell us about the methodology behind who was chosen to receive the survey or how they were chosen.

There are two things that are important regarding evaluating the survey of mostly closed-ended questions: (1) We are given no information regarding the representativeness of the census or sample, and (2) This survey is undertaken by a partisan group, which can be problematic. So going into the survey, we should be somewhat careful in evaluating the reliability and validity of the questions.

In looking at some closed-ended questions from the survey, several issues arise. Here are some of the questions:

"Do you believe the huge, costly Democrat-passed stimulus bill has been effective at creating jobs or stimulating America's economy?"

1. Yes
2. No
3. No opinion

---

[1] See Propublica at www.propublica.org for the document.

SPECIAL NOTICE: You have been selected to represent Republican voters in New Jersey's 8th Congressional District Enclosed please find documents registered in your name.

2010 CONGRESSIONAL DISTRICT CENSUS

DELIVER EXCLUSIVELY TO:

Commissioned by the Republican Party

Census Document Registered To:

Representing Congressional District: New Jersey #8

Census Tracking Code: #A10PD177

Please Respond By: February 11, 2010   551880571

0123456

Your Participation is Greatly Needed and Appreciated:

Strengthening our Party for the 2010 Elections will take a massive grassroots effort. As a key facet of our overall campaign strategy, the Republican Party is conducting a Census of Congressional Districts all access America. The opinions registered in this document will be used to help ensure that our Republican leaders and candidaters are specifically addressing those issues most important to voters in your area.

Instructions: Please answer all questions to the best of your ability. All individual responses will be kept confidential and only survey tallies will be shared with our Congressional leaders and candidates. When finished answering your Census, please return it along with your generous contribution in the enclosed postage-paid envelope. Thank you.

SECTION I

POLITICAL PROFILE

1. Do you generally identify yourself as a:
   - ☐ Conservative Republican   ☐ Moderate Republican   ☐ Liberal Republican
   - ☐ Independent Voter who leans Republican   ☐ Other

2. Do you traditionally vote in all elections?
   - ☐ Yes   ☐ No

3. Do you plan to vote in the 2010 elections?
   - ☐ Yes   ☐ No   ☐ Unsure

4. If you vote in the 2010 elections, are you more likely to vote for the Republican or Democrat candidate?
   - ☐ Republican   ☐ Democrat   ☐ Other   ☐ Unsure

5. What age category below applies to you ?
   - ☐ 18–29   ☐ 30–44   ☐ 45–59   ☐ 60+

6. How close do you think your views are to other voters in your community?
   - ☐ Very Close   ☐ Somewhat Close   ☐ Not Very Close   ☐ No Opinion

7. From what media source do you regularly receive your political news? (Check all that apply)
   - ☐ NBC/CBS/ABC   ☐ Friends   ☐ Radio   ☐ Other_____
   - ☐ CNN/MSNBC   ☐ Fox News   ☐ Twitter
   - ☐ News Websites   ☐ Facebook/MySpace   ☐ Candidate Websites
   - ☐ Local Newspaper   ☐ Internet Blogs   ☐ National Magazines

8. How much does it concern you that the Democrats have total control of the federal government?
   - ☐ Very Much   ☐ Not Too Much   ☐ Not A Concern   ☐ No Opinion

Figure 12.1. The 2010 Congressional District "Census."

"Do you think the record trillion dollar federal deficit the Democrats are creating with their out-of-control spending is going to have disastrous consequences for our nation?"

1. Yes
2. No
3. No opinion

"Do you trust the Democrats to take all steps necessary to keep our nation secure in this age where terrorists could strike our country at any moment?"

1. Yes
2. No
3. No opinion

"How much does it concern you that the Democrats have total control of the government?"

1. Very much
2. Not very much
3. Unsure

"Do you think things in this country are generally going in the wrong direction, or do you feel things are starting to improve?"

1. Wrong direction
2. Starting to improve
3. Unsure

When you answer questions on a survey, how much time do you spend reading through and thinking about their construction? Probably not too much, since we mentioned that one of the strengths of closed-ended questions on surveys is their efficiency. Being efficient partly means that respondents can read through questions quickly. Sometimes surveyors count on this. In reading the previous questions, several concerns present themselves.

Biased Language

Reread these questions. Do you find a pattern in terms of what the "correct" answers are? The questions are written in a way as to bias the answer. Here is another clue that this is not a scientific survey. Each of these questions uses **biased language,** or language that directs us to respond in certain ways.

Consider the first question:

"Do you believe the *huge, costly Democrat-passed* stimulus bill has been effective at creating jobs or stimulating America's economy?"

The words *huge, costly,* and *Democrat-passed* lead respondents to answer "No." Since this is a partisan survey, using words like *Democrat-passed,* along with *huge* and *costly* paints the issue in a negative light, predisposing respondents to subsequently answer "No."

Look again at the other questions. Can you spot the embedded biases in the language being used?

> "Do you think the record trillion dollar federal deficit the Democrats are creating with their out-of-control spending is going to have disastrous consequences for our nation?"

> "Do you trust the Democrats to take all steps necessary to keep our nation secure in this age where terrorists could strike our country at any moment?"

> "How much does it concern you that the Democrats have total control of the government?"

> "Do you think things in this country are generally going in the wrong direction, or do you feel things are starting to improve?"

Phrases like "record trillion dollar federal deficit" "out-of-control spending" "disastrous consequences," and "have total control of the government" make assumptions that predispose respondents to answer the questions in specific ways. Look at the juxtaposition of "the Democrats" and "our nation" or "our country". What is the underlying supposition here? As a partisan question, this supposes that "they" (the Democrats) do not understand "our" (the Republicans) nation/country, setting up conditions for the survey respondents to give "stock" answers.

Anchors

Can you spot the bias in the last question? It seems to be fairer, giving options within the stated question and the answer categories that include both "wrong direction" and "starting to improve." The question, however, *anchors* the response categories. By using "things in this country are generally going in the wrong direction," the starting point of the question again biases respondents to select "Wrong Direction" as the preferred choice. **Anchor questions** distort the range and preferred selection of categories. When we quickly read through questions, we simply do not stop to analyze each question word for word. When we do not have a specific opinion that is contradicted by the wording of the question, the anchors move us to answer in specific ways.

Answers to questions can also be manipulated through the answer categories. Forgetting for the moment that the questions themselves use biased language, look at the category options given: "Wrong Direction," "Starting to Improve," and "Unsure."

"Do you think things in this country are generally going in the wrong direction, or do you feel things are starting to improve?"

1. Wrong direction
2. Starting to improve
3. Unsure

Balanced Categories

Do these questions provide exclusive, exhaustive, or balanced categories? Again, in a quick pass these survey items may seem fair, but they are neither fair nor balanced. The two substantive categories assume that the country is going or has been going in the wrong direction (whether or not they are "starting to improve"). Answer categories for survey items must be **balanced categories,** or categories that have the same *degree* of responses. If the item's beginning category is "Wrong Direction," there should also be a "Right Direction," rather than "starting to improve." As we discussed in Chapter 5 on measurement, these category responses are also not exclusive or exhaustive.

The objective researcher does not want to pick out one political party as being any better than another political party; all are partisan and seek to cast issues is specific ways. This is an important point. We have stressed that people in general do not understand the process of science, especially as it relates to the social sciences. Partisan and special interest groups may actually count on this, seeking to craft questions with language that biases responses toward support of their own issues. Any survey or study undertaken by a partisan or special interest group should be viewed cautiously with respect to possible biases. Scientists are interested in answering research questions. The strength of a scientific survey is to measure the world as it is, not to manipulate the findings to support our biases. It takes a keen eye and a clear understanding of social location to write a good survey. Be skeptical of studies that you hear about; check to see who sponsored the study and how the study was conducted (there is a difference between being skeptical and being cynical; cynicism throws doubt on subjects without regard to weighing the evidence, whereas skepticism brings a healthy regard to considering all of the available evidence to come to an informed conclusion).

While the faux Congressional Census highlights the problems with biased language, anchors, and unbalanced category responses, there are some additional issues that neophyte survey writers should beware of when creating good surveys. These issues include having double-barreled questions and having the appropriate variation within the item response categories.

Double-Barreled Questions

Double-barreled questions are items that are measuring two ideas. Each survey question should be written as clearly and concisely as possible. Sometimes, however, we put too much into one question. An example of this would be an original item McKinney wrote for her "Clergy Connections" survey:

"What is your congregation's approach to interpreting and teaching the biblical meaning of Christian faith and the Church? (Circle one.)"
1. There is one best or true interpretation and our congregation or denomination comes closest to teaching it.
2. There is one best or true interpretation, but no church can legitimately claim to be closer to it than another.

3. There are probably many interpretations which are equally valid, so many churches may be correctly teaching Christian faith.

When this item was pretested, several respondents noted in the margins that the question was confusing. They were right. Not only was the question itself confusing, the response categories were also confusing. The first issue was the double double-barreled question. Double-barreled questions are questions that appear to be only one question, but they really have two or more questions embedded in them. There are actually four possible questions embedded in the previous question (thus the "double" double-barrel):

What is your congregation's approach to interpreting the biblical meaning of the Christian faith?

What is your congregation's approach to teaching the biblical meaning of the Christian faith?

What is your congregation's approach to interpreting the biblical meaning of the Church?

What is your congregation's approach to teaching the biblical meaning of the Church?

Because denominations have historically distinctive doctrines, they see Christian faith differently. Respondents did not know how to answer the question because they did not know if it was asking about how their congregations interpret (or teach) the meaning of Christian faith, or if the question was asking how a particular church interprets (or teaches) the meaning of the Church. This is why pretests are so important. In reflecting on what the question had originally intended to indicate, it was revised for the actual survey (test) to read:

"What is your congregation's approach to interpreting and teaching the biblical meaning of Christian faith? Circle the number on the continuum that represents where you believe your congregation falls."

1 = Our congregation teaches the one true interpretation of Christian faith.

7 = Our congregation teaches that there are many interpretations of the Christian faith.

          1   2   3   4   5   6   7

By changing the question, it is much clearer what the researcher is asking, making the response categories easier for respondents to understand. With the second iteration of the question, respondents had one item to answer with a clearer meaning. Even professionals make mistakes. That's why we have checks and balances, colleagues, and committees to read through surveys, critiquing them with fresh eyes. But for researchers who know their subject well, sometimes it takes the pretest and the eyes of the population being studied to see some of the faults of individual items.

Variation in Survey Analyses

Ensuring greater variation among questionnaire items yields more valid and reliable statistical results (e.g., see the correlation section discussing restricted range). In addition, response categories can truncate responses and in such a way actually restrict or bias the respondent's intent. Critics of survey analyses often cite the problem that results are an artifact of the way questions are asked. Survey researchers are also interested in forcing variation, or making sure that the response categories for an item have the appropriate variation. We talked earlier about the fact that closed-ended questions by nature actually suppress variation. There are ways to make sure that you maximize variation even when using closed-ended questions.

One of the things we see on surveys is premature collapsing of answer categories. Many people have answered survey questions that have items like this:

"What is your age?"

1. Less than 18 years old
2. 18–24 years old
3. 25–39 years old
4. 40–64 years old
5. 65 years old or older

This seems like a perfectly good question with perfectly good answer categories (okay, so some represent approximately 20 years while another represents 6 years). But you know you've seen these questions. In order to force variation, rather than asking for specific categories (making this an ordinal variable), would not it be just as efficient (and better for analysis) to ask:

"What is your age? _____"

With the second question, we maximize variation because every respondent puts down his or her current age. Once we have all the data, we can look at the distribution of all of the data and if need be collapse the data into meaningful categories based on the full distribution of responses. What if we used the first question and then after the fact realized that we needed to look at the answers of teens? Since the second category includes two years of teens, we would never be able to learn which respondents were 18 to 19, rather than 20 to 24. Be thoughtful as you construct response categories to think through what you are asking. Does the question lend itself to a wider array of data? If possible, always ask the question in a way that allows for maximum variation (several of the questions in the GSS have collapsed categories; the difference is that in the original survey most of these questions asked for the maximum variation and were collapsed after the fact to make some analyses easier, e.g., cross-tabs).

Another way variation can be maximized is by making sure response categories are balanced. In looking at the Congressional District Census, we noted that some of the categories made certain assumptions. For example, what is the assumption behind the following answer categories (question adapted from the 2010 GSS)?

"Do you think the amount of money the government spends on the space exploration program is:"

1. Too little
2. About right
3. Slightly too high
4. Moderately too high
5. Far too high

The **unbalanced categories** favor respondents answering that the cost of this program is some degree of "too high." Notice that there are three degrees of "too high" and only one for "too little." When using categories, you must have equal degrees for each:

"Do you think the amount of money the government spends on the space exploration program is:"

1. Far too little
2. Slightly too little
3. About right
4. Slightly too high
5. Far too high

The question is asked clearly so changing the response categories to have two degrees of "too little" and two degrees of "too high" makes the item balanced.

Socially Acceptable Answers

One last thing to think about when writing response categories: selecting socially acceptable answers. We talked about the fact that respondents develop a rapport with interviewers, in the case of face-to-face interviews. This may exert unconscious pressure on them to give responses they feel are more socially acceptable. For example, excessive drinking in our cultural context is considered irresponsible (unless you're living in a fraternity house, which makes it seemingly necessary to social relations). Determining how often people frequent a bar or tavern (an old GSS standard) may make some respondents hesitate and select an answer that underestimates how often they visit a bar or tavern. You can, however, rank the response categories in a way that reframes what's normal. Notice in the bar/tavern item how the responses are ranked:

"How often do you visit a bar or tavern?"

1. Almost every day
2. Once or twice a week
3. Several times a month
4. About once a month

5. Several times a year
6. Once a year
7. Never
8. Do not know

Rather than beginning with the more socially acceptable answers:

1. Never
2. Once a year
3. Several times a year . . .

Ordering the categories in the reverse sequence allows respondents to consider the full range of response categories before they select one—resulting in them giving more accurate responses, rather than quickly choosing one that appears to be "higher" or "better" on the list.

Biased language, anchors, balanced categories, double-barreled questions, variation, and socially acceptable responses are some of the issues involved with item construction for surveys. We hope you recognize how complicated creating a good survey can be. Survey methodologists must have an eye on the big picture of data collection, as well as an eye for the minute details of item wordings and item response categories. Even when good surveys are created, there are a few things to keep in mind, including the reliability and validity of the survey items.

RELIABILITY AND VALIDITY

Each research design will have unique issues regarding the reliability and validity of the design itself. For surveys, much of the reliability and validity lies in the construction of the survey items. Even for well-written items, reliability can be a problem (reliable survey items would yield similar results over time).

Respondent Knowledge

Some questions are difficult for respondents to answer, not because they do not have experience with them, but they simply cannot remember or just do not know what an accurate answer would be. Think back to last year. How many meals did you eat out over the last 12 months? I'm sure you've eaten out over the last 12 months, but can you give an accurate account of how many times (and do you count grabbing a sandwich at a convenience store or sitting down in a restaurant)? This is an example of a question that cannot be answered (accurately). In the "Clergy Connections" survey, one question that seemed innocuous enough was:

"How many hours a week do you work? _____ "

The question is clear. The question allows for maximum variation. The question for pastors, however, was almost unanswerable. Since the work of pastors bleeds into many areas of life, it was difficult for some of the pastors to give an answer other than 168 hours. A total of 168 hours? How is that possible? If you work 24 hours a day, 7 days a week, that's 168 hours. Pastors were clearly letting McKinney know that the nature of their job was impossible to quantify, except to say it was a job that consumed them 24/7. Clearly, for some pastors, this was a question they felt they could not answer. There are also questions that are not answerable because they are not clear (for example, double-barreled questions, questions respondents feel are biased, etc.).

Respondent Refusal

Some questions simply cannot be answered, and there are also questions for which people have clear answers but they just will not answer them. While collecting data for the "Clergy Connections" survey, McKinney received calls from respondents concerned that the nature of some questions was too personal, including a question asking about the respondent's political preferences and respondent's annual household income. Although the callers had political preferences and annual household incomes that they were very aware of, they did not want to answer those questions. McKinney's advice was to the respondents was to feel free to leave those questions blank if they felt the questions were too personal, but to please complete the rest of the survey. Since social scientists do not track or identify individual respondents, McKinney does not know if the people who called her ever returned their surveys. When entering the data, though, she noticed that several people did leave certain questions blank.

Some questions are clear and have clear answers, but people feel they are too personal and either will not answer at all (skipping those questions but completing others) or they do not answer them honestly. There are patterns to these questions. Questions regarding age, income, or sexual behaviors tend to be ones that respondents are least likely to answer, either at all or honestly. Gender is an independent variable here. While men tend to inflate their age (older men are seen culturally as "wiser"), income (men's income is equated to their success) and sexual behaviors (men are more likely to embellish the number of sexual partners they've had), women tend to deflate these same variables—age (younger women are thought to be more attractive), income (culturally women are supposed to privilege relationships over career), and sexual partners (women are more likely to deemphasize how many sexual partners they've had).

Nonresponse in Questionnaire Data

Between questions that respondents cannot or will not answer, another reliability issue is the amount of nonresponse within items. We have talked about how random samples need certain percentages of responses in order for the study results to be generalizable to the population from which they are drawn. Similarly, in the case of survey items you need to make sure that you have a good response rate per item, or else you cannot use that item in your analyses (you can tabulate them as frequencies, but if the response to an item is too low, then the item cannot be used as a variable).

One of the components of the "Clergy Connections" survey was to ascertain pastors' religious beliefs. With so much controversy in mainline denominations regarding ordination of homosexual pastors, it seemed a good idea to include an item measuring belief in homosexual practice. This item was included in a religious orthodoxy scale:

"Would you please think about each of the religious beliefs listed below and then indicate how certain you are that each is true."

　　　　1 = Completely true
　　　　2 = Probably true
　　　　3 = Probably not true
　　　　4 = Definitely not true

a. There is a life beyond death.	1	2	3	4
b. Jesus was born of a virgin.	1	2	3	4
c. Jesus walked on water.	1	2	3	4
d. Only those who believe in Jesus Christ can go to heaven.	1	2	3	4
e. Jesus was both fully human and fully divine.	1	2	3	4
f. The Scriptures have binding authority on all people; they are the sole rule of doctrine and conduct.	1	2	3	4
g. Homosexual practice is a sin.	1	2	3	4
h. God really exists; there is no doubt about it.	1	2	3	4

After cleaning and entering the data from the survey and running reliability checks on the data, the item measuring belief regarding homosexual practice had a very high rate of nonresponse. In fact, this question had 10 times the amount of nonresponse as any other question in the scale. People answered the questions above and below this one item, but they specifically left this question blank (sometimes people skip whole questions or turn pages too quickly and miss whole pages; it is rare for them to circle answers in an eight-part question, skipping only one). The nonresponse was too high for the item to be used as a variable in any of the analytic models (or combined in a scale). Clearly, beliefs about homosexual practice were sensitive for many of these clergy.

When we turn to issues in survey research with validity, the question is whether your items measure what you think they are supposed to measure. Sometimes we can confuse items that measure beliefs with items that measure behavior and vice versa. These are not the same concepts, and beliefs should not be used to extrapolate to behaviors, whereas behaviors should not be used to measure what people think. Social scientists often note that attitudes can be the worst predictors of behavior. Often the best predictor of future behavior is past behavior (not an attitude about behavior). Even questions that ask what someone might do in a given situation are not good measures of what they end up doing in that situation (as we noted in a previous chapter, while most people agree that helping those in need is important, not many of them actually helped those in need). When assessing the validity of survey items, make sure you are clear about what the item actually measures.

BIAS IN SURVEYS

Once satisfied that you've constructed good survey questions, the survey should be evaluated one last time to check on two issues that may impact your respondents: conformity bias and response bias. We discussed how question response categories need to navigate the tendency of some respondents to lean toward a more socially acceptable answer. Reordering your categories from highest to lowest (e.g., "How often do you go to a bar or tavern?") mitigates some of this and can avoid **conformity bias,** or the tendency to answer questions in socially acceptable ways. The survey questions themselves also need to be worded in such a way that they are clear yet avoid any kind of social judgment. Be aware of the population; know that questions worded in some ways may come across differently than what you intend.

Another bias that may result from respondents is the **response set,** a tendency to fall into a pattern of responding in particular ways. In the previous discussion of item reliability, we noted one item that fell in a religious beliefs section of the "Clergy Connections" survey. The survey actually asked two sets of beliefs questions. Being well aware of the danger of a response bias, the second set of beliefs responses were reversed. Here is how they appeared:

"How important is each of the following to you personally?"

        1 = Not important at all
        2 = Not very important
        3 = Important
        4 = Most important

a. Knowing that God loves me	1	2	3	4
b. Helping others who are in need	1	2	3	4
c. Accepting Jesus Christ as my personal savior	1	2	3	4
d. Knowing the distinctive doctrines of the [church]	1	2	3	4
e. Working to bring about social justice	1	2	3	4
f. Telling others about Jesus	1	2	3	4
g. Belonging to this particular denomination	1	2	3	4
h. Socializing with persons with similar values	1	2	3	4

"Would you please think about each of the religious beliefs listed below and then indicate how certain you are that each is true?"

        1 = Completely true
        2 = Probably true
        3 = Probably not true
        4 = Definitely not true

a. There is a life beyond death.	1	2	3	4
b. Jesus was born of a virgin.	1	2	3	4

c. Jesus walked on water.	1	2	3	4
d. Only those who believe in Jesus Christ can go to heaven.	1	2	3	4
e. Jesus was both fully human and fully divine.	1	2	3	4
g. The Scriptures have binding authority on all people; they are the sole rule of doctrine and conduct.	1	2	3	4
g. Homosexual practice is a sin.	1	2	3	4
h. God really exists, there is no doubt about it.	1	2	3	4

For the first eight items, responses range from 1 to 4 on a scale of "Not important at all" to "Most important." Recognizing that one goal of survey research and closed-ended questions is efficiency, respondents can quickly go through questions, simply circling their response. We do this unthinkingly, not stopping to read the questions thoroughly. To mitigate this, the second set of religious belief questions reversed the 1 to 4 continuum, running from "Completely true" through "Definitely not true." Interestingly, when the surveys were collected and cleaned, there was a pattern among several respondents that was inconsistent. These respondents had given "Most important" to all of the first set of eight beliefs, but they gave "Definitely not true" to the second set of eight beliefs. This was totally inconsistent. In reviewing the rest of the surveys and creating a measure that combined other theological belief questions from the surveys to test for reliability, it seemed that the response set had occurred regardless of how the items had been constructed to avoid it. Once we get into a pattern of responding to questions in a survey, particularly where items are grouped, we tend to tune out and just keep doing what we did originally. In hindsight, it would have been better if all 16 beliefs had the same ordering of responses.

STUDYING CHANGE WITH SURVEYS

Social research is fun. It is exciting. Getting the tools to think about how to find out about human interactions and experiences is a wonderful thing. As a research design, surveys also offer us a look into the past and future. Many research questions deal with change over time. There is a general idea as each generation comes of age that things have changed precipitously. As much as things change, they also stay the same. But how do we know when things actually change?

When social scientists sample populations to survey, they have a cross-section of that population. **Cross-sectional studies** are those that take a snapshot of a respondent group at one particular time and place. The attempt is to measure what exists (in terms of beliefs, behaviors, etc.) at that time. Researchers note that these findings cannot be considered particularly valid or reliable, since they do not necessarily reflect the overall dynamic nature of individual belief and behavior. However, cross-sectional studies allow the researcher to sample individuals' belief and behavior in order to obtain larger patterns. The general hope is that these sample responses are normative across respondents. Once researchers have data from a cross-sectional study, they are able to measure change using cohort, trend, panel, or longitudinal studies.

USING TIME IN SURVEY STUDIES

Cohort Studies

Over time, researchers can survey new samples of the same population, replicate survey items from an earlier survey, and compare how whole populations have changed over time. The first GSS was given in 1972. Since then the GSS has replicated approximately 250 items on each of the subsequent items that were given to a representative sample of Americans in 1972. We can use multiple years of the GSS to trace how the general attitudes, behaviors, and experiences of Americans have changed or perhaps not changed. The GSS continues to add new survey questions, using modules of topics in each new GSS to add to the overall array of questions that can be studied as trends over time.

Trend Studies

When using two or more cross-sectional studies from the same population, you have a **trend study.** Drawing on independent samples of the same population is important. You may survey a random sample of a different population, but then you cannot compare to see how either population has changed (since you're not comparing samples of the same population). Once you have a random sample of the same population as a previous cross-sectional study, you also need to make sure that the surveys use at least some of the same questions; otherwise you have no items that you can compare to see how they may have changed. A good example of a trend study using the GSS is the study by Jeni Loftus (2001) of how attitudes toward homosexuality have changed in America over time. Drawing on GSS data from 1973 to 1998, Loftus traced the responses to four items consistently included in the GSS measuring attitudes toward homosexuality:

> "What about the sexual relations between two adults of the same sex—do you think it is always wrong, almost always wrong, wrong only sometimes, or not wrong at all?"

> "And what about a man who admits that he is a homosexual? Suppose this admitted homosexual wanted to make a speech in your community. Should he be allowed to speak or not?"

> "Should such a person be allowed to teach in a college or university or not?"

> "If some people in your community suggested that a book he wrote in favor of homo-sexuality should be taken out of your public library would you favor removing this book, or not?

In examining how American attitudes regarding homosexuality had changed over time, Loftus found that while Americans had become increasingly negative about the morality of homosexuality through the year 1990, they had then become increasingly less negative over the next eight years. Loftus also found that during the same 25-year period, Americans had become less and less willing to restrict the civil liberties of homosexuals.

Panel Studies

Sometimes when you want to study change over time, you find that you need a large amount of data. But people will only spend so much time in an interview or filling out a questionnaire. One way to ameliorate this is to do a panel study. Whereas trend studies rely on multiple surveys given to different samples of the same population, **panel studies** rely on the *same respondents*. Recognizing that respondents can only sit so long and answer so many questions, panel studies select a random sample of respondents and send them multiple surveys over short periods of time. A short period of time could be anywhere from a few months to a few years (generally no longer than a three- to five-year period).

The Presbyterian Church, U.S.A. (PCUSA) has a research group that implements a regular denominational panel study (you may have run into them when looking at studies on The ARDA, the Presbyterian Panel Studies). Each study targets specific topics, with each topic its own survey. Multiple surveys are sent to the sample over the course of study. The responses to the surveys are collated into one large data set, spanning thousands of variables over the same respondents. The data are in panel studies can be very rich, since there are so many variables for each case (person).

Longitudinal Studies

Panel studies solve the problem of not enough data per case (or too few variables for each person). But sometimes we do not expect people to change significantly over such short periods of time. One of the most interesting ways to study change over long periods of time is to use **longitudinal studies.** By following along with panel studies in using the same respondents over time, we yield a great deal of data for one sample. The difference between the two, however, is that the surveys are given over long periods of time, perhaps five years and longer. Some longitudinal studies take data from the same people over the course of their entire adult lives. The data gleaned from these studies are amazing.

Longitudinal studies encounter some unique issues, however, because of the unique way they study change over time. When begun, the sample is usually selected using probability sampling. Over the course of time, however, that sample ceases to be representative of the population, sometimes because of attrition, or having people drop out. Over long periods of time members of the sample may lose interest and stop responding to the surveys, they may move and lose touch with the project, and/or in some cases they may die (the issue of mortality). We've made clear how important probability samples are to generalizing the results of the data. We do not want to lose generalizability when the data are such a rich resource of studying change. Researchers have developed strategies to mitigate these issues. As attrition occurs, researchers are able to continue to add cases (people) who are representative of the population. An example of a longitudinal study is the National Survey of Youth and Religion (NSYR) (http://youthandreligion.org/). Using random-digit dialing, researchers sampled nationally representative households with youth age 13 to 17 to learn about the religious and spiritual lives of adolescents. The telephone interviews consisted of a 40-minute questionnaire for the youth and a 30-minute questionnaire for their parents. Since the first *wave* of surveys was completed in 2002, two additional waves of surveys from these respondents have been collected (the second wave in 2005 and the third wave between 2007 and 2008). The NSYR data are particularly compelling because each wave of telephone surveys has been followed up with in-depth interviews with a sample of each group of respondents. Together, these data provide a picture of how religion changes within the life course of adolescents and young adults.

Censuses and Sampling

As we've gone over surveys as a research design, one thing we want to reemphasize is the importance of the census and sampling techniques discussed in Chapter 8. Survey results are inextricably tied to how the target population is selected. Even with good survey design, without a clear population census or sample, the results of the survey are not generalizable to any meaningful population. The point of survey research is generalization; thus the sampling must be done using probability sampling (unless a whole population is being used for the data collection). The biases inherent in sampling—nonresponse, selective availability, and areal bias—will impact the generalizability of your survey results, even apart from the specific biases that directly relate to survey construction (poor item construction, conformity bias, and response set). Surveys, censuses, and samples are inextricably linked; all must be conducted methodically and thoughtfully. This is the essence of social science.

Exercise: Survey Research in Action

Ghosts, UFOs, and Other Paranormal Beliefs[2]
We usually think of fairly traditional beliefs when considering religion, such as belief in God or heaven. However, many individuals in the United States hold supernatural or paranormal beliefs that do not fit into those traditional categories. This learning module will walk you through survey data examining some of these beliefs to see how many people hold them and who they are.

Open your Web browser and go to the home page for the Association of Religion Data Archives (www.theARDA.com).

---

[2] Used with permission from the Association of Religion Data Archives (ARDA) (http://www.thearda.com/learningcenter/learningactivities/module18.asp).

Let us begin by looking at belief in ghosts. Search for "ghosts" in the search box in the upper right-hand corner. The results will be organized by the area of the ARDA's Web site in which they were found.

1. In what areas of the ARDA's Web site were results for "ghosts" found? (Hint: Look for the green icons that mark each area of results. The name of area will be to the right of these icons.)

   _____

   You should have found some results in the "ARDA Data Archives-Questions/Variables on Surveys" section. Look through these results and find the question from the Baylor Religion Survey (2005) that asks whether a person believes that ghosts exist.

2. What is the label for this question? (Hint: It will be hyperlinked and to the left of the survey's name. While they seem strange, these labels are simply shorthand names for survey questions.)

   _____

3. If you had to guess, what *percentage* of the U.S. population do you think "absolutely" believes that ghosts exist?

   _____

   Click on the name of the question. This will take you to the overall responses that people gave to this question.

4. What *percentage* of the U.S. population say that they "absolutely" believe ghosts exists?

   _____

   Let us explore this in more detail. Click on the "Analyze Results" link below the responses to the question. You will see a page with a pie chart and the same results you looked at on the previous page. Below this, though, are several tables looking at people's responses broken down by various social and demographic categories.

5. Do you think men or women are more likely to believe in ghosts? Why do you think there might be a difference?

   _____

   _____

   Find the table for the gender of the individual and fill in the following table with the percentage of male and female respondents who gave each response.

	Male	Female
Absolutely not		
Probably not		
Probably		
Absolutely		

(Continued)

6. How does a person's church attendance relate to their belief in ghosts? Fill in the following table and briefly discuss the findings.

	Never	<1 Month	<1 a Week	Weekly+
Absolutely not				
Probably not				
Probably				
Absolutely				

_____
_____
_____

7. Look at the other tables and describe some of the patterns you see in beliefs in ghosts.

_____
_____
_____
_____
_____

Let's move on to another supernatural belief. As you did earlier, search for "UFO" in the upper right-hand corner. (Note: UFO stands for "unidentified flying object" and is usually associated with visitors from space.) Find the question that asks whether the person has "ever read a book, consulted a Web site, or researched the following topics . . . UFO sightings, abductions, or conspiracies."
After you find it, click on the name of the question.

8. What *percentage* of people have done research on UFO-related issues?

_____

9. Do responses differ by gender? How does this compare to what you found earlier concerning ghosts? If there is a difference, why do you think this is?

_____
_____

10. How does a person's church attendance relate to their interest in UFO-related topics?

_____
_____
_____

    You should have a good idea of how to search and move around within the ARDA's Web site. Now it is time for you to do some independent research. Feel free to explore different surveys or different areas of the ARDA.

Search for some other nontraditional beliefs. Here are some examples you can use: New Age, astrology, horoscope, Bigfoot, Loch Ness monster, dreams, haunted, psychic. If you search for something but do not find any results, try any synonymous terms. If you still do not find anything, it is possible that it has not been asked on any surveys.

Choose one of the questions you find and do some analysis. Be sure to address the following issues: How many people overall believe or practice the thing you are examining? How do responses vary by other social or demographic characteristics? How does it relate to more traditional religious beliefs or behaviors? **Print out any relevant tables or results you find and attach them to this module.**

11. Discuss your findings below.

_____

_____

_____

_____

_____

_____

12. Given what you have seen in this learning module, what seems to be the relationship between traditional religious beliefs and practices and such nontraditional beliefs?

_____

_____

_____

_____

_____

_____

_____

13

AGGREGATE RESEARCH

Sometimes individual-level data are not appropriate for answering research questions; we need data from **aggregates,** or units that are composed of many individuals, such as cities, states, nations, schools, clubs, or churches. Each of these units (states, clubs) is composed of hundreds, if not thousands, of individuals. Aggregate-level data, as opposed to survey-level data, relies on a smaller number of cases, but each case is made up of thousands of smaller units. Say, for example, you had a data set made up of variables collected by the U.S. Census Bureau that reported data on the 50 states. How many cases would the data set have? The cases, or units of analysis, would be each individual state, so there would be 50 cases where each one case is composed of millions of individual units (people). The SPSS screen shown in Figure 13.1 is an example of an aggregate database in which there are many variables for each state (cases).

Apart from census data, another example of aggregate data would comprise variables taken from a congregation, where variables may include total number of members, number of baptisms, number of deaths, value of churches/equipment/property, or debt on buildings/property. These examples of aggregate data stand in stark contrast to the data collected for the General Social Survey (GSS) 2010. This GSS has 4901 cases or individuals. The discrepancy in the number of cases we deal with depending on the type of data we have will cause us to look at some unique aspects of aggregate data.

Understanding and Applying Research Design, First Edition. Martin Lee Abbott and Jennifer McKinney.
© 2013 John Wiley & Sons, Inc. Published 2013 by John Wiley & Sons, Inc.

STATE	NUMBELOW POV08	LOWPOV08	NUMFAMBE LOWPOVE	PERCENTB RTY08	PERCENTF AMBELOWP OV08	SSIREC08	SSIPAY08	TANFREC08	TANFFAM08	TANFTOT08	TANFASSIS T08	MEDINCHS0 8	MEDINCFA M08	POP09
United States	39108	7252		13.2	9.7	7520	43035.00	2355.00	1622.00	28130.00	10047.00	52029.00	63366.00	307007.00
Alabama	713	146		15.7	12.0	167.00	875.00	41.00	29.00	143.00	46.00	42066.00	54270.00	4709.00
Alaska	56	9		8.4	5.7	12.00	61.00	8.00	10.00	63.00	42.00	68460.00	79541.00	698.00
Arizona	939	154		14.7	10.3	103.00	554.00	78.00	10.00	349.00	123.00	50958.00	60547.00	6596.00
Arkansas	481	98		17.3	13.0	99.00	511.00	19.00	20.00	144.00	14.00	38815.00	47648.00	2889.00
California	4778	825		13.3	10.0	1272.00	6980.00	1217.00	46.00	6687.00	3750.00	61021.00	70029.00	36962.00
Colorado	553	95		11.4	7.8	60.00	319.00	21.00	65.00	231.00	45.00	56993.00	70164.00	5025.00
Connecticut	315	59		9.3	6.7	55.00	303.00	37.00	22.00	496.00	107.00	68596.00	85344.00	3518.00
Delaware	85	15		10.0	6.9	15.00	79.00	12.00	11.00	68.00	15.00	57989.00	68745.00	885.00
D.C.	97	15		17.2	13.7	23.00	131.00	12.00	35.00	161.00	21.00	57936.00	66722.00	600.00
Florida	2371	433		13.2	9.5	445.00	2321.00	63.00	3.00	948.00	175.00	47778.00	57455.00	18538.00
Georgia	1381	263		14.7	11.1	213.00	1123.00	38.00	7.00	615.00	116.00	50861.00	60268.00	9829.00
Hawaii	115	18		9.1	6.0	24.00	139.00	14.00	5.00	229.00	49.00	67214.00	78659.00	1295.00
Idaho	188	37		12.6	9.4	25.00	129.00	2.00	33.00	35.00	6.00	47576.00	54695.00	1546.00
Illinois	1532	283		12.2	9.0	266.00	1500.00	55.00	14.00	1013.00	63.00	56235.00	68958.00	12910.00
Indiana	806	158		13.1	9.6	108.00	601.00	84.00	116.00	308.00	75.00	47906.00	59380.00	6423.00
Iowa	335	58		11.5	7.3	45.00	232.00	39.00	24.00	173.00	73.00	48980.00	61663.00	3008.00
Kansas	307	57		11.3	7.7	42.00	230.00	31.00	2.00	176.00	68.00	50177.00	62402.00	2819.00
Kentucky	721	147		17.3	13.1	187.00	988.00	59.00	82.00	193.00	124.00	41538.00	51729.00	4314.00
Louisiana	744	146		17.3	13.4	165.00	867.00	22.00	8.00	173.00	48.00	43733.00	53963.00	4492.00
Maine	158	29		12.3	8.6	34.00	176.00	24.00	19.00	127.00	96.00	46581.00	57719.00	1318.00
Maryland	443	75		8.1	5.4	101.00	565.00	46.00	50.00	405.00	113.00	70545.00	64415.00	5699.00
Massachusetts	627	112		10.0	7.1	182.00	1058.00	91.00	12.00	915.00	293.00	65401.00	81569.00	6594.00
Michigan	1410	265		14.4	10.5	233.00	1332.00	166.00	8.00	1230.00	361.00	48591.00	60615.00	9970.00
Minnesota	491	83		9.6	6.2	81.00	438.00	48.00	15.00	435.00	71.00	57288.00	71817.00	5266.00
Mississippi	602	128		21.2	17.0	122.00	624.00	23.00	3.00	91.00	25.00	37790.00	49908.00	2952.00
Missouri	768	148		13.4	9.7	124.00	662.00	85.00	53.00	332.00	114.00	46867.00	56068.00	5988.00
Montana	140	23		14.8	9.7	16.00	82.00	8.00	52.00	39.00	18.00	43654.00	56820.00	975.00
Nebraska	187	31		10.6	6.8	24.00	123.00	17.00	5.00	94.00	23.00	49693.00	62067.00	1797.00
Nevada	290	49		11.3	7.9	37.00	198.00	18.00	3.00	85.00	31.00	56361.00	64910.00	2643.00
New Hampshire	97	17		7.6	5.0	16.00	84.00	9.00	.00	85.00	31.00	63731.00	76710.00	1325.00
New Jersey	741	135		8.7	6.2	160.00	877.00	79.00	29.00	955.00	244.00	70378.00	85761.00	8708.00
New Mexico	333	61		17.1	12.6	58.00	298.00	36.00	52.00	129.00	58.00	43508.00	52172.00	2010.00

Figure 13.1. Aggregate-level data in data view.

NATURE OF AGGREGATE DATA

Aggregate data refers to individual data that have been summed up to larger units (e.g., like individual patient data across different hospitals being summed to county- or state-level reporting). Formal organizations often use these data because you can convey several general patterns embedded in individual responses. School districts, hospitals, businesses, and other such social units aggregate data to summarize a great deal of information and to protect against identifying specific individual responses. Conclusions made in studies using aggregate data should be made at the highest level of aggregation to avoid the *ecological fallacy,* or trying to generalize to people when the data are taken at the aggregate level (see Chapter 10).

Areal and Social Units

There are two types of aggregate units: areal units and social units. **Areal units** are cases made up of geographically defined boundaries (e.g., counties, states, nations, etc.). **Social units** are derived from social boundaries (e.g., churches, clubs, sports teams, etc.). Statistically both types of units are treated the same in analysis. One unique feature of aggregate-level data collection is that since the units are composed of thousands of smaller units, they are not usually data that researchers can collect on their own. For example, researchers cannot go door to door asking every state resident how many murders he or she has committed that year in order to determine the state's murder rate. Since aggregate measures are composed of counts of individuals based on areal or social units, social scientists generally rely on some type of official agency to collect these statistics so that they can be utilized in social research.

At the beginning of this book, we talked about the need for the process of science to help establish a baseline for how the world really is, rather than what we tend to think it is or should be. Aggregate data allow us to describe phenomena on a broader level. In the book's introduction, we noted the common American belief that religious participation has been in decline since the birth of the nation. Much debate exists among researchers who study religion about individual church attendance patterns; people can exaggerate their religious participation in national surveys (for example, saying they go to church every week, when it's closer to once or twice a month; see Hadaway and Marler 2005). Individual reports of religious participation cannot tell us about the *overall* picture of American church growth and decline patterns, only aggregate data can to that.

In their book *The Churching of America 1776–2005*, Finke and Stark (2005) used aggregate data over 200 years of American history to illustrate patterns of church growth and decline. What they found was a complex picture of individual denominations growing or declining, but what was so surprising is that their data show an overall growth pattern in American religious adherence over time, noting that at the beginning of the Revolutionary War, 17 percent of Americans were "churched." By the mid-1860s, 37 percent of Americans were churched. By the mid-1920s, religious adherence had reached 56 percent, and then by the 1960s the adherence rate stabilized at 62 percent.

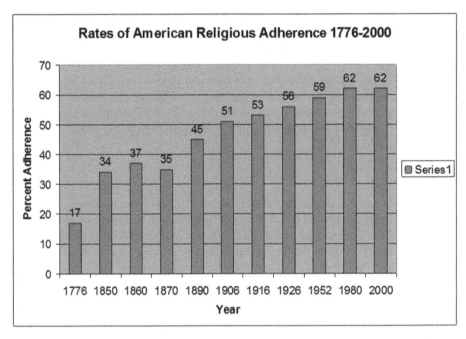

Figure 13.2. The growth in rates of American religious adherence 1776–2000.[1]

Figure 13.2 shows the rates of American religious adherence from 1776 to 2000. As you can see, the rates of adherence increase steadily across these years. The last years (1980–2000) show some stabilization of the growth pattern. Current research may suggest reasons for this plateau, although it may simply overrepresent a dynamic on the continued growth of the phenomenon. As we noted in Chapter 2, much has been made about religious "nones" and how that dynamic impacts the overall growth of religion. At this point, there is no clear reason these two ideas—continued religious growth and the growth of "nones"—are contradictory. With a 62 percent adherence rate holding steady, that leaves 38 percent of Americans who do not regularly participate in religious organizations. Many of these 38 percent claim a religious affiliation, even though they are not formal adherents. At this point it appears the 15 percent of American religious "nones" accurately describe a portion of the percentage of America's nonadherents.

Finke and Stark's (2005) study could only have been undertaken using aggregate data. In their book, Finke and Stark describe the data sources they used for the study. One of the hallmarks of the study is their use of U.S. Census data ranging from 1890 to 1936. During these decades the U.S. Census collected data on every American religious body including total membership, Sunday school enrollment, number of congregations, and many other variables (see Finke and Stark 2005, Chapter 1, for a description of their data).

[1] Adapted from Finke and Stark 2005.

Using a variety of aggregate data was appropriate for Finke and Stark's research question, which hypothesized that religion operates as a religious economy and that in countries with a relatively unregulated religious market, religious organizations will thrive. What that means is that living in a country where people are free to practice any kind of religion, anyone can build their own firm (church)—or participate in a competitive religious market in order to win souls/adherents. The aggregate data (200 years' worth) support Finke and Stark's hypothesis that when religious regulation remains relatively low, religious participation will be relatively high (a 62 percent religious adherence rate is quite high).

Rates

Aggregate data are also unique in that they rely on different types of numerical values. Since aggregate data rely on people's actual behaviors/characteristics (murder rate, property crime rate, percentage of the population with a college degree, median family income in states, etc.), the numerical values tend to be interval- or ratio-level measures (the hallmark of both measures is that they have equal distance between units; see Chapter 5 on measurement to review the levels of measurement). In comparing areal or social units, we need to make sure that the measures are standardized so that we can compare variables across units (another key is to make sure to compare similar units, e.g., nations compared to nations, states compared to states, Girl Scout troops compared to Girl Scout troops, congregations to congregations, etc.).

Units that are comparable may not be the same size (for example, look on a map at California and Rhode Island—can you even find Rhode Island?). If you take just the raw number of some variable to compare across states, the total number of anything within a state will be skewed toward the size of the population of that state. For example, if you had a variable that measured the number of people receiving food stamps in a state, chances are the states with the highest number of food stamp recipients would be California, New York, and Texas. In looking at these states, do you notice anything that they have in common? They are the largest three U.S. states. It should make sense that they would have the highest numbers of people receiving food stamps, since they have the highest populations. But do these states have a higher *proportion* of their population receiving food stamps compared to other, smaller states? Unless we take into account some common denominator, we cannot compare which states may have a higher *percentage* of food stamp recipients.

In order to compare phenomena across states, we need to standardize raw numbers by using rates and percentages. **Rates** are calculated using a common base for each unit so we can compare variables across units. Remember when you learned fractions in elementary school? You learned to add and subtract things that had different bases (or denominators):

$$\tfrac{1}{2} + \tfrac{1}{4} = \tfrac{2}{6}$$

Remember when you did that? What? Is that not how it works? Oh, that's right, we need to make each fraction that we're using *equivalent* or comparable. We do this by finding a common denominator (base) for each fraction. For the previous equation, the common denominator between one-half and one-fourth is fourths:

RANK	CASE/STATE	VALUE
1	California	2259
2	Texas	1636
3	New York	1627
4	Florida	991
5	Illinois	923
6	Pennsylvania	907
7	Michigan	772
8	Ohio	734
9	Georgia	632
10	Tennessee	538
11	Louisiana	537
12	North Carolina	528
13	Alabama	427
14	New Jersey	425
15	Kentucky	412
16	Missouri	411
17	Virginia	397
18	Washington	362
19	South Carolina	333
20	Mississippi	329

Figure 13.3. Number of persons (in thousands) receiving food stamps.

$$\frac{1}{2} = \frac{2}{4}, \quad \text{therefore } \frac{2}{4} + \frac{1}{4} = \frac{3}{4}$$

Creating a common base for a rate often relies on the type of aggregate unit you have. If we're looking at data taken on the 50 U.S. states, a commonly used base is "per 100,000 population." By using the state's population as a common base, we can compare which states have the highest or lowest rates of some variable because each variable takes into account the state's total population. Let's go back to the example of the variable for "number of persons (in thousands) receiving food stamps" in a state. This variable gives the raw number (or total number) of people within each state receiving food stamps "in thousands." If we look at the states with the highest number of food stamp recipients, indeed, California, Texas, and New York are at the top of the list: California with 2259 food stamp recipients, Texas with 1636 food stamp recipients, and New York with 1627 food stamp recipients (see Figure 13.3).

What do the numbers in Figure 13.3 tell us? The variable description explains that the numbers given for each state are "in thousands"; therefore we need to multiply the number by 1000 to see the number of food stamp recipients for each state (California with 2,259,000 food stamp recipients, Texas with 1,636,000, and New York with 1,627,000).

Again, having these particular states with the highest number of food stamp recipients is logical because these are the states with the largest populations. If we want to compare which states have the highest *proportion* of their populations receiving food stamps, we need to calculate using rate. Rates allow us to compare across states to see which has a higher proportion or percentage of their populations receiving food stamps.

In order to calculate a rate, we take the raw number of food stamp recipients (the numerator) and divide by a common base. For this example we divide the total number of food stamp recipients by the total population of the state:

Number of persons receiving food stamps in state

(numerator/raw number)

———————————————————————————————

Total state population

(denominator/common base)

When we take each state's total number of people receiving food stamps (the raw number) and divide it by the state's total population (common base), we can then see the states with the highest *percentage* of their populations receiving food stamps, shown in Figure 13.4.

The three states with the highest percentage of food stamp recipients from Figure 13.4 are West Virginia with 14.85 percent of their population receiving food stamps, Louisiana with 12.29 percent of their population receiving food stamps, and Mississippi with 11.95 percent of their population receiving food stamps. Where do California, Texas, and New York fall on percentage of population receiving food stamps? New York is ranked 11 out of the 50 states, with Texas ranked 14 out of 50 states, and California ranked 25 out of 50 states (California not shown on Figure 13.4).

Rates allow us to standardize variables in order to compare the variables across aggregate units. When calculating and evaluating rates we want to be aware of what is being used as the common base. Look at the following variables. How do they differ?

RANK	CASE/STATE	VALUE
1	West Virginia	14.85
2	Louisiana	12.29
3	Mississippi	11.95
4	Kentucky	10.47
5	Hawaii	10.23
6	Arkansas	10.09
7	New Mexico	10.07
8	Tennessee	9.91
9	Alabama	9.81
10	Maine	9.24
11	New York	8.95
12	South Carolina	8.68
13	Oklahoma	8.60
14	Texas	8.28
15	Georgia	8.27
13	Michigan	7.86
17	Vermont	7.78
18	Illinois	7.66
19	Missouri	7.56
20	Pennsylvania	7.56

Figure 13.4. Percentage of the population receiving food stamps.

BEER = gallons of beer consumed per person 16 and over

%BEER = percentage of alcoholic beverages consumed that was beer

While both variables measure beer consumption, they are calculated differently. The first (BEER) is calculated using total gallons of beer consumed; the other (%BEER) is calculated using total percentage of alcoholic beverages consumed that was beer. Variables that address the same overall concept (beer consumption) can be calculated quite differently depending on the numerator (raw number) and denominator (common base). When evaluating rates, we want to be aware that depending on the common base used to calculate the rate, we could have very accurate statistics that tell very different stories. Let's explore this more using the example of divorce rates.

It's important to evaluate a rate by knowing how the rate is calculated. What story is being told by the use of a particular rate? One way to think about evaluating a rate is by asking what is being used as the common base—is it the most relevant base? A relevant base can give you a more precise measure. For example, most Americans can tell you that the divorce rate is 50 percent. The description of this statistic tells you something about how it is created. Since it is a rate, we know it's an aggregate-level measure. Since aggregate-level measures/rates take a raw number and divide it by a common base, what is the common base used here to calculate the divorce rate of 50 percent? Do you know how this rate is calculated? Based on our previous discussion, you might guess that the raw number of divorces per year would be the numerator, and perhaps the total population the denominator for this rate. That's not how this particular rate is calculated. We should also add that the 50 percent divorce rate has become an urban legend. The rate has a very real beginning, but is it the best measure of divorce in America?

Divorce rates are a good example of how you can have *accurate* statistics that differ from one another. There are a number of ways to calculate the divorce rate—all of them accurate, yet all of them different. Depending on the common base, we can come up with a variety of ways to describe the divorce rate including the ratio of annual marriages to annual divorces, the crude divorce rate, or the refined divorce rate. Let's look more in depth at these three measures.

The 50 percent divorce rate seems to be the rate that has won the popular debate. This is the highest possible estimate for divorce and receives the most media attention. How is it calculated? Generally this rate is calculated by taking the raw number of divorces granted in a given year (numerator) and dividing by the raw number of marriages performed in a given year (the denominator). In 1981 there were 1.2 million divorces and 2.4 million marriages. Using annual divorces divided by annual marriages for 1981 yields a divorce rate of 50 percent.

$$\frac{\text{Total number of divorces granted}}{\text{Total marriages performed}} = \frac{1,200,000 \text{ divorces}}{2,400,000 \text{ marriages}} = \frac{50 \text{ percent}}{\text{divorce rate}}$$

Consider how this rate is calculated. How many marriages performed in a given year are the same ones ending in divorce that same year (although there are certainly a few, it is probably not a significant number). So what does this rate mean? Although accurate, we are not comparing the same dynamic in the numerator and denominator.

While marriages were performed in a given year, the divorces are the result of marriages of varying length over the course of several years. There is another problem with the precision of the variable. What happens if the number of divorces stays the same but the number of marriages declines? The divorce rate will rise. Again, this is an accurate statistic; it is clear how it is calculated. The 50 percent divorce rate, however, is probably not the best estimate of divorce in America.

We could look at the total number of divorces granted per year to give a different picture to see how divorce changes over time. But what should we use as a common base? We cannot simply compare the raw numbers of divorces per state—they would be skewed toward the states with the largest populations having the highest numbers of divorces (and we would not be able to compare the raw numbers across states). An alternative measure is the crude divorce rate, which takes the total number of divorces and then divides them by the total population of each unit (state). Think about this rate's common base. Who is included in the total population? Is the total population of a state at risk for divorce (children and the unmarried)? No, not all members of a state's population share an equal risk for divorce. While calculating the crude divorce rate is an *accurate* way to measure divorce rates, it still may not be the best way to measure divorce rates (it will be a significantly lower rate than the 50 percent rate, running somewhere around 30 percent).

The most precise divorce rate, the refined divorce rate, calculates the number of divorces per 1000 married women over the age of 15. This rate is considered the best estimate of divorce because it counts the total number of divorces (numerator) and divides the number by the total number of women eligible for divorce (the denominator). The refined divorce is the best indicator of divorce because the common base is the most **relevant base,** or the base that takes the most accurate population. By taking only married women over 15 years old as the common base, we arrive at a rate that is more precise in describing divorce. These data can be more difficult to collect; thus social researchers often rely on individual level data regarding divorce to see a picture of divorce in America. Nationally representative surveys like the GSS ask respondents, "[Excluding those who have never been married] Is respondent or has respondent ever been divorced or legally separated?" When asked if a respondent has ever been divorced, the number responding affirmatively hovers around 42 percent. Although 42 percent may not seem very far away from 50 percent, it is significantly less.

Aggregate-level data give us a fuller picture of a social unit (e.g., states or nations), than individual survey data can for some measures. All of the measures of divorce we've discussed represent *accurate* measures of divorce. This is something to keep in mind. When we talked about valid and reliable measures for survey data, we noted that partisan or special interest groups will phrase items in such a way as to bias the outcome. This can also happen with the reporting of rates. Depending on the specific interests of a group, rates can be used to selectively support those interests. As you read about rates, think about what type of group is reporting the data. Does the group reporting the data have a stake in how people think about the issue? How is the rate calculated? If how the rate is calculated is not included in the report, be cautious. If the rate calculation is described, consider what the base tells you; is it the most relevant base? Does the base yield a lower (crude divorce rate) or higher (divorces granted in a year divided by marriages performed that same year) estimate of the phenomenon?

Outliers

One last unique feature of aggregate-level data is the problem of outliers. An **outlier** is an extreme *case* that distorts the true relationship between variables, either by creating a correlation that should not exist or suppressing a correlation that should. While aggregate level measures are composed of hundreds, thousands, or millions of individuals, the aggregate units themselves have significantly fewer units. The United States has over 310 million residents, but for data taken at the state level (aggregate level) those 310 million people are divided into just 50 states. When analyzing 50 cases, if even one case is significantly different from the others, that case may throw off any statistical calculations (creating a significant correlation when there is not one or creating an insignificant correlation when there really is a significant relationship). Therefore, we need to test for outliers—cases that distort the true relationship between variables.

Look at Figure 13.5, which shows a scatterplot between percentage white in a state and the number of astrology offices per 100,000 population. We hypothesize that whites are the most likely to visit astrology offices.

The scatterplot in Figure 13.5 illustrates a flat cluster of the states with an $R^2 = 2.601E-4$! Notice at the top of Figure 13.5, however, the lone dot that is far from the regression line (line of best fit). It appears that there is an outlier—an extreme case that distorts the true relationship between percentage white and astrology offices. When data are arrayed in this manner, we need to think about what is so unique about that case that it would be so far afield of the other data points. In going back to SPSS and sorting the cases, we find that the outlier is the data for the whole United States. In our census data file, each state, the District of Columbia, and the United States are included

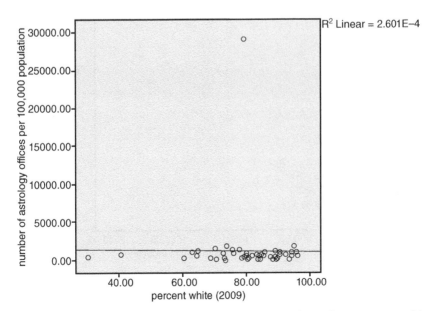

Figure 13.5. Number of astrology offices per 100,000 population by percentage white.

as cases. Clearly, we need to "select cases" and take out the United States before we can come to any conclusions about the relationship between our variables. Because aggregate data use fewer cases, we want to make sure that our analyses do not contain any cases that are significantly different (for example, Washington, D.C., may also serve as an outlier when compared to the states; rather than being a state, it is more equivalent to a large urban city. States and cities are not comparable units, so Washington, D.C., may need to be removed from the data set).

As we've illustrated, aggregate data help social scientists look at the big picture; social patterns that are beyond the scope of individuals. How the data are collected, how rates are calculated (and interpreted), and how one case can distort the relationships within the data demonstrate the unique features of this research design. In order to explore more about aggregate data, we have provided two practice exercises. The first exercise deals with how rates are calculated. The second exercise comes from the ARDA (www.theARDA.com) and walks you through the Learning Module for Exploring Congregations in America.

Exercise: Putting It Together[2]
Let's look at an example of a scatterplot using aggregate data, with two variables, number of abortions in a state, and number of Social Security recipients (women) 65 and older in a state. Using the scatterplot function in SPSS, we look at the relationship between these two variables:

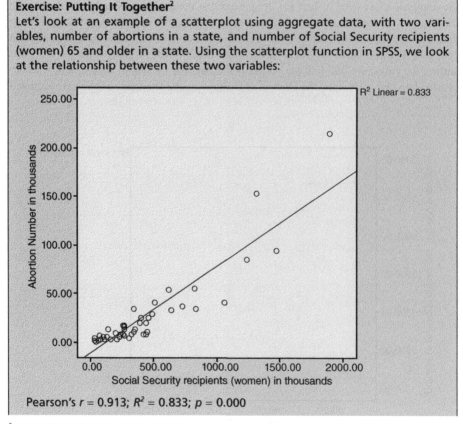

Pearson's $r = 0.913$; $R^2 = 0.833$; $p = 0.000$

[2] Adapted from Corbett and Roberts 2002.

The analysis shows a Pearson's r equal to 0.913 and a probability level equal to 0.000, which tells us there is a strong and significant relationship between Social Security beneficiaries and abortions in states.

What explains the strong relationship between these two variables?

_____

_____

_____

_____

_____

_____

_____

Although we can try to find the "missing link" between these two variables, there is one clear reason why these variables have such a high correlation. Look at each variable:

What does the description tell you about how number of abortions is calculated?

_____

What does the description tell you about how the number of the value was calculated?

_____

What do the variables have in common? Both variables are simply measured as raw numbers. When looking at aggregate data, we tend to jump to the conclusion that the variables are rates. That is not the case. If these variables were measured as rates, the description would say, "Social Security beneficiaries *per* 1000 population" or "Abortions *per* 1000 population." Instead, each is measured "in thousands." If we look sort the cases by descending order in SPSS, we can see that California, New York, and Florida have the highest numbers of both Social Security recipients and abortions. The high correlation between these two variables is a function of state size. When aggregate data are not transformed into a rate or percentage, we cannot compare them. *The results we get will be a function of the size of the unit, not a function of a potential relationship between variables.*

If you take these variables and calculate a rate, you will find very different results. Using variables that measure Social Security recipients (woman) 65 and older per population and abortions per population" (ABORTRATE1), we see the following results:

(*Continued*)

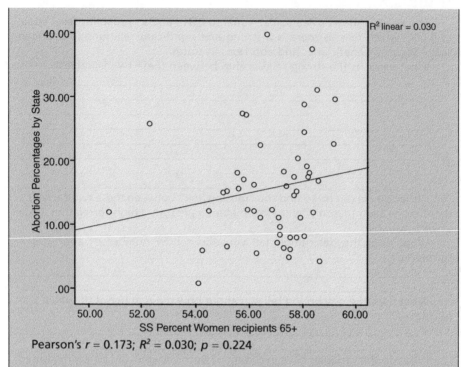

Pearson's $r = 0.173$; $R^2 = 0.030$; $p = 0.224$

The analysis shows a Pearson's r equal to 0.173 and a probability level equal to 0.224. When using variables calculated as percentages or rates, there is no relationship between percentage Social Security beneficiaries and abortion rates. As we stressed in Chapter 5 on measurement, how data/variables are measured is important. Our results rely on our ability to think through how data are measured. When using aggregate data, this is very important. How rates are calculated and that they are calculated is important, impacting both the reliability and validity of your findings.

Exercise: Exploring Congregations in America[3]
Congregations have been called the "basic unit" and "bedrock" of American religious life (Chaves 1999; Warner 1994). As one group of sociologists summarized (Chaves et al. 1999),

> [Congregations] are the primary site of religious ritual activity, they provide an organizational model followed even by religious groups new to this country, they provide sociability and community for many, they offer opportunities for political action and voluntarism, they foster religious identities through education and practice, and they engage in a variety of community and social service activities.

---

[3] This exercise is printed by permission from the Association of Religion Data Archives (http://www.thearda.com/learningcenter/learningactivities/module8.asp).

In this learning module, you will explore a nationally representative survey of congregations in the United States to learn more about this central feature of American religion.

Open your Web browser.

If you find yourself lost at any point, look to the instructions on the right.

Open your Web browser and go to the home page for the Association of Religion Data Archives.

Go to www.thearda.com

Studying congregations can be surprisingly difficult because there is not a comprehensive list of every congregation in the United States. This makes simple sampling techniques impossible. The National Congregations Study, however, solved this issue using innovative methods to create the first nationally representative sample of congregations.

Click on "Data Archive" in the main menu.

Let us begin by locating this file in the Data Archive.

Click on the "Browse Alphabetically" tab.

1. The National Congregation Study has been conducted twice. When was the first National Congregations Study, or "wave" completed? When was the second wave completed?

Find "National Congregations Study, Cumulative Dataset"

_____

_____

2. According to the "Collection Procedures," what individual within congregations typically answered the survey? (Hint: this person is often called the key informant.)

_____

Let's now begin exploring the findings of the National Congregations Study. You can find the full list of questions included in the survey by clicking on the "Codebook" tab. You can search this codebook by clicking on the "Search" tab and entering keywords.

Click on the "Search" tab.

Many congregations are affiliated with a larger network of congregations united by historical or theological tradition. These networks are called denominations.

Enter keywords to locate the question asking about denominational affiliation.

3. According to the National Congregations Study, what percent of congregations are "formally affiliated with a denomination, convention or some similar kind of association"?

Hint: The shorthand name for this question is HAVEDEN.

_____

If you click on "analyze results" under the question, you can look at some patterns of denominational affiliation by a select number of other questions.

Click on "analyze results."

4. What percent of "white conservative, evangelical or fundamentalist" congregations are formally affiliated with a denomination? What about "white liberal or moderate" congregations?

Find the table for "Religion (I-RELIGION)."

(_Continued_)

_____

_____

White conservative, evangelical, or fundamentalist: _____

White liberal or moderate congregations:
_____

Let's examine the leadership of congregations in the United States.

5. What percent of congregations are led by a female religious leader?

_____   **Click on the "Search" tab.**

6. What types of congregations are most likely to have a female leader?

_____   **Enter keywords to locate the question asking whether the religious leader is male or female (CLERGSEX).**

_____

_____

7. What percent of congregational leaders graduated from a seminary or theological school?   **Click on "analyze results" and examine the tables.**

_____

8. Fill in the following table to show the percent of seminary-trained leaders by the religious tradition of the congregation.   **Click on "analyze results" and examine the tables.**

Has the religious leader "graduated from a seminary or theological school"?   **Click on the "Search" tab.**

	Roman Catholic	White conservative, evangelical fundamentalist	Black Protestant	White liberal or moderate	Non-Christian
Yes					

Enter keywords to locate the question asking whether the religious leader graduated from seminary (CLERGRAD).

Let's now examine some of the features of congregations' worship services. Some features are found in many congregations, while others are rarer.   **Click on "analyze results" and fine the "Religion (I-RELIGION)" table.**

9. Find the percentage of congregations with worship services that included the following features:

Silent prayer or meditation (MEDITATE): _____

Worshippers greeted one another by shaking hands (GREET): _____

Part of service directed specifically at children (KIDTIME): _____   **Click on the "Search" tab.**

"Amen[s]" or expressions of approval voiced (AMEN): _____

People heard speaking in tongues (TONGUES): _____

Incense used (INCENSE): _____

10. What percent of congregations participated in a joint worship service with another congregation?

_____

11. Many congregations engage in social and political issues at the local, state or national level. Find the percent of congregations that report the following activities:

Organized or participated in a demonstration or march (MARCH): _____

Organized or participated in efforts to lobby officials (POLOPPS): _____

Organized voter registration groups or meetings (VOTERREG): _____

Distributed voter guides to congregation (VOTRGUID): _____

12. Describe how responses to these political engagement activities differ by the religious tradition of the congregation. What tradition(s) are most likely to participate in these activities? What tradition(s) are least likely? Do these patterns differ depending on the specific type of activity?

_____

_____

_____

_____

Because the National Congregations Study has been conducted twice, we can explore how congregations have, or have not, changed between the two editions of the study. Let's look back of some of the findings above in light of this.

13. Some sociologists have pointed to an apparent growth in independent or non-denominational congregations. Does the National Congregations Study show any growth in the percentage of congregations that do not have a formal affiliation with a denomination (Hint: Analyze results for HAVEDEN)? Discuss your findings below.

_____

_____

_____

Enter keywords like "prayer" or "amen" to locate these questions.

Click on the "Search" tab.

Enter keywords to locate the joint service question (JOINTWOR).

Click on the "Search" tab.

Enter keywords like "march" or "voter" to locate these questions.

Go back and click on "Analyze Results" for the questions above and examine the tables.

Click on the "Search" tab.

(Continued)

Although less than 10 years separate the two editions of the National Congregations Study, many changes have occurred within society. One such change is the growth in Internet access and use.

14. What percent of congregations reported having a Web site in 1998 (WEBSITE)? In 2006?

1998: _____

2006: _____

15. What percent of congregations used electronic mail to communicate with members in 1998 (EMAIL)? In 2006?

1998: _____

2006: _____

Enter keywords to locate the question asking about denominational affiliation.

Click on "Analyze Results" and find the "Year (I-YEAR)" table.

Click on the "Search" tab.

Enter keywords to locate the question asking about the congregation's Web site and e-mail.

Click on "Analyze Results" and find the "Year (I-YEAR)" table

14

EXPERIMENTS

Probably the most notorious example of a scientific experiment is Stanley Milgram's obedience experiment. A social psychologist at Yale in the 1950s and 1960s, Milgram wanted to design an experiment to help him explain the horrors of Nazi Germany. He wondered what would influence people to obey authority to the degree that they could willingly allow and/or participate in the extermination of millions of other human beings. Milgram designed his study to test how far an individual would go in obeying an authority figure when the results of this obedience could hurt another person.

Milgram's experimental design was set up as a learning and memory exercise. Each subject was joined in the experiment by a **confederate**, a person who is in on the experimental design but pretends to be another subject. In the experiment, both men (confederate and subject) came to Milgram's lab and "randomly" drew a piece of paper out of a hat. While both pieces of paper had the word "Teacher" written on them, the confederate (a middle-aged accountant) pretended that he received a piece of paper labeled "Learner." The three men (experimenter, subject/Teacher, and confederate/Learner) then walked into a separate room where they saw a chair attached to electrodes. The experimenter explained that the Learner would be strapped to the chair

Understanding and Applying Research Design, First Edition. Martin Lee Abbott and Jennifer McKinney.
© 2013 John Wiley & Sons, Inc. Published 2013 by John Wiley & Sons, Inc.

while the Teacher, in another room, would read word pairs aloud over an intercom to the Learner. If the Learner answered incorrectly to the word cues, the Teacher would administer an electric shock to the Learner. The Teacher was then strapped into the chair to experience, for himself, the electric shocks.

The Teacher then watches the experimenter strap the Learner into the chair. The experimenter ushers the Teacher into an adjacent room equipped with a "shock-generating" machine. Bedecked with switches, relays, and lights, the machine allowed the Teacher to read word pairs to the Learner through an intercom. When the Learner responded to the Teacher's cues incorrectly, the Teacher was instructed to flip a switch that administered the electric shock to the Learner. With each wrong answer the Teacher was instructed to increase the shock by 15 volts. The voltage gauge on the shock generator ranged from 0 to 450 volts, marked in 15-volt increments. At the 435- and 450-volt level the gauge was labeled "XXX" for severe shock.

The experiment begins. At 75, 90, and 150 volts the Teacher hears the Learner grunt with the administration of the shocks. At 120 volts the Learner shouts that the shocks are getting painful. At 150 volts the Learner cries out, "Experimenter, get me out of here. I won't be in the experiment anymore. I refuse to go on." If the Teacher stops to ask the experimenter if he should continue, the experimenter simply says, "Please continue."

At 270 volts the Learner screams in agony and tells the Teacher he wants out of the experiment. Again, the Teacher is instructed that the experiment requires him to continue to read the word pairs and administer the shocks for each wrong answer. At 300 and 315 volts the Learner screams his refusal to answer the questions. "It is absolutely essential that you continue," the experimenter tells the Teacher. At 330 and 350 volts the Teacher is greeted with silence. Silence is considered a wrong answer, and the experimenter instructs the Teacher, "You have no other choice; you must go on."

When Milgram began his study, he did not anticipate the results; 63 percent of his subjects (the Teachers) obeyed the command to continue the shocks through 450 volts—even though each heard the same responses from the Learner beginning at 150 volts, asking for the Teacher to stop the shocks in the experiment. While the Learner (the confederate) never experienced any actual shocks, the Teachers acted in such a way as to expose the power of obedience to authority.

Experiments are the most powerful research design, showing us—in action—how people react to a stimulus (in Milgram's case, how pervasive obedience to authority is even in situations where the outcome is not trivial). At their base level, **experiments** involve taking an action and observing the effects of that action. The action is the independent variable (often termed *stimulus*), and the result or effect is the dependent variable.

Experiments are unique in a number of ways. While surveys or aggregate data are designed for descriptive purposes, experiments are designed for explanatory purposes (theoretical generalizing, rather than statistical generalizing). And while they are powerful, experiments can also be quite limited. Experiments have a limited number of independent variables in play at a time. By contrast, the survey and other approaches we've looked at may include multiple causes (independent variables, called *factors* in experiments). In their simplest form, experimental stimuli have two categories: those who receive the stimulus and those who do not. There are multiple variations to be seen in the execution of experiments.

EXPERIMENTAL DESIGNS

The primary advantage of experiments is that they lend themselves to making *causal attributions*. For example, when experiments are done with precision, limiting all extraneous influences and randomly selecting and assigning subjects, the conditions are set for comparisons in which the treatment group (or experimental group) is as similar to the comparison group (or control group) in every way except being exposed to changes in the independent variable. In these cases, since the only thing that changes for either group is that one group (treatment group) is exposed to the changes in the treatment condition, researchers can attribute any changes in the outcome of the experimental study (i.e., the difference in the outcomes of the treatment and comparison groups) to the treatment itself. It is the only thing that changes, so any difference in outcomes is therefore assumed to be due to the nature of the treatment variable itself.

This is what researchers mean when they say that experiments have the potential to make causal conclusions. This is not the case with nonexperimental designs (which we discussed previously), nor is it the case with badly designed experiments. In 1963 Donald Campbell and Julian Stanley published *Experimental and Quasi-Experimental Designs for Research*, which became a touchstone work for experimental design. In their book Campbell and Stanley describe three pre-experimental designs, three true experimental designs, and ten quasi-experimental designs ("not-quite" experimental designs created when full control cannot be maintained over the experiment). Their attempt was to show how the various types of experiments were different, leading to the differential ability to make causal conclusions.

A question that can be tested with an experimental design and that often comes up in class is the issue of sex education in schools. Does a sex education program make a difference in the attitudes and behaviors of students? Experiments allow you to test the efficacy of a school sex education curriculum in impacting student perceptions and attitudes regarding sexual behavior and information.

For example, a middle school decides to offer a sex education program to their sixth graders over a nine-week period. Students are sent home with permission slips for parents to sign, allowing them to participate in the program (or not). When students obtain parental permission, they are put into the program. Experimental designs need several major components to be effective. As we walk through the next experimental designs we also report on each component that impacts the efficacy of each design.

PRE-EXPERIMENTAL DESIGNS

Pre-experimental designs do not have all of the features of a true experimental design (i.e., independent/dependent variables, pre-/posttests, experimental/control groups, or randomization). Researchers often use pre-experimental designs because they do not have access to the features that would improve the design and must create at least the primary features of a more comprehensive experimental design. This section discusses three common pre-experimental designs.

Design 1: The One-Shot Case Study

In the one-shot case study design, we select the students who have been given permission to participate in the sex education program and then enroll in the nine-week program. At the end of the program we give them a **posttest** to measure how their attitudes regarding sex behaviors and attitudes may have changed (we use a questionnaire as the posttest in this example). Here is the design:

Subjects:	independent variable →	dependent variable
(students	*(the sex education program)*	*(posttest consisting of*
with permission		*a survey of attitudes*
to enroll in the		*regarding sex)*
sex education program)		

Let's evaluate this design. The first major component of experiments is that the experiment has an independent variable and a dependent variable that is measured after the subjects have been exposed to the independent variable/stimulus. This design meets the first criteria. You may ask, however, if we are measuring the dependent variable as a *change* in previous attitudes toward sex behaviors: *How do we know what our students' attitudes were before they participated in the program*? We do not currently have any information regarding their previous knowledge/attitudes toward sex.

The hallmark of a good design is having an understanding of the nature of the subjects. If an experimenter has the ability to randomly select a group of subjects, they can be reasonably assured that their subjects at least are comparable to the population from which they were drawn. In the preceding example, however, the group (only one) was not necessarily chosen randomly. The glaring problem with this design is that it violates the major sociological injunction of asking, "Compared to what?" If the experimenter does not know what the attitudes of the subjects were before the intervention, at the very least they could compare this group to a similar group whose members did not take the sex education course. This comparison group is known as a **control** or **comparison group** because it provides the experimenter the opportunity to see how much of the eventual (outcome) change is due to the experimental treatment and how much is due to the nature of the experimental subjects. Thus the one-shot case study is a pre-experimental design. We need to know, before we begin the experiment, what our students attitudes are—one way to do this is to add a pretest measuring their attitudes and knowledge about sex before the intervention begins.

Design 2: One Group Pretest-Posttest Design

In order to measure subject knowledge and attitudes regarding sex we give the subjects a **pretest,** which measures their current knowledge and attitudes toward sex (we use a questionnaire as a pretest in this example). Once the subjects have answered the questionnaire, they are sent into the nine-week program.

Subjects: pretest → independent variable → dependent variable

When the nine-week sex education program has concluded we give the students a second survey (the posttest). If their attitudes have changed from the pretest measure, we will conclude that having the students complete a sex education course has impacted their attitudes or knowledge of sex. This design contains an independent variable and a dependent variable, as well as a pretest and a posttest.

Evaluate this design. Although we can now compare previous attitudes to attitudes post sex education program, we have to critique the design. Suppose one of the most popular shows middle-school students watch is MTV's *16 and Pregnant*. While trying to illustrate the perils of teenage and unplanned pregnancy, the program can also serve to glamorize teenage parenthood. Suppose our middle school students are learning more about sex from watching the current season of *16 and Pregnant*, rather than from our sex education program? How would we know the difference between our sex education program and what subjects are learning from a television show?

When we are only looking at the effects of an independent variable within one group, we do not yet have a true comparison. During the course of the sex education program, there could be things going on that have nothing to do with our program—and yet may be influencing the outcome of our dependent variable. There are numerous "effects" that can impact the measurement of our dependent variable apart from our independent variable, including the effects known as "history," "maturation," "attrition," "testing," and/or "instrument effects."

History Effects. **History effects** can impact your subjects through events that happen simultaneously with the independent variable. For example, if *16 and Pregnant* becomes a focus for students to talk about, they could be learning about sex from the television show. When we see a change from the pretest to the posttest, we may mistakenly attribute the change to the sex education program, when in fact the change originated by watching the television program.

Maturation. It could also be that sixth grade students are going through a period of **maturation,** or growing into or out of a particular phase. Sixth graders are beginning to go through puberty and may be suddenly more aware of themselves as sexual beings and/or about significant others that may sensitize them to issues of sex (or even learning about sex when their parents realize it's time to have "the talk"). A change in the student attitudes toward sex may be due to their maturation, rather than our sex education program.

Testing Effects. Apart from what could be going on simultaneously with the students, two effects may come from the experiment itself: testing effects and instrument effects. When we institute a pretest, a potentially necessary measure, it could be that students become tipped off to the aim of the experiment. When they take the pretest they may then be primed to recognize that the experiment seeks to impact their knowledge/attitudes about sex and then respond to the posttest based on their understanding of what happened with the pretest. Thus the pretest and posttest themselves may have introduced **testing effects,** influencing how the students responded, rather than the sex education program.

Instrument Effects. Slightly different from testing effects are the instrument effects. Instruments are the pre- and posttests—the questionnaires we give to measure attitudes about sex before and after students take the program. **Instrument effects** occur when the tests we use to measure attitudes toward sex may be flawed or biased, not actually measuring student attitudes reliably. Therefore any generalizations we make from the measurements might not reflect how the sex education program affects the students.

Attrition. Apart from these other effects, the impact of subject attrition may impact the dependent variable. Over the course of a nine-week program, students who participate in sports and/or who experience illness may miss significant portions of the program and would then be excluded from the experiment, resulting in **attrition.**

We need a more robust design to alleviate some of these effects. The way we do that is by adding another group to the experiment: a control group. Two groups that comprise the third type of experimental design are the control group and experimental group. The **control group** is not exposed to the stimulus, whereas the **experimental group** is. For our example that would mean we need a group of students who are not participating in the sex education program to take a pretest. This becomes design 3, or the static group control design.

Design 3: The Static Group

Experimental: pretest → independent variable → dependent variable

Control: pretest →   → dependent variable

Adding a control group allows us to mitigate the effects of history or maturation. Theoretically, if something occurs for the experimental group, it's also occurring to the control group (everybody who is anybody is watching *16 and Pregnant*). So we would expect that if our sex education program has an impact on knowledge and attitudes toward sex, then the posttest for the experimental group will show a significant difference from the control group scores, regardless of any simultaneous event or any process of maturation.

Evaluate this design. This design has the independent variable, a pretest and posttest (dependent variable), an experimental and control group. This design seems to have taken care of any concerns about the execution of an experiment.

Selection Bias

In looking at our control group, it seems that a simple solution to obtaining students who do not participate in the sex education program might be to put the students who did not return their permission slips into the control group. This is problematic. Can you think of any differences between students whose parents sign permission slips and students whose parents do not? Perhaps parents who do not want their students to participate in a school sex education program may have very different values from those who do want their children to participate. This introduces a selection bias: we do not know if the students in the experimental or control groups differ significantly from each other. This bias leaves us with the next major component of experiments: random assignment.

TRUE EXPERIMENTAL DESIGNS

Design 4: The Classic Experiment

In the classic experiment, the experiment includes the independent variable, the pretest and posttest (the dependent variable), and a randomized experimental and control group. This is now a true experimental design. By randomly assigning students who have permission slips to take the sex education program, we have randomized their basic characteristics, making the groups equivalent to each other.

		Random Assignment	
Experimental:	pretest →	independent variable →	dependent variable
Control:	pretest →	→	dependent variable

We have now taken care of almost everything by adding the components to make a true experimental design. There are two remaining issues, however: the effects of testing or of the instruments and the manipulation of the independent variable. Whereas the static group comparison takes care of the effects of history and maturation, we still have been dealing with the possible effects of testing (the pretest and posttest tipping off students as to what the "right" answers are) or the instrument effects (the pre- and posttests not actually measuring the variables). These last two effects can be addressed by tweaking the classic experiment design, creating a fifth design: the Solomon four-group design.

Design 5: Solomon Four-Group Design

In the Solomon four-group design, two additional groups are incorporated into the classic design, a second experimental and control group:

Random Assignment

1. Experimental$_1$:	pretest →	independent variable →	dependent variable
2. Control$_1$:	pretest →	→	dependent variable
3. Experimental$_2$:	pretest →	independent variable →	dependent variable
4. Control$_2$:	pretest →	→	dependent variable

The first experimental group gets a pretest, is exposed to the independent variable, and then gets the posttest. The first control group gets the pretest and then the posttest. The third group is an experimental group that does not get a pretest but then is given the independent variable and the posttest. The fourth group is given neither the pretest nor the independent variable but is given a posttest. If the pretest tips off the subjects and/or does not measure the variables, the second two groups will control for this. We expect that if our sex education program actually impacts attitudes and knowledge regarding sex that groups 1 and 3 will have significantly different scores than groups 2 and 4. By adding groups 3 and 4, any interaction between the testing and the stimulus should be ruled out.

QUASI-EXPERIMENTAL DESIGNS

The last designs we talk about are those known as **quasi-experimental designs,** since they do not quite have all the elements necessary to be considered a true experimental design. Typically, these designs do not use full randomization (usually because the conditions of real-world research prevent this) and therefore they do not yield clear-cut outcomes that can be attributed only to the treatment variable.

Although there are several types of these designs, Campbell and Stanley (1963) note that they are used extensively in educational research when the experimental and control groups are not individually randomized from a common population, but they are formed through naturally assembled collectivities, for example classrooms. Which collectivity is exposed to the independent variable is randomly selected by the experimenter.

Design 6: Pre-Post Nonequivalent Control Group Design

In this design, using our sex education program example, the classes we use may already be formed (which is typically the case) and as researchers we may only be able to randomize which class is designated the treatment group and which is to be the control or comparison group. We have therefore introduced some randomization, but only for entire classes. The problem with this is that the classes have been created already according to some criteria that yield groups of individuals who are not necessarily equal on all individual characteristics. Whatever processes that were used to create the groups in the first place are already built in to the groups. The outcomes of the experiment are thus made a bit cloudier, since we cannot measure how much the individual differences among the groups affects the outcomes of the study *beyond the exposure to the treatment variable.*

Partial Random Assignment

Experimental: pretest →   independent variable →   dependent variable

Control: pretest →   →   dependent variable

The pre-post nonequivalent control group design has the positive features of the classical design that can help control certain problems in the experiment, like the inclusion of a control group for comparison sake and perhaps the positive contributions of pretesting. However, the lack of complete randomization is a thorny issue that prevents the researcher from making truly causal attributions. For this reason, many experimental social scientists consider this a useless test (at worst) or simply equivalent to a post facto test (at best). In the real world of research, however, this design is often the best alternative. Even though it may never reach the level of purity of the other designs, it nevertheless can yield very strong conclusions, nearly causal in nature depending on the level of control possible by the researcher.

FIDELITY OF EXPERIMENTAL DESIGN

These pre-, experimental, and quasi-experimental designs highlight the important components to consider when designing an experiment (independent and dependent variables, pretest/posttest, experimental and control groups, and random assignment). There is, however, one other component to consider: the manipulation of the independent variable.

Under careful conditions, the outcome of an experiment can yield reliable and valid results. However, where the treatment is not consistently applied, or when different research settings do not ensure that the treatment is applied in a standardized way, the outcomes may not point to the treatment itself but to differences in the treatment. **Fidelity** is the concept that captures the extent to which researchers and research settings apply the treatment in a consistent, standardized way. This last component, the manipulation of the independent variable, should be clearly evaluated when the experiment is put together. Here is an example of an experiment that failed to take into account this component of experiments.

In Chapter 4 on ethics, we discussed the study *Pygmalion in the Classroom* by Rosenthal and Jacobson (1968). These researchers were interested in how teacher expectations impacted student performance. While the experiment introduced us to the *paradox* (where researchers have the ability to give or deny potential benefits to subjects), there is another useful concept to be learned from the experiment. Rosenthal and Jacobson replicated the study, slightly tweaking their design (which is standard practice when replicating experiments). Recall that in the first study the researchers randomly selected students who would be going through a fictitious intellectual growth spurt over the course of a school year.

To manipulate the variable, Rosenthal and Jacobson needed to make sure teachers were aware of the "spurters." Toward this end, the teachers were sent a series of letters explaining that they would have these special students in their classrooms. In the new iteration of the experiment, teachers were sent a letter over the summer listing the next year's "spurters." Yet at the end of the school year the researchers found no difference in the test scores between the "spurters" and the "non-spurters." How could this be?

It turns out that most teachers did not remember receiving the letter alerting them to the study, since they only received the one letter over the summer when they were not teaching. The critical letter in the initial study was the one that came after the beginning of the school year. Teachers were able to place faces to the names of the students, which was a crucial piece in manipulating the independent variable. When teachers realized they had two groups of students, spurters and non-spurters, only then was the variable manipulated (by having two conditions—students the teachers knew to be "spurters" and students the teacher knew to be "non-spurters"). When conducting an experiment, researchers must make sure that the independent variable is manipulated, taking on at least two values. Otherwise, there is no experiment at all.

EXPERIMENTAL SETTINGS

This latter experiment highlights another issue that researchers face when doing experiments: where should the experiment take place? Most of us think of experiments happening in laboratories that are painted white, with men in white coats and clipboards watching our every move. Laboratory experiments are highly controlled, giving power to the results of the experiment. Labs, however, are not the only place experiments are conducted. Like the Rosenthal and Jacobson experiment, many experiments are taken out into the field or are done through surveys, and even in the wake of natural disasters.

Laboratory Experiments

The power of the experiment is exemplified in the **laboratory experiment,** where experimenters have complete control over their environments. This control helps to eliminate exogenous factors that may impact the relationship between the independent and dependent variables. Some of our classic experiments, like the Milgram shock experiment, illustrate how labs help the process. Social psychologists Latané and Darley (1968) have a series of experiments that test the bystander syndrome (where people witnessing emergencies do not intervene to help). In one variation of these experiments, a group of university men are asked to fill out a questionnaire. In one condition of the experiment, the men fill out the questionnaire alone in the room. In the second condition there are three men in the room. While the students fill out the forms, a staged emergency is put in place; smoke begins to pour into the room through a wall vent. Students who were alone in the room to complete the questionnaire tended to glance around the room idly and when they did, they noticed almost immediately the smoke pouring into the room. Once they noticed the smoke, the students who were alone went to report the problem. Interestingly, when there were three students in the room, it took them significantly longer (almost four times as long) to even notice the smoke. For this experimental condition, in only one of eight groups did anyone get up and leave the room to report a problem. Most of the students in this condition stayed in the room until it was filled with smoke. Interestingly, the subjects would begin to explain why the room was filling up with smoke but never left to see about it. This experiment illustrates the way that in ambiguous situations, we look to others to see how we should respond. Myers (1994) notes that we try to understand situations that are unfamiliar by watching how others are responding. Since they seem to be calm or indifferent, we assume that we have nothing to worry about. The problem is that the others in the room were doing the same thing, each appearing calm while they assessed how the others were reacting. Thus we each interpret the emergency or ambiguous situation as not an emergency.

This example of the smoke-filled room experiment illustrates the power of the lab; within their own labs, experimenters have complete control over the environment (thus in a six-minute experiment, Latané and Darley could see the influence of others in how we interpret and react to events). *The strength of experiments is in the amount of control that researchers can impose on their design.* This strength, however, also acts as a weakness. How often are people in the real world going to encounter a room filling with smoke or be asked to administer electric shocks to others? The lab creates a space of artificiality that may impact how people react. Would they do the same in the real world? Labs are also sometimes impractical; how many people can be brought into a lab space, for example? These limitations in a lab setting move researchers out into the field so that they can observe real-world constraints on people.

Field Experiments

Field experiments allow researchers to see how people react to an experimental treatment or independent variable in natural settings and other real world interactions, rather than in a laboratory. For example, can you imagine having a conversation with someone only to find that the person who began the conversation is not the same person who finished the conversation? It seems fairly ridiculous to think we would not notice the change in our conversation partner. Research studying the effect of "change blindness," however, illustrates that changing conversation partners mid-stream may not be easily detectable.

Change blindness is "the inability to detect changes to an object or scene" (Simons and Levin 1997:261). Although we experience a detailed visual world, we often miss obvious changes to our vision field. Simons and Levin (1997, 1998) noted that change blindness occurred in a variety of lab settings. In one lab experiment, for example a man was placed behind a reception desk that had a sign above it labeled "Experiment." He distributed forms to study participants and then took the forms back as the subjects completed answering them. He then walked out of their view to "file" the form. A second experimenter returned to the reception desk, continuing the conversation and handing subjects a packet of information. Approximately 75 percent of subjects did not notice the change in the men, even though the two men had different faces, hair, and shirt colors. Could this same change blindness occur in the "real world"?

Simons and Levin (1998) created a field experiment to test the effect of change blindness on everyday interactions. The experimental design involved having an experimenter carrying a campus map asking directions to a nearby building from people on a college campus. Within 10-15 seconds of the beginning of the interaction, two additional experimenters carried a door, passing "rudely" between the pedestrian and the first experimenter. As the door interrupts the conversation (for approximately one second), the first experimenter switches places with one of the door carriers. The second experimenter stays behind, continuing to ask for directions. These two experimenters wore different clothing, were different heights, and had "clearly distinguishable" voices. After the interaction (which lasted less than five minutes), the second experimenter explains to the pedestrians that the psychology department is doing a study on "the sorts of things people pay attention to in the real world. Did you notice anything unusual at all when that door passed by a minute ago?" (Smith and Levin 1997:645). Despite the clear differences in the two experimenters, more than 50 percent of the subjects did not notice the change. Although more subjects detected the change in conversation partner in the context of a field experiment than in the similar lab experiment, the fact that more than 50 percent of experimental subjects did not detect the change illustrates the substantial effect of change blindness.

Conducting experiments in a laboratory setting is beneficial because more sources of error (i.e., extraneous variables) can be controlled, which lends more confidence to the conclusion that the intervention is really the reason for any results. On the other hand, the conclusions of a laboratory experiment are less likely to be generalized to the world outside the laboratory because the setting is not something the subjects encounter in real life. The term for these differences that must be examined is **ecological validity**. This refers to the extent to which the laboratory conditions (setting, experimenters, procedures, etc.) are similar to those encountered in the natural settings and therefore the extent to which the conclusions can transfer or generalize to the world outside the lab. In the change experiment, perhaps one of the differences in outcomes between lab and natural setting is that the former was conducted in an artificial setting in which the subjects may not have used the same cues they would in everyday life governing these kinds of interactions. In the natural situation, which would be more in line with how they use ordinarily use cues to negotiate interactions, they were a bit more aware. It is hard to say if that is the only difference, but experimenters have always been aware that lab experiments have more challenges in transferring or generalizing results to the real world.

Natural/Disaster Experiments

Being out in the field lends itself to another type of experiment, the **natural/disaster experiment.** Since researchers cannot ethically wreak havoc on the everyday lives of their subjects, they have to take advantage of nature and/or of events that we simply cannot engineer. For example, when two airplanes hit the twin towers of the World Trade Center in New York on September 11, 2001, emergency personnel were the first to the scene. Upon their heels were a host of social researchers. Social scientists assessed the social damage that began on that day, measuring, for example, the psychological stress of such an event. In a special 2011 issue of the journal *American Psychologist*, researchers found that psychologists overestimated the number of people who would experience emotional trauma due to the attacks and that some may have been harmed by the crisis intervention of counselors (Carey 2011). Since events on this scale cannot be designed, natural disaster researchers measure the intervention strategies after the event.

Another example of a naturally occurring "experiment" was the 2010 earthquake in Haiti and the cholera epidemic that followed about nine months later. The tiny nation was violently disrupted by the earthquake event, and aid workers rushed to help. Before the aid workers arrived, there was a measureable incidence of cholera. Some few weeks later, however, cholera was rife, centered mainly in the center of the country around the Artibonite River. Resulting outbreaks of cholera occurred in the center and north of the country, but not in the south. Epidemiologists and other health workers later determined that UN peacekeepers from Nepal (host) brought the cholera organism (agent) to the country, and the environmental conditions were such (primarily sanitation problems and proximity to the waterways) that cholera spread north and west. The southern areas, unconnected to the same waterways, were not affected. As a natural experiment, therefore, the cholera agent affected part of the country ("experimental group"), but not the other part ("control"). Public health researchers used this information to isolate the cause of the outbreak and create solutions.

Survey Experiments

One other place that experiments take place is in the context of surveys. When doing surveys you should know both your population and your topic quite well. Recognizing that certain words or phrases impact question responses, survey researchers experiment with word choices to see if measuring the same concept with different wordings garners a similar response. The General Social Survey (GSS) does this often, randomly assigning one question to one survey form and an alternative question to the other survey form. One of their perennial questions looks at how Americans interpret the word "welfare" as opposed to the phrase "assistance to the poor." Recognizing that these two wordings measure the same concept, but that "welfare" carries pejorative connotations, the 2010 GSS asks the following questions (from the ARDA).

Welfare. "We are faced with many problems in this country, none of which can be solved easily or inexpensively. I'm going to name some of these problems, and for each one I'd like you to name some of these problems, and for each one I'd like you to tell me whether you think we're spending too much money on it, too little money, or about the right amount. First (READ ITEM A) . . . are we spending too much, too little, or about the right amount on (ITEM)? K. Welfare"

0. Inapplicable
1. Too little
2. About right
3. Too much
8. Do not know
9. No answer

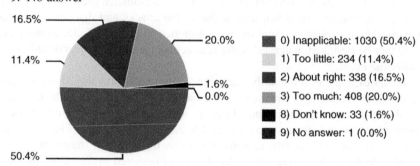

Assistance to the poor. "We are faced with many problems in this country, none of which can be solved easily or inexpensively. I'm going to name some of these problems, and for each one I'd like you to tell me whether you think we're spending too much money on it, too little money, or about the right amount. First (READ ITEM A) . . . are we spending too much, too little, or about the right amount on (ITEM)? K. Assistance to the poor"

0. Inapplicable
1. Too little
2. About right
3. Too much
8. Do not know

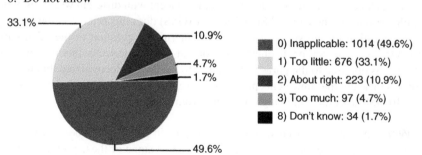

You can see that half of the respondents received only one form of the question (denoted by category 0) Inapplicable). Interestingly, however, the percentage of those who believe the government is doing "too little" changes significantly. Only 11.4

percent believe that the government spends too little on welfare, whereas 33.1 percent believe the government spends too little on assistance to the poor.

ETHICS

Lab and field experiments deal directly with people, which means taking extra precautions regarding our ethical considerations for the study. As we noted earlier in the book, ethics are an integral part of social science. The three very specific issues that need to be addressed within the context of experiments are the paradox, the use of deception, and the debriefing. As we noted in Chapter 4, experimenters do not know if the independent variable will impact people; that is why we test relationships through experiments. The paradox is the giving of potential benefits and the denying of potential benefits. Note that the word *potential* indicates that we simply do not know, beforehand, the impact of the independent variable on the dependent variable. The general rule for experimenters is that as long as the subjects are no worse off than if they had not taken part in the experiment, then the assumption is that there is no harm with which to contend—for the group that receives the independent variable or the group that does not.

The other concerns, deception and debriefing, are related in experiments. It is not unusual for experimenters to use deception. If subjects knew what was being measured, they may try to make sure to meet the experimenter's expectations and/or they may try to do exactly the opposite of the experimenter's expectations. Therefore, deception is used to draw the subject's gaze away from the true aim of the study. When deception is used ethically, we assume that the potential increase in knowledge far outweighs any possible study risks (and the deception will have to be approved by a human subjects committee/institutional review board). When deception has been employed in the study, experimenters must debrief the subjects following the end of the subject's participation. The debriefing allows the experimenter to give the subject the full explanation of the true aim of the study.

RELIABILITY AND VALIDITY

As we close the chapter on experiments, there are two other issues to keep in mind: the reliability and validity of the research design. Specifically related to experiments is the problem of subject bias, where subjects are aware they are being observed. The Hawthorne effect may impact some subjects (not responding naturally because they are being observed). The use of deception can help to alleviate those issues, since subjects are not aware of what experimenters are truly watching or measuring. There may also be a place for experimenter bias within the context of the study. Experimenters are aware of the study's true aim and may unconsciously alter their behavior, impacting the results of the dependent variable.

In terms of the validity of the experimental study (are the researchers measuring what they say they are measuring), experimenters need to make sure that they have at least two conditions (or values) of the independent variable. Like the second Rosenthal and Jacobson study we discussed, teachers were not given enough information to think about their students being in two different states of intellectual growth (spurters and non-spurters); thus the independent variable was not manipulated (there were not two conditions). Another issue with the validity of experiments is the problem of realism. Are the results we get from experiments realistic or because the design can have high artificiality, can we trust in the outcomes? Since experiments can be replicated, having more studies with different designs offering the same outcomes helps us to interpret if the results are truly generalizable.

One last concern we want to address is the fallacy of unmanipulated "causes," or the mistaking of subjects' individual characteristics for impacting the dependent variable. While researchers observe how an independent variable impacts a dependent variable, they may also notice other differences between groups and attribute that difference as a causal factor. For example, in Chapter 3 we described an experiment with three conditions. One randomly assigned group of four-year-olds watched nine minutes of *SpongeBob SquarePants,* another group watched nine minutes of *Caillou,* and another group simply drew pictures over the course of the nine-minute period. After each child had finished the nine-minute tasks and took the tests of cognitive ability, researchers may have (hypothetically) noticed that the boys who watched *SpongeBob SquarePants* scored even lower than the girls on the tests of cognition (dependent variable). The researchers might then try to conclude that gender played a part in determining how well a child did on the test. This conclusion is an example of the fallacy of unmanipulated causes. In an experiment, we can only make causal statements about the independent variable that is measured. Since gender was not the variable being tested (exposure to *SpongeBob*, *Caillou*, or drawing was being tested), we cannot conclude that gender impacted the results of the dependent variable.

Exercise: Experiments in Action

As we noted previously in this chapter, one setting for experimental research is in surveys. The GSS 2010 has a variety of items testing word choices, to see how (or if) different words or phrases bias respondent answers. In the text we looked at the example of whether or Americans felt the government spent too much, too little, or about the right amount on "assistance to the poor" versus "welfare." We noted the significant differences between the two responses (11.4 percent of Americans believe that the government spends too little on welfare, whereas 33.1 percent believe the government spends too little on assistance to the poor).

Here we have listed the other experimental questions from the 2010 GSS. Look at each pair of questions and run Frequencies in SPSS (from the Analyze | Descriptive Statistics menu). Notice the differences between the responses for each pair of questions. Also make sure to look at the "missing" or "total" responses to see that only half of the sample answered each question (each was randomly selected). Follow the example for the variables "natcity/ natcityy":

1. What are the two variable descriptions?
   a. natcity: _Solving the problems of big cities_
   b. natcityy: _Assistance to big cities_
   c. What differences are implied by the phrases? _Natcity seems to be more value-laden because it uses the phrase "solving problems" implying that big cities have big problems and we shouldn't help. Natcityy seems to be more polite because it offers "assistance" to big cities._
   d. Which version of the question do you expect to be more biased? natcity
   e. Compare the frequencies for the three categories of each:

	natcity	natcityy
Too little	19.2%	8.7%
About right	16.5%	18.6%
Too much	8.8%	15.7%

   f. What do you notice comparing the different frequencies? _Wow, 19.2 % of people think the government spends to little on "solving problems" of cities, while only 8.7% think too little is spent on "assisting" cities. This is the opposite of what I thought would happen._
   _____
   _____

2. What are the two variable descriptions?
   a. natdrug: _____
   b. natdrugy: _____
   c. What differences are implied by the phrases?_____
   _____
   _____
   _____
   _____

   d. Which version of the question do you expect to be more biased?
   e. Compare the frequencies for the three categories of each:

Too little		
About right		
Too much		

   f. What do you notice comparing the different frequencies?_____
   _____
   _____

(Continued)

_____
_____

3. What are the two variable descriptions?
   a. natcrime: _____
   b. natcrimey: _____
   c. What differences are implied by the phrases? _____
      _____
      _____
      _____
      _____

   d. Which version of the question do you expect to be more biased? ____
   e. Compare the frequencies for the three categories of each:

Too little		
About right		
Too much		

   f. What do you notice comparing the different frequencies? _____
      _____
      _____
      _____
      _____

4. What are the two variable descriptions?
   a. natenvir: _____
   b. natenviry: _____
   c. What differences are implied by the phrases? _____
      _____
      _____
      _____
      _____

   d. Which version of the question do you expect to be more biased? ____
   e. Compare the frequencies for the three categories of each:

Too little		
About right		
Too much		

   f. What do you notice comparing the different frequencies? _____
      _____
      _____
      _____
      _____

5. What are the two variable descriptions?
   a. natspac: _____
   b. natspacy: _____
   c. What differences are implied by the phrases?_____
      _____
      _____
      _____
      _____

   d. Which version of the question do you expect to be more biased?_____
   e. Compare the frequencies for the three categories of each:

Too little		
About right		
Too much		

   f. What do you notice comparing the different frequencies?_____
      _____
      _____
      _____
      _____

15

STATISTICAL METHODS OF DIFFERENCE: *T* TEST[1]

There are many types of statistical procedures that attempt to detect statistical differences between and among research conditions. We have already looked at chi square, which examines whether the *frequencies* of the categories of one variable are distributed equally across the categories of another variable. Statistical procedures that use a higher level of data, namely interval data, also detect differences among variables.

Independent *t* tests determine whether the two categories of a predictor (independent) variable result in statistically different mean values of an outcome (dependent) variable. Thus, do men and women postal workers (independent variable) indicate different levels of job satisfaction (dependent variable)? This procedure assumes that the predictor categories are "independent," or unrelated to one another. Other *t* tests can accommodate for the two predictor categories being related or dependent (e.g., when the same people are tested twice as in a pre-post test), but they use different formulas.

ANOVA (analysis of variance) tests do essentially the same thing, but instead of comparing two categories of a predictor variable (or factor), they compare three or more categories. Thus, which of three different instructional methods leads to greater math achievement scores? The added complexity is that the procedure uses special drill-down

[1] Parts of this section are adapted from M. L. Abbott, *Understanding Educational Statistics Using Microsoft Excel® and SPSS®* (Wiley, 2011), by permission of the publisher.

methods to determine the differences among each of the pairs of categories if the overall test is significant. Thus, if the overall test is significant, are there significant differences between method A and method B, between method A and method C, or between method B and method C? As you can see, the procedure is specific, and the results can get quite involved depending on how many categories there are to compare. Like the "dependent" *t* test, there are also ANOVA procedures that use related categories and must therefore use special formulas to accommodate for the dependencies (as in within-subjects ANOVA).

INDEPENDENT AND DEPENDENT SAMPLES

We need to be careful to point out, however, that when these procedures are used (in experimental or post facto designs), the researcher needs to ensure that the groups they choose to compare are independent of one another. *Independent samples* mean that choosing subjects for one group has nothing to do with choosing subjects for the other groups. Thus in experimental situations that involve two groups, if we randomly select Bob and assign him randomly to group 1 (experimental group), it has nothing to do with the fact that we choose Sally and assign her randomly to group 2 (control group). In post facto procedures with two groups, if we compare the job satisfaction of "workers with more than five years on the job" to "workers with five or fewer years on the job," it is important that there are no connections between the subjects chosen (close relatives, spouses, etc.). The independence assumption is important because it assures the researcher there are no built-in linkages between subjects. The power of randomization will result in the comparability of the two groups in this way.

Dependent samples would consist of groups of subjects that had some structured linkage, like using the same people twice in a study. For example, we might use pretest scores from Bob and Sally and *compare them with their own posttest scores* in an experimental design. Using dependent samples affects the ability of the randomness process to create comparable samples; in such cases, the researcher is assessing *individual* change (before to after measures) in the context of the experiment that is assessing *group* change.

Who we measure, and how affects the nature of our research design. Experiments study the differences that can be attributed to a change in the independent variable, while post facto designs measure the differences after a change has already taken place. Experimental designs that analyze whether *entire groups* demonstrate unequal outcome measures are called **between-group designs**. **Within subjects designs** are those in which researchers analyze differences *among* subjects in a group, or matched groups, over time.

INDEPENDENT *t* TEST

The independent *t* test is a powerful but common statistical procedure because it conforms nicely to a lot of what researchers are interested in doing. The *t* test is so common because it allows us to perform a very basic function in statistics and common practice: *compare*.

Perhaps you have heard an advertisement like the following: "New Boric Acid mouthwash is 30 percent better!" This statement begs a series of questions including "Compared to what? Old Boric Acid mouthwash? Used dishwater? Other brands of mouthwash?" In order to find out whether this claim has merit, we must compare it to something else to see if there is really a difference.

With the independent *t* test, we can assesses whether two samples, chosen independently, are likely to be similar or sufficiently different from one another that we would conclude they do not even belong to the same population. The versatility of the procedure is shown in the fact that it can be used in experimental designs (i.e., comparing control and experimental group performance on some outcome) or post facto designs (i.e., comparing two categories of an outcome variable to see whether there are differences in their outcome measures).

Example of Experiment

When Abbott was studying experimental psychology as an undergraduate student, he performed an experiment on the effects of noise on human learning. He randomly selected students and randomly assigned them to either a high or low noise condition (by using a white noise generator with different decibel levels). He then gave students in both groups the same learning task and compared their performance. The learning task (the outcome measure) was simple word recognition. Figure 15.1 shows the research design specification for this experiment.

Note some features of the experiment shown in Figure 15.1:

- Randomly selected students were randomly assigned to the two treatment groups.
- No control group was used (the "absence of the treatment") but rather a second level or condition of the treatment variable to yield two treatment groups.
- The experiment did not include a pretest of subjects' word recognition before exposing them to different experimental treatment conditions.

Research treatment variable (noise)			Dependent variable (word recognition)
Random selection and assignment	No pretest	Experimental group (high noise level)	Outcome test scores (# of recognized words)
Random selection and assignment	No pretest	Control group (low noise level)	Outcome test scores (# of recognized words)

Figure 15.1 The experimental research design specification: *t* test with two groups.

- The word recognition test (outcome measure) was administered to both groups after exposing them to different experimental treatment conditions.

Students were randomly selected and assigned, which allowed Abbott to assume they were equal on all important dimensions (to the experiment). He exposed the two groups to different conditions, which he hypothesized would have differential effects on their learning task. Thus, if he had observed that one group learned differently (either better or worse) than the other group, he could attribute this difference to the different conditions to which they were exposed (high or low noise). If their learning was quite different, he could conclude, statistically, that the groups were now so different that they could no longer be thought to be from the same population of students he started with. That is the process he used for testing the hypothesis of difference. Specifically, he used the *t* test with independent samples to detect difference in posttest scores.

By the way, the short answer as to whether or not he observed statistical differences between the high and low noise outcome measure is no. This did not mean that noise does not affect learning; it just gave him a way to look at the problem differently. As we will see, this example shows several features of the theory testing process as well as the *t* test procedure.

Post Facto Designs

Post facto designs compare group performance on an outcome measure after group differences have already taken place. These designs can be correlational or comparative depending on how the researcher relates one set of scores to the other (i.e., using correlation or difference methods, respectively). A post facto design compares conditions with one another to detect differences in outcome measures. Thus, for example, rather than perform an experiment to detect the impact of noise on human learning, we might ask a sample of students to indicate how loud their music is when they study, and ask them to record their grade point average (GPA). Then we could separate the students into two groups (high and low noise studiers) and compare their GPA measures.

In this design, therefore, we would not manipulate the noise measure; we would simply create groups on the basis of *already existing differences* in noise conditions. If the outcome measure (GPAs) was different between the groups, we would conclude that noise would *possibly* be a contributing factor to GPA. We could not speak *causally* about noise, since many other aspects of studying may have affected GPA (for example, sleep deprivation, studying in groups, caffeine consumption, etc.). Figure 15.2 shows how the post facto design might appear using the noise research question. We would simply use an independent *t* test to compare the GPA measures for high and low noise studiers.

This same design could use *dependent* samples if we deliberately stacked the two samples to be the same on some issue. For example, we might equate the numbers of women and men students as well as ensure equivalent numbers of freshmen, sophomores, juniors, and seniors in both noise groups. If we did this, we would be ***matching*** the groups and therefore creating dependent samples. Under these circumstances, we would need to use the dependent *t* test.

Studying under high noise	Studying under low noise
GPA scores	GPA scores
GPA scores	GPA scores
.	.
.	.
Mean GPA$_{High}$	Mean GPA$_{Low}$

Figure 15.2 The post facto comparison for independent *t* test.

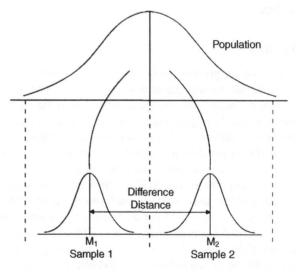

Figure 15.3 The independent *t* test process.

INDEPENDENT T TEST: THE PROCEDURE

In the independent *t* test, the researcher compares a pair of samples to see whether these can be said to belong to a common population. The experimental and post facto designs that we discussed earlier would both yield sample data for two samples. Figure 15.3 shows how the two sample process works.

The chief concern with this test is the difference in the means of the two samples. If the samples are chosen randomly, by chance the means will both be close to the actual population mean (the value of which is unknown). By chance alone, the difference between the means should be fairly small.

If we chose two sample groups (or, in an experiment, if we randomly chose a group and randomly assigned them to two groups), we would expect the group means to be similar. In research, we start with this assumption but observe whether the two sample means are *still equal after an experimental treatment or if the group means are different when we compare different conditions of the research variable.*

Using our previous post facto example above, this would be our reasoning:

- We have two groups of students, some who study under high noise and some who study under low noise.
- *We assume the groups of students were equivalent before they developed their habits of studying under different amounts of noise.*
- Our task is to determine whether now that they have developed their habits of study, they are still equivalent or different on a word recognition task. If we reject the null hypothesis, that will indicate they no longer belong to the same population of students.
- If they are different now, then we can say that the different noise *may have affected* their ability to recognize words. However, there were surely other influences that led them to develop their study habits, so we cannot say the different word recognition ability is caused only by the noise.

But how large does this difference have to be before it could be said that a difference that large could not reasonably be explained by chance and therefore the two groups do not represent a single population? That is the nature of the *t* test process that we will examine using data from the General Social Survey (GSS).

INDEPENDENT *T* TEST EXAMPLE

As an extended example of the independent *t* test, we will use data from the 2010 GSS database. We can use questionnaire items that related to the issue of general health and work. One of the theoretical questions that we have pursued in several places in this book is whether individual health is in any way connected to the structure and experience of work.

In this example, we use two GSS items to address the research question of whether a person's perceived health is affected by their judgment of the "fairness" of their earnings. Obviously, the larger research question can be answered in many ways. We are choosing to operationally define the variables (health and fairness) by the GSS respondents' answers to two questionnaire items. Different studies might use other methods of operationalization, perhaps obtaining an objective measure of health (e.g., electro-cardiogram reading) from workers who have either publicly voiced their concern over earnings fairness or not. Providing confidence in the larger issue of concern (health and fairness) can therefore come from a variety of research studies. We are choosing this method to describe the independent *t* test and how to interpret the findings.

The Variables Used in the Example[2]

The GSS contains two questions that we will use as the basis of our example:

1. HEALTH1—"Would you say that in general your health is excellent, very good, good, fair, or poor?" We recoded[3] this variable into a new variable, 'GenHealth', so that higher values (1–5) indicate better health.

2. fairearn—"How fair is what you earn on your job in comparison to others doing the same type of work you do?" (Relevant response categories: Much less than you deserve, Somewhat less than you deserve, About as much as you deserve, Somewhat more than you deserve, and Much more than you deserve). We recoded this variable into a new variable "deserveearn" that has two categories (1) Earnings are Less Than Deserved and (2) Earnings are As Much As or More Than Deserved.

The SPSS Results

In order to use SPSS to assist us with the t test for the example, we use the Analyze button in the top menu ribbon and specify Compare Means and Independent Samples T Test as shown in Figure 15.4. When we make this choice, another window appears as shown in Figure 15.5.

The choices for the t test in Figure 15.5 show that GenHealth is the Test Variable (outcome) and deserveearn is the predictor variable with two categories (1 and 2). You can choose the variables for the analysis from the window on the left side of the panel by using the arrow buttons in the middle of the screen. With this test example, you can simply choose OK at this point, and SPSS will complete the t test analysis.

Figures 15.6 and 15.7 show the SPSS results of the t test analysis of the example variables where the mean of GenHealth is compared between categories of deserveearn. The first panel shown in Figure 15.6 indicates the GenHealth group means for both categories of deserveearn. As you can see, the Less Than Deserved group indicates a lower GenHealth mean than the As Much As or More Than Deserved group. The group mean differences (0.2331) do not appear to be large, but the statistical analysis will determine whether the difference is a significant (nonchance) difference.

Figure 15.7 shows the statistical analysis of the difference in GenHealth means between the two categories of deserveearn. As you can see (shaded), the T ratio of

---

[2] Some researchers use these kinds of data (ordinal questionnaire items) routinely in survey research, but the procedure is contested by other researchers. The former argue that (even single) items like these from the GSS are commonly used and exhibit many of the properties of interval data. The latter researchers might argue that the questionnaire data are ordinal data and should not be used with parametric procedures. While researchers should be cautious, we take the former position that GSS data have been used meaningfully in similar research and are appropriately used when additional procedures support their use in this fashion. In this example, a separate nonparametric test (Mann-Whitney U *test*) confirms the parametric (t test) results that are presented and discussed below.

[3] See Data Management Unit B, "Using SPSS to Recode t Test," for an explanation of how we used SPSS to recode variables.

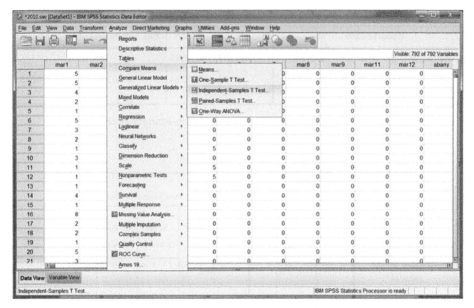

Figure 15.4 Using SPSS specification windows for the independent *t* test.

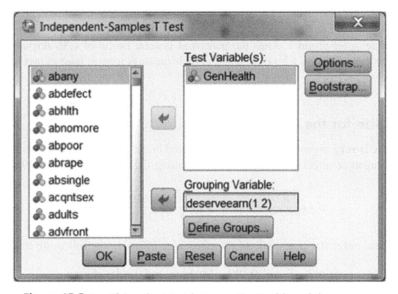

Figure 15.5 Specifying the *t* test between GenHealth and deserveearn.

Group Statistics

deserveearn		N	Mean	Std. Deviation	Std. Error Mean
GenHealth	1.00 Less than deserved	515	3.4893	1.01611	.04478
	2.00 as much as or more than deserved	616	3.7224	1.06346	.04285

Figure 15.6 The t test group means for the example problem.

Independent Samples Test

		Levene's Test for Equality of Variances		t-test for Equality of Means						
		F	Sig.	t	df	Sig. (2-tailed)	Mean Difference	Std. Error Difference	95% Confidence Interval of the Difference	
									Lower	Upper
GenHealth	Equal variances assumed	1.024	.312	-3.746	1129	.000	-.23308	.06223	-.35517	-.11099
	Equal variances not assumed			-3.761	1109.097	.000	-.23308	.06197	-.35468	-.11148

Figure 15.7 The t test omnibus findings for the example problem.

−3.746 is significant as indicated by the 0.000 figure in the Sig. (2-tailed) column. The T ratio is a negative figure because SPSS entered the value for group 2 (the higher mean value) after group 1. Thus the perceived general health of GSS respondents is higher for those workers who indicated that their earnings were at least as much as they deserved.

Effect Size for the Example

Effect size is very important, as we have indicated in earlier sections. There are several ways to calculate effect size for the t test, including the following (Cohen 1988):

$$d = t\sqrt{\frac{n_1 + n_2}{(n_1)(n_2)}} \quad d = -3.746\sqrt{\frac{515 + 616}{317240}} \quad d = .224$$

We can judge the magnitude of the effect size (d) using the following criteria:

0.200—Small, 0.500—Medium, and 0.800—Large.

In this example our effect size is judged to have a small effect. Therefore, even though the perceived health is different between groups of deserveearn, it is a small difference. In the world of research, however, even a small effect size difference can be meaningful.

Additional *t* Test Considerations

If you are interested in developing a working knowledge of using SPSS for independent *t* tests, consult a more comprehensive statistical work (e.g., see Abbott 2011). This pursuit will include discussions of the assumptions for the test, how to use SPSS to provide information for the test assumptions, and alternative procedures to use when the assumptions have been violated (e.g., **Mann-Whitney *U* test**).

16

ANALYSIS OF VARIANCE[1]

Analysis of variance (**ANOVA**) is a statistical method that allows the researcher to compare several different groups, rather than the *t* test that only compares two groups. The scores of a dependent variable are compared within groups of the independent variable. In our discussion of the *t* test, for example, we compared word recognition scores (dependent variable) within *two* groups of the independent variable "noise" (high noise versus low noise). Using ANOVA allows us to compare *three or more* independent variable groups. For example, we might want to compare subjects' word recognition scores depending on their group membership in low, medium, or high noise conditions.

There are several variations of this test, but we focus primarily on the "one-way" ANOVA in this section. One-way ANOVA refers to the number of independent variables. *In research, independent variables are known as factors.* Therefore, if we have a research problem that has several groups of one independent variable, we can use one-way ANOVA to detect any differences on the dependent variable measure among the groups.

We do not have the space in this book to talk about variations of ANOVA, but there are several. Each kind is particularly designed to handle multiple independent

[1] Parts of this section are adapted from M. L. Abbott, *Understanding Educational Statistics Using Microsoft Excel® and SPSS®* (Wiley, 2011), by permission of the publisher.

Understanding and Applying Research Design, First Edition. Martin Lee Abbott and Jennifer McKinney. © 2013 John Wiley & Sons, Inc. Published 2013 by John Wiley & Sons, Inc.

Research treatment variable (noise)			Dependent variable (word recognition)
Random selection and assignment	No pretest	Experimental group I (90 decibels)	Outcome test scores (No. of recognized words)
Random selection and assignment	No pretest	Experimental group II (60 decibels)	Outcome test scores (No. of recognized words)
Random selection and assignment	No pretest	Experimental group III (30 decibels)	Outcome test scores (No. of recognized words)
Random selection and assignment	No pretest	Control group (no noise level)	Outcome test scores (No. of recognized words)

Figure 16.1 The four groups in the noise–learning experiment.

variables (e.g., factorial ANOVA), designs that introduce control variables (ANCOVA), those that include multiple measures of the same subjects (within-subjects ANOVA), and those that include more than one dependent variable (MANCOVA). In factorial ANOVA, for example, we introduce a second (or more) independent variable to our research analysis. *Factorial* refers to the fact that we have more than one factor in the procedure. In our noise group example, we might introduce sex groupings in the design so that we could see whether two independent variable groups (sex groupings and noise groupings) have independent and joint effects on word recognition.

THE NATURE OF THE ANOVA DESIGN

In Chapter 15 on the *t* test, we discussed an experiment on the effects of noise on human learning. Since there were only two groups to compare in that example (high versus low noise), we showed how the independent *t* test could be used to test the null hypothesis. Figure 16.1 shows a design of the experiment using four groups (of the independent variable) rather than two.

As you can see, there are four groups, three of which used different levels of noise and one control group in which no noise was present. The noise level in group III, 30 decibels, is comparable to quiet conversation; 60 decibels is equivalent to normal conversation; and 90 decibels is equivalent to heavy truck traffic. (Sustained exposure to 90-decibel noise could result in hearing loss.)

There is (still) only one independent variable (noise) in the experiment, but now there are groups in four levels of the independent variable. A *t* test would be inappropriate because we would need to conduct six *t* tests to compare all the group results, as shown in Figure 16.2.

Even if we conducted all these *t* tests independently, we would have no idea of the whole test result. We use the ANOVA test for this, since it conducts *all the comparisons*

Group comparisons
Exp. group I versus exp. group II
Exp. group I versus exp. group III
Exp. group I versus control group
Exp. group II versus exp. group III
Exp. group II versus control group
Exp. group III versus control group

Figure 16.2 The paired group comparisons in the experiment.

within the same procedure at the same time. We need one omnibus answer to the question of whether noise affects human learning. If this result rejects the null hypothesis, we could then examine each of the pairs of groups to see which are different from the others and therefore which may be responsible for the overall omnibus finding.

If the mean differences between the paired groups are markedly dissimilar to one another (in a statistically significant sense), we could reject the null hypothesis and conclude that at least one of the group means differs from the others. If we are able to reject the (omnibus) null hypothesis, ANOVA will not indicate which mean (or means) is different from the others. In that event, we would perform a separate test, called a "post hoc comparison," in order to identify which of the means was significantly different from one another.

THE COMPONENTS OF VARIANCE

Recall that variance is a measure of the dispersion of scores in a distribution. We used variance measures in our discussion of the *t* test, and we can look at these measures again in ANOVA. Variance is a more global measure of variance, so we use it instead of the standard deviation (another measure of dispersion) as a way of determining how sample groups differ from one another. Look at Figure 16.3 to see how our experimental groups might be represented, using the noise experiment we introduced earlier.

Figure 16.3 shows the sample groups arrayed from left to right on an x axis of noise groups that show different "Number of Learning Errors," which is the dependent variable for this experiment (represented on the y axis). Remember that the experimental question is whether noise affects human learning. We can operationally define learning (the dependent variable) as the number of simple words recalled during a memorization task. The experiment thus would provide information about whether different amounts of noise would decrease the number of words recalled.

Figure 16.3 shows the three sources of variance in an ANOVA analysis. The four *sample groups have their own distributions, so all their scores are spread out around their own group means. This is known as **within variance**, since the variance is measured within each sample distribution.* Figure 16.3 also shows a shadow distribution that represents a total distribution if all the individual scores from all the sample groups

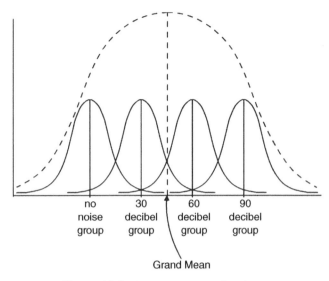

no
noise
group

30
decibel
group

60
decibel
group

90
decibel
group

Grand Mean

Figure 16.3 The components of variance.

were thrown into one large distribution. If we were to measure *the variance of that large composite distribution, this would constitute the **Total Variance***. As you can see, the composite distribution has a mean value, indicated by the dashed line. This total mean is known as the "Grand Mean." ***Between variance*** *is the variation between each sample group mean and the grand mean.*

- Total variance is the total variation of all individual scores in all sample groups *together*.
- Between variance is the variation between the sample means and the grand mean.
- Within variation is the variation of individual scores within their own sample groups.

THE PROCESS OF ANOVA

In effect, the ANOVA process determines *whether the sample means vary far enough away from the grand mean that they could be said to be from different populations.* Figure 16.4 shows two possible results that illustrate the process of ANOVA. The bottom panel shows the results of an experiment in which the groups (designated by G) are so squished together that the group variances are intermingled. In the top example, however, the groups are far enough apart that individual group variances do not have extensive overlap. That is, the variances *within* each sample group do not confuse the distance *between* the groups in the top example. If we had an actual result

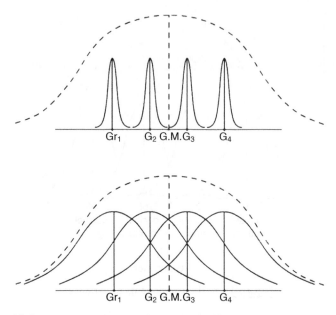

Figure 16.4 ANOVA possibilities of groups of different within group variances.

like the top example, we would be likely to determine that the groups are statistically different from one another. This might not be the case in the bottom example.

Formally stated, ANOVA seeks to determine whether the variance between measures *would be large relative to the* variance within measures as appears to be the case with the bottom example of Figure 16.4. In the bottom example, the within variance muddles the picture of how the groups relate to one another. It is for this reason that within variance is often referred to in research (and in SPSS) as *error variance*. If we want to understand how far apart the group means are from the grand mean, the error variance gets in the way. When there is less error variance, then the distance between the group means and the grand mean (i.e., the between variance) becomes easier to distinguish, as in the top panel.

CALCULATING ANOVA

The calculation of ANOVA compares between variance to within variance. When between variance is *large* relative to within variance (as it appears to be in the top panel of Figure 16.4), then there is a greater likelihood that result will be statistically significant. We can suggest a conceptual equation that expresses this relationship:

$$F = \frac{Between\,Variance}{Within\,Variance}$$

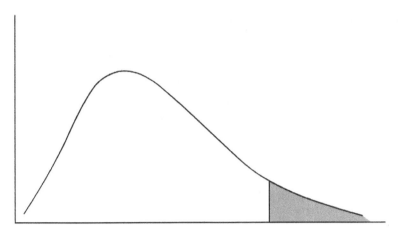

Figure 16.5 The F distribution with shaded exclusion area.

TABLE 16.1. An Example ANOVA Results Table

Source of Variance	SS	df	MS	F ratio
Between	257.5	3	85.83	32.70
Within	31.5	12	2.625	
Total	289	15	–	

The F ratio is the calculated relationship between the between variance and the within variance, according to the formula shown earlier. This is the value that is compared to the F distribution to see whether the calculated F ratio falls into the exclusion range and therefore is considered a significant (nonchance) finding.

ANOVA uses the F distribution as a comparison distribution for the relationship of the between-to-within variance. The F distribution is used as a benchmark to establish the boundaries of exclusion that indicate whether a calculated F ratio should be considered significant (not by chance) or not.

The preceding equation thus expresses the relationship of between-to-within variance as a point on the F distribution to determine whether the sample group means are far enough from the grand mean that the calculated F ratio would fall in the exclusion area of the comparison (F) distribution. The F distribution is shown in Figure 16.5 with a shaded portion in the tail (exclusion area). If a researcher obtained an actual calculated F ratio from a study like the top panel in Figure 16.4, there would be a greater likelihood that the (calculated) F ratio would fall in the exclusion area than if the top example was obtained. If so, it would be declared significant.

Table 16.1 shows an example of an ANOVA table with values that have been calculated from actual data. This is only an example, so we will not go through all the calculation steps. In a later section, we demonstrate an ANOVA example using SPSS so that you can learn how to recognize the appropriate information from the table and interpret the results.

If the values in Table 16.1 were actual results of a study, the F ratio (32.70) would be compared to an F "exclusion" value of 3.49. When the F ratio shown in the results table exceeds this exclusion value (obtained from a table of values), the researcher can conclude that the findings are significant. The findings in Table 16.1 would therefore be considered significant. This is an example of the **F Test**, the statistical test comparing the between and within variances in an ANOVA test. The calculated ratio of between to within variance is compared to the exclusion values of the F distribution.

If the example values in Table 16.1 represented the data in the experimental study discussed earlier, we could conclude that the noise group differences (between variance) were large relative to the within variance measure. This would mean that, taken together, the means of the group noise levels vary so greatly around the grand mean that the groups represent different populations of people.

EFFECT SIZE

As with our other statistical analyses, we can create an effect size that shows the impact of the independent variable on the dependent variable. In the case of our noise-learning example, the effect size concern could be asked as, "How much does the *grouping on the independent variable* affect the outcome measure"? We would be trying to determine how much noise impacts learning by measuring how far apart the learning group means of the specific noise categories (0, 30, 60, and 90 decibels) were from one another.

The ANOVA procedure determines the significance of group differences by examining variance, so we can use a variance measure to express effect size. The symbol that represents effect size in ANOVA is **eta squared**, or η^2. It is the Greek symbol for eta. *It refers to the proportion of variance in the outcome measure explained by the grouping on the independent variable.* SPSS reports the effect size in some applications, but it is easy to calculate from the ANOVA table using the following formula:

$$\eta^2 = \frac{SS_{Between}}{SS_{Total}}$$

$$\eta^2 = \frac{257.5}{289}$$

$$\eta^2 = 0.89$$

Thus 89 percent of the variance in learning errors is due to assigning subjects to our four different noise levels. The effect size is very large (only 100 percent is possible, obviously), but this is probably due to the fact that we have only groups of size four, and this is a hypothetical example.

How large is large? That is, how large does η^2 have to be before you would say it is meaningful? That is a question for which there are many answers. The reason it has many answers is that it depends on the sample sizes, number of groups, and so on. We alluded to power analysis in earlier chapters. Statisticians have constructed tables that take into account factors that would have an impact on the size of the effect size value. For example, there are tables for the 0.05 level of significance and the group size as considerations in judging effect size.

There are therefore many standards offered by statisticians and researchers for judging effect size meaningfulness. We tend to take two approaches to this question. First, we can suggest the following as benchmark comparisons: 0.010 (small), 0.060 (medium), and 0.150 (large). These are ballpark figures because the next approach is the most crucial.

The second approach to the question is allowing the researcher to judge the meaningfulness of the effect size by the nature of the problem studied. It would therefore have a lot to do with the subject of the research question. If we were studying a hypothetical research question with very small sample sizes (as we did in the hypothetical noise-learning example), we might not be very excited about a very large effect size (and in our study, 0.890 far exceeded the 0.150 guideline for large effect size). However, if we were studying a new drug that could lessen the death rate from AIDS, we would be ecstatic to find an effect size of 0.030 even though it would be judged small by the earlier benchmarks.

One way to visually understand effect size as explained variance is to use Venn diagrams. Look at Figure 16.6. The top circle represents the outcome measure of learning errors. The bottom circle represents the noise groups. Effect size is represented in the portion of the top circle with diagonal lines. The "89%" refers to our η^2 value of 0.890. If the top circle represents all the variation of learning errors, we can reduce that variance by 89 percent just by knowing the different noise conditions. There are many things that probably result in learning errors (e.g., aptitude, previous training, poorly executed experiment, etc.), but in our small study, we would conclude that the different noise conditions reduced our lack of knowledge by 89 percent.

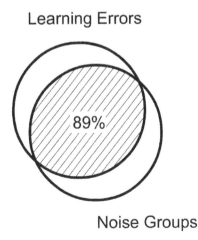

Figure 16.6 Venn diagram showing effect size.

POST HOC ANALYSES

If a researcher rejects the null hypothesis with the ANOVA procedure, the next question is what the differences between the groups mean. This question identifies the extent to which individual group means differ from the grand mean. The larger the difference, the more likely that these group differences may be responsible for the overall differences that led to the omnibus finding. Some of the group means will be (statistically) different from one another when the overall null hypothesis is rejected.

It remains for the researcher to analyze all the paired differences (like those listed in Figure 16.2) to see which mean differences are excessive. These tests of paired group means are known as **post hoc analyses**, since they are conducted after the omnibus test is performed. In effect, they are like individual *t* tests of the paired groups. If any of these paired means are statistically significant, it indicates that the specific group means are considered too far apart to occur by chance. As you might expect, there are specific statistical procedures to help us analyze the difference between these paired means. The many types of such procedures go well beyond our treatment of ANOVA, but you might refer to a statistical guide for further information. We present an example and discuss one such procedure, the **Tukey's HSD** (honestly significant difference) that compares all possible pairs of means referencing a distribution of values to determine significant differences.

Figure 16.7 shows two possibilities of group differences that might produce a significant F ratio in a test with four independent variable categories. The top panel shows that the first three groups are similar in their means on the dependent variable measure, but the subjects in the fourth group are much more likely to produce a different group mean (in this case, larger than the other three). The bottom panel of Figure 16.7 shows that the first two groups are similar, but both are very different from the third and fourth groups (which are similar to one another). There are many other possible combinations of findings.

ASSUMPTIONS OF ANOVA

As with the other statistical tests we discussed, the researcher needs to assess whether the conditions of the data are appropriate for the ANOVA test. Here are the primary assumptions, although ANOVA is also somewhat robust with respect to slight variations of assumptions:

1. Population is normally distributed. Do the different independent variable groups show normally distributed dependent variable distributions?
2. Population variances are equal. This assumption refers to the variance of the sample groups. Make sure that the groups have equal variance on the outcome measure. (This is known as the homogeneity of variance assumption.)
3. Samples are independently chosen.
4. Interval data are on the dependent variable.

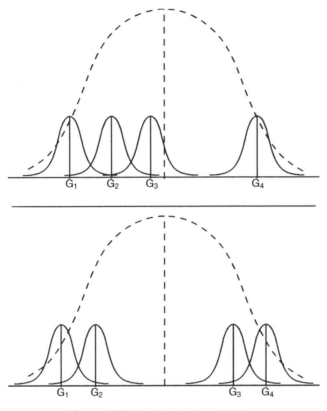

Figure 16.7 Post hoc test possibilities.

ADDITIONAL CONSIDERATIONS WITH ANOVA

ANOVA, like the *t* test, can be used in experimental contexts (where we consciously change the value of an independent variable to see what effects this has on a dependent variable), and for post facto situations (where we simply try to determine if there are existing differences among several sample groups without changing the value of the independent variable). In the earlier hypothetical example, we used ANOVA in an experimental context, since we directly "manipulated" or changed the value of the independent variable to see if this affected the dependent variable. The following real-world example shows how to use ANOVA with post facto designs.

A REAL-WORLD EXAMPLE OF ANOVA

In the section on *t* test, we examined the differences in perceptions of general health between those respondents who judged their earnings to be fair and those who believed their earnings were not fair. We found that there was a difference between the groups such that those believing their earnings were fair had significantly better perceptions of their general health. If we had introduced three independent variable categories (e.g., those believing earnings were unfair, those who believed earnings were appropriate, and those who believed earnings were more than deserved), we would have needed to use ANOVA, since three independent variable groups would require three separate *t* tests.

Shifting to a different example of ANOVA, we return to our Washington state database (2011 data) showing schools with various percentages of (fourth grade) students passing the reading assessment. In the extended example that follows, we use SPSS to analyze the question, "Does the percentage of low-income students in a school have an effect on school-based achievement levels?" This is an important question, since family income, while important, has not been the subject of a great deal of scrutiny as a potential factor in influencing school-based achievement.

You will note that we are also shifting to aggregate data in this example, as we have on several occasions in the book. Since our school database cannot provide individual-level student data, we cannot analyze the individual impact of family income on students' achievement. However, we can study the question when it is viewed at the school level. In the latter case, we will use the 2011 database from the state of Washington that shows the percentages of students who pass the reading assessment and the percentage of students who qualify for free/reduced price meals (FR). As we have noted in past sections, researchers use the FR variable as a way (often, the only way) to indicate low income.

We use a data set that is randomly chosen from all the schools in the state with fourth grade classes ($N = 1232$). We gathered a random sample of 100 schools resulting in 88 schools reporting scores on reading achievement (i.e., Reading Percent Passing Without Previous Pass) and on Free or Reduced Price Meals (Percent Free or Reduced Price Meals).

In order to show how to use ANOVA with this example, we categorized Free or Reduced Price Meals into a variable called "FR3Group" with three groups: low, medium, and high, which indicate schools with different percentages of students qualified for free or reduced price meals. (See the section "Data Management Unit B: Using SPSS to Recode for *t* Test," on using SPSS to transform variables to see how to categorize variables.)

USING SPSS FOR ANOVA PROCEDURES

Before conducting any statistical analysis, we would check to see whether the assumptions were met for the specific procedure we are using. In this case, we would examine the data to make sure the assumptions for ANOVA we outlined earlier were met. (We refer you to the Descriptive Statistics section for how to use SPSS to assess the normal distribution assumption.) In what follows, we point out the parts of the SPSS output that addresses the assumptions.

1. Do the Different Independent Variable Groups Show Normally Distributed Dependent Variable Distributions?

In order to assess this assumption, we can use SPSS to create descriptive information on the schools' reading achievement within categories of the FR groups. Figure 16.8 shows the SPSS specification for the "Means" procedure that will provide the descriptive information we need to examine this assumption.

Once this choice is made, SPSS provides another menu window in the Means procedure in which you can further specify which descriptive outcomes you wish. As you can see in Figure 16.9, we have called for descriptive output on the three FR groups (in the Independent List window) using schools' reading achievement as the dependent variable (in the Dependent List window). At this point, we can select the Options button to specify the outcomes we need. This list appears in Figure 16.10.

Figure 16.10 shows the numerical output that we can use to assess the normal distribution assumption. As you can see, we specified several statistical measures to help us in the Cell Statistics window.

Figure 16.11 reports the descriptive information on the study variables that we can use to assess whether the assumptions have been met. As you can see, the figure shows

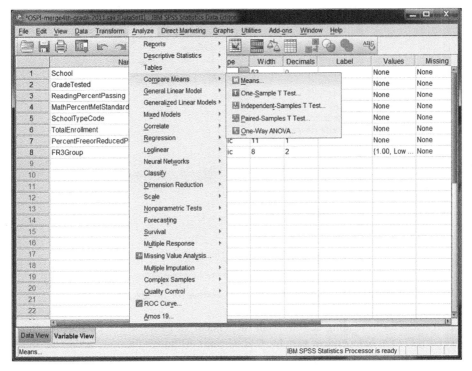

Figure 16.8 The SPSS specification for comparing group means within FR categories.

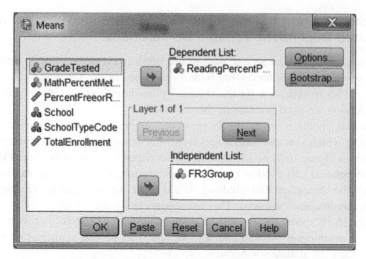

Figure 16.9 The "Means" specification window for the example.

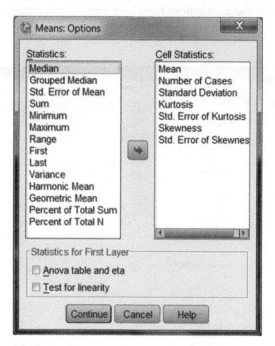

Figure 16.10 The Means: Options window for the example study.

the distribution of reading achievement (dependent variable) within each of the three categories of the independent variable (FR3Group). Although there are additional measures to help assess the normal distribution assumption, you can see from Figure 16.11 that the group distributions of reading achievement indicate relatively normal distributions, with one exception. The Low FR group shows a negative skew (most likely due

Report

ReadingPercentPassing

FR3Group	Mean	N	Std. Deviation	Kurtosis	Std. Error of Kurtosis	Skewness	Std. Error of Skewness
Low FR	78.88	27	11.421	2.876	.872	-1.344	.448
Medium FR	68.62	31	11.028	-.678	.821	-.191	.421
High FR	55.26	30	15.485	-.713	.833	.035	.427
Total	67.21	88	15.908	-.386	.508	-.512	.257

Figure 16.11 The descriptive information on the study variables.

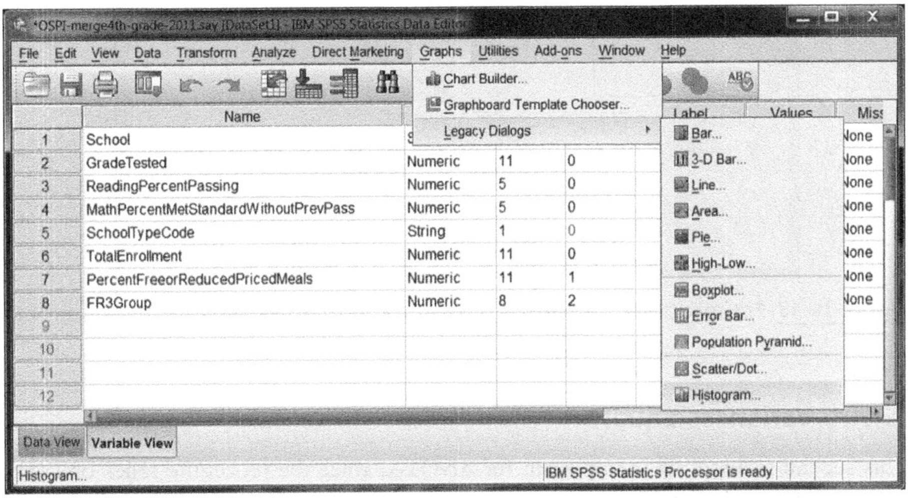

Figure 16.12 Using the SPSS Graphs menus to create group histograms.

to one school) and is slightly leptokurtic. We can watch this in our continuing analyses to see if this affects our outcomes.

Another way to understand these values is to look at the visual descriptions available through histograms. SPSS provides these through the Graphs menu. Figure 16.12 shows the specification for histograms. Once this choice is made, SPSS provides the window in which you may create the histogram you need for the analysis. This is shown in Figure 16.13. As you can see, the reading assessment variable is placed in the variable window, with FR3Group in the Rows window so that SPSS will create three reading assessment histograms, one for each of the FR3Group categories. Note also that we checked the Display Normal Curve box so that our group histograms will show the superimposed normal curve.

Figure 16.14 shows the histograms of these group distributions so you can confirm the numerical analyses. Visually, it appears that the histogram for Low FR is negatively skewed, probably due to the one extreme score on the left.

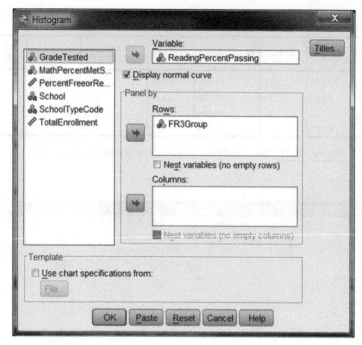

Figure 16.13 The specification window for creating group histograms of reading assessment.

2. Are Variances Equal?

This second assumption is important for ANOVA, since the procedure relies on comparisons of variance measures. The primary way to check this assumption is to use the **Levene's Test** available in the SPSS ANOVA procedure. We will point this out as we encounter it in the ANOVA analysis.

3. Are Samples Independently Chosen?

This assumption is met by virtue of schools being randomly chosen and not placed in groups due to any structurally linking criteria.

4. Are Interval Data on the Dependent Variable?

This assumption is met. Percentages are interval data.

SPSS PROCEDURES WITH ONE-WAY ANOVA

There are two procedures for creating ANOVA results in SPSS, both of which produce the same overall results. The first way is only used with ANOVA problems in which

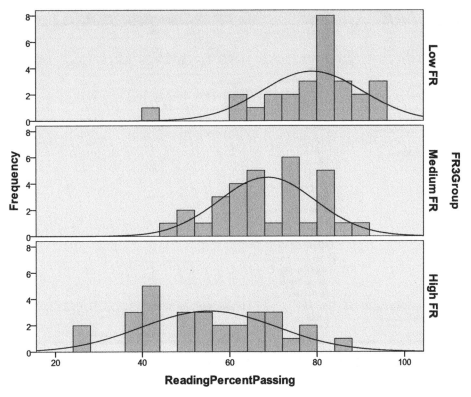

Figure 16.14 The reading achievement histograms for FR groupings.

there is one independent variable, as there is in this example. We will use this procedure for our example study, but we will point out the other way at the end of the discussion.

The one-way ANOVA procedure in SPSS is easy to create and thorough in what it reports. Figure 16.15 shows the menu choices that creates the one-way ANOVA procedure. As you can see, it is in the same menu group that includes the single sample t test and the independent samples t test that we discussed in other sections. This menu also includes the Means procedure we used earlier to create the descriptive data for the three FR groups.

When you select the one-way ANOVA procedure, the window shown in Figure 16.16 appears that allows the researcher to specify the analysis. As you can see, we specified the Reading achievement variable for the "Dependent List" and "FR3Group" as the grouping variable ("Factor"). The three buttons on the top right of the screen can be used for further specification.

Figure 16.17 shows the choices available from selecting the Post Hoc button from the one-way ANOVA menu. As you can see, several post hoc procedures are available for inclusion in the analysis, from which we selected the Tukey procedure, perhaps the

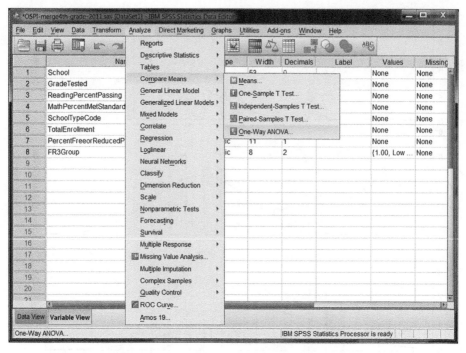

Figure 16.15 The SPSS menu options for accessing one-way ANOVA.

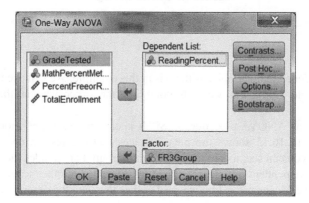

Figure 16.16 The one-way ANOVA specification windows.

most commonly used post hoc method. Notice in Figure 16.17 that the post hoc procedures are grouped according to whether equal variances (of the FR sample groups) is assumed or not. There are a few post hoc procedures available if you find the variances to be unequal.

When you select the Options button from the one-way ANOVA menu, you can select several important procedures that we can use in our analysis. Figure 16.18 shows these

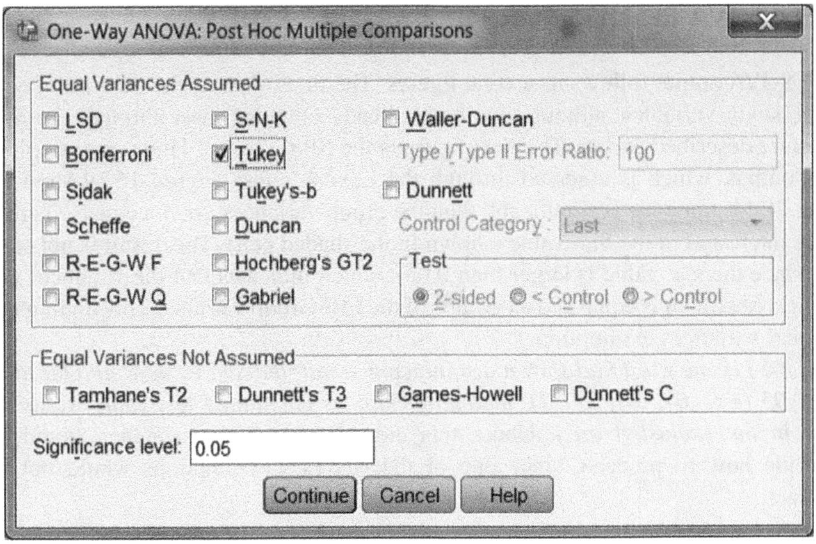

Figure 16.17 The post hoc choices from SPSS one-way ANOVA.

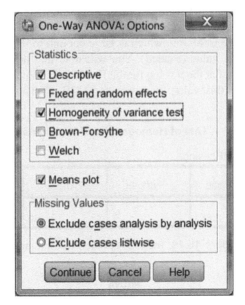

Figure 16.18 Options for SPSS one-way ANOVA.

choices. As you can see, we chose "Descriptive" procedures to derive the group means and other data, and the "Homogeneity of variance test" option that will produce the Levene's test for equality of variance. This will allow us to assess the equal variance assumption for one-way ANOVA that we discussed earlier. We also called for the "Means plot" that will provide a visual representation of the numerical descriptive results.

SPSS ANOVA RESULTS FOR THE EXAMPLE STUDY

The ANOVA output follows in several figures. The first output is the descriptive values for the study variables, although we have already obtained these through the Means procedure described earlier. The next figure is the SPSS Test of Homogeneity of Variances output, which is assessed through the Levene's test. Figure 16.19 shows this output. If this outcome is significant, then the group variances *are not* equal. Therefore, we are interested in the Sig. value (shown in the shaded cell). This result is not significant (since the sig. value is larger than 0.05), which indicates that the variances of the reading assessment is equal across groups of the FR3Group variable. This finding meets the equal variances assumption.

If the Levene's test had shown a significant result, the Sig. value would be smaller than 0.05 (e.g. .04, .01, .0032), indicating that its calculated test value would have fallen in the exclusion area. Under this circumstance, the researcher would need to decide how to proceed, since one of the ANOVA assumptions would not have been met.

The ANOVA Table

The next output is the ANOVA summary table similar to the one we described earlier. Figure 16.20 shows the overall, or omnibus, results of this study.

The reported F ratio is 24.369, which is large enough to be considered significant as indicated by the Sig. value (shaded). You will note that the reported significance value of 0.000 indicates that the p value for the F ratio is extremely small. If you change the decimal places for this value, it shows "4.3113E-9". This is scientific notation

Test of Homogeneity of Variances

ReadingPercentPassing

Levene Statistic	df1	df2	Sig.
2.661	2	85	.076

Figure 16.19 The SPSS Levene's test findings.

ANOVA

ReadingPercentPassing

	Sum of Squares	df	Mean Square	F	Sig.
Between Groups	8023.683	2	4011.841	24.369	.000
Within Groups	13993.741	85	164.632		
Total	22017.424	87			

Figure 16.20 The SPSS one-way ANOVA summary table.

indicating an actual significance of 0.000000004. Thus there is a very small likelihood that the different group means of reading assessment by FR3Group are different by chance.

Effect Size

SPSS does not routinely produce η^2 through this one-way ANOVA procedure, but you can confirm the effect size by dividing the appropriate values from the ANOVA summary table. If you divide the Sum of Squares-Between Groups by the Sum of Squares-Total (shown as shaded values in Figure 16.20), you will find that the effect size is 0.364 and has the effect size value expressed as follows:

$$\eta^2 = \frac{8023.683}{22017.424} = 0.364$$

The value of 0.364 would be considered large by the criteria we discussed earlier. In terms of interpreting its value, we could say that 36.4 percent of the variation in school reading achievement is explained by the three groupings on the FR3Group variable. The differences in the FR3Group categories result in meaningful differences in reading achievement.

The Post Hoc Analyses

The omnibus (F test) and effect size calculations indicate significant differences among the different FR3Groups. We can determine which of the groups might be different from which others, in terms of reading achievement, by using the Tukey's HSD analysis. Figure 16.21 shows the results of this test.

We highlighted some of the values in the first row to explain the output. The first two columns indicate which two FR group means are being compared. According to

Multiple Comparisons

ReadingPercentPassing

Tukey HSD

(I) FR3Group	(J) FR3Group	Mean Difference (I-J)	Std. Error	Sig.	95% Confidence Interval	
					Lower Bound	Upper Bound
Low FR	Medium FR	10.262*	3.378	.009	2.21	18.32
	High FR	23.621*	3.404	.000	15.50	31.74
Medium FR	Low FR	-10.262*	3.378	.009	-18.32	-2.21
	High FR	13.359*	3.286	.000	5.52	21.20
High FR	Low FR	-23.621*	3.404	.000	-31.74	-15.50
	Medium FR	-13.359*	3.286	.000	-21.20	-5.52

*. The mean difference is significant at the 0.05 level.

Figure 16.21 The Tukey HSD analysis findings.

TABLE 16.2. The Significant FR3Group Differences on Reading Achievement

Groups	Paired Difference	Interpretation
Low FR and High FR	23.621 (78.88-55.26)	Study schools with High FR (percentages of students eligible for free or reduced lunch) show an average 23.62 percent less students passing the reading achievement assessment than study schools with Low FR.
Medium FR and High FR	13.36 (68.62-55.26)	Study schools with High FR (percentages of students eligible for free or reduced lunch) show an average 13.36 percent less students passing the reading achievement assessment than study schools with Low FR.

the shaded values, the first comparison is between Low FR and Medium FR. The difference in the group means is 10.262, which you can recreate by subtracting the two means reported in Figure 16.11 (10.262 = 78.88 − 68.62). This value and the associated asterisk indicate that the reading achievement of the study schools is significantly different depending on whether the FR percentage of students is Low or Medium. The group means indicate that reading achievement is higher when the FR percentage of students in schools is low (by over 10 percent). The significance of this finding is confirmed by the (shaded) value in the Sig. cell (0.009).

Table 16.2 shows the interpretations of the other paired tests. As you can see from Figure 16.21, all the pairs are significant (see values in the Sig. column), so the interpretation focuses on the magnitude of the differences between groups.

In our study, the group means indicate percentages of students in the study schools who passed the reading assessment. Therefore, positive values indicate stronger math performance. The Low FR thus outperforms the other two groups while the Medium FR group outperforms the High FR group. Our measure of family income (FR3Group) appears to have a strong effect on school-based reading achievement. We will extend our analysis with this variable in subsequent analyses.

17

FIELD RESEARCH

Have you ever been in a public space, like a coffee shop, restaurant, doctor's office, or department store, and found yourself listening to someone else's conversation? Have you ever watched others try to navigate an awkward social situation? Have you ever waited in a line, surreptitiously surveying everyone else? If you are majoring in a social science, chances are you love observing other people. Many of us are natural observers, watching others and making suppositions as to why they act in certain ways. This is the lure of observational research, or field research. **Field research** is primarily a qualitative research design where researchers go into the field, or a natural setting, to observe people, collecting very detailed information about individuals, groups, or interactions.

For professional (and armchair) social scientists, we are constantly taking in data wherever we are. Most of us practice "studied nonobservance," or pretending not to notice what others are doing, in public spaces. Natural observers use this to their advantage to notice others in their natural habitats. In fact, while reviewing classic field studies in preparation for writing this book, McKinney did a little informal field observation of her own when she pulled out Laud Humphreys's classic study, *Tearoom Trade*, at the airport waiting for a flight.

As she read from *Tearoom Trade*, McKinney realized something was happening. It was clear that several other airport patrons were interested in something around her.

Understanding and Applying Research Design, First Edition. Martin Lee Abbott and Jennifer McKinney.
© 2013 John Wiley & Sons, Inc. Published 2013 by John Wiley & Sons, Inc.

She tried to subtly observe the area around her to see what was causing a stir but saw nothing unusual. She finally realized what was happening. In this very public space, with children as well as adults whiling away the time before flights were announced, people were concerned about what she was reading. *Tearoom Trade* has a plain white cover with the title and subtitle of the book scrawled across the simple cover in large black letters: *Tearoom Trade: Impersonal Sex in Public Places.* Mystery solved. A good guess would be that people were surprised by the "impersonal sex in public places" part of the title, and that they feared McKinney was reading an inappropriate how-to guide rather than a classic sociological study.

This natural interest in those around us is the core of social science—how do people relate to each other and interact with each other? Most social science students are introduced to their disciplines by a substantive course that allows them to explore the eccentricities of societies, groups, and/or individuals. For example, in our Sociology Department we have a colleague that teaches sociology of deviance. Inevitably we hear students in our classes discussing what they are learning in the deviance course (where they cover the research in *Tearoom Trade*). Learning interesting things about people and groups seems to be more direct when doing field research (like how Humphreys discovered that tearooms utilize a third person to serve as a lookout, or "watchqueen"). Whereas surveys and aggregate data are important tools for describing people and populations, and experiments allow us to generalize to theoretical principles, field research puts researchers front and center in observing people in their natural habitats.

Field research is quite broad in what can be studied as well as how it is studied. Researchers Lofland and Lofland (1995) looked at several elements of life that field research can illuminate including practices (behaviors), episodes (events like crime, divorce, illness), encounters between people, role associations with the positions people occupy, relationships, groups, organizations, or subcultures. These elements of life are often difficult to quantify using survey or aggregate data. At the same time, these elements lend themselves to careful study by field researchers because they deal with the *process* of social interactions. Field researchers spend time watching how interactions within groups, relationships, or organizations take place. Field research is unparalleled in bringing to the fore a rich, descriptive account of how these things happen.

Because many people see themselves as natural observers, it is sometimes difficult to explain why it takes skill to do field research. What is there to learn about observing others when you have been doing that your whole lives? Isn't observation intuitive? When deciding to do field research, there are a variety of hurdles to overcome just in selecting a topic and entering into the field—much less how data is collected once in the field and then how the data are organizaed and prepared for analysis. We begin with how to think about preparing to do field research by selecting an appropriate topic.

SELECTING A TOPIC

Because field research takes place in a setting teeming with a variety of people and interactions, it seems to carry a more glamorous status as a research design than something like aggregate data. Studies on rock music groupies, street gangs, wives of NBA players, impersonal sexual activity, and/or *Star Trek* devotees also seem to make field research more trendy than scientific. While these topics are certainly interesting, there are only so many studies you can do looking at these types of groups.

Relevance

As social scientists, one of the questions that should be asked prior to beginning field research is how *relevant* the topic is. Is the topic being selected because it is glamorous, rather than because the topic contributes to the larger scientific literature? For example, a graduate student decides that she and her husband have been part of the "swingers" subculture (a group of husbands and wives who meet to switch sexual partners) and decides that studying swingers will make a good field research project. Although the graduate student may have unprecedented access to a group of swingers, dozens of studies have been done on the swingers' culture. Is she choosing this topic because it will add to the scientific literature, or is she choosing this project simply because it is convenient? Of course a second question would be whether or not she could maintain her objectivity as someone studying her own group. Can she report on this subculture without her own biases getting in the way?

Selecting a relevant research topic requires thoughtfulness. Has anyone done previous scientific study on this particular group/organization/interaction/relationship/episode/subculture? Topics can be chosen responsibly by selecting research questions that illuminate an element of life we need to know more about. There is a balance to be struck in doing this. Many studies have been done on new religious movements (NRMs), for example, yet what if there is a burgeoning religious movement that seems to be doing something different from previous movements? Is studying this movement redundant, or will studying the movement add to current scientific knowledge? In looking at the academic literature pertaining to the group, we find that little to nothing has been written regarding this group, except in the popular press or by the group's proponents and detractors. Studying this NRM would then be a relevant choice. Drawing from previous studies on NRMs, there would be a sense of the appropriate strategies to use for the research. At the same time, however, how the NRM uniquely adds to present religious knowledge makes the study relevant.

The key to selecting a relevant topic for field research is in clearly defining the topic when going into the field. Once in the field a clear topic guides what will be observed, as well as how it will be observed. It is not uncommon for field researchers to enter the field and be overwhelmed by the sheer amount of data they are observing. Without clearly defined reasons for going into the field, researchers can quickly become lost in the sea of interactions they are observing. Going back to the research plan keeps the research on track, ensuring that the data coming out of the field are the data that were intended to be collected.

Accessibility

Another hurdle in selecting a topic is assessing the accessibility of the individuals, groups, or process being studied. Many people, groups, sites, and/or activities are not available for scientific observation; sometimes this is counterintuitive. In religion research, the expectation is that traditional American religious groups would be the most open to being studied, whereas NRMs would be the least likely to grant access to researchers. The opposite, however, is often more true; mainstream religious organizations sometimes refuse to grant permission for outsiders to study their organizations. Often NRMs give social researchers unprecedented access to study them (see Barker's 1984 study on the Unification Church in Britain).

As we noted in Chapter 4, getting permission from the group is usually the most ethical way to proceed with research. Overt observation (being out in the open about being a researcher) allows the researcher to observe in good conscience—everyone knows that they are being observed. In some cases, however, it is permissible to seek degrees of covert observation, where the group is not aware of the researcher's true aim to study the group. One example of this is when Lofland and Stark (1965) decided to study religious conversion taking place (see also Lofland 1966; Stark and Finke 2000). After sifting through a number of "deviant" religious groups in the San Francisco Bay area, Lofland and Stark chose a small group that had recently relocated from Oregon to San Francisco. The religious group was led by a former professor of religion from Seoul, South Korea. The Reverend Kim and her group of followers were the first members in the United States of the Unification Church (also known as the "Moonies").

Not knowing how long the process of conversion would take, Lofland and Stark did not want to disrupt the group by introducing themselves as social researchers. The researchers sought permission from Reverend Kim to conduct field research without alerting the other group members to their status as social researchers. They were given permission to study the group and subsequently published one of the first scientific studies of the process of religious conversion. Lofland and Stark found that religious conversion was a product of social network ties; those most likely to convert to the new religious group were those most socially connected to the group. The first converts to the group had all been friends prior to joining the church. When Lofland and Stark began studying the new group, they found the group had not yet succeeded in attracting any strangers—of the people drawn to the group, the only ones who joined were those whose social network ties to the group outbalanced those without (Stark and Finke 2000). By choosing a different degree of covert research, Lofland and Stark discovered a process of religious conversion that has stood the test of time (see also Kox, Meeus, and t'Hart 1991; Stark and Finke 2000).

When permission has been obtained for the field research, a determination needs to be made as to how much access a researcher is given. A group being observed may initially want to steer the researcher to particular members of the group, hoping that the researcher observes the best parts of the group. This, of course, will give a biased view of the group. Researchers must make sure that they have access to the full variety of individuals, rituals, practices, interactions, and so on, within the group in order to get the broadest possible view of the group. Researchers need to *maximize their access*. When studying religious groups, it is helpful to get a list of those who have left the group, as well as those who participate, in order to understand mechanisms through which members choose to leave. By having observed both current and former members of a group, a fuller picture of the group emerges. Having access to the group being studied is necessary; it can also be something more elusive than expected.

Several years ago McKinney was involved with a research group where members reviewed each other's work in progress. While reviewing one of the research projects, an interesting conundrum presented itself. Another participant had proposed an exciting project dealing with an unusual religious group that blended traditional native religious practices with traditional Christian practices. While the researcher did some background work on the group by observing the group's public festivals, he also gained knowledge of the group by way of an **informant** (someone who had insider information about the group). Based on the researcher's interest in studying the group, the background research he had done (through public news reports, etc.), along with what he had learned from the informant, he began to set up the project. Several months later, however, the researcher discovered that the group had previously approved guidelines for people seeking to study them. The researcher's informant had been unaware of these guidelines, leaving the researcher to wait for permission to get full access to the group. At the research group's last meeting, the researcher had to report that he had still not been able to get permission, and thus access, to the group he had hoped to study.

Censorship

Having negotiated for full access to the group is a beginning point to field research. One pitfall sometimes encountered in doing field research is censorship. Oftentimes unwittingly, those within the group may seek to censor what a researcher reports. It is imperative for social researchers to walk the line between being *part* of the group and yet being *apart* from the group, keeping in mind at all times his or her place in the group as a researcher. Even in overt research, over time those that are being observed come to see the researcher as a part of the group—in effect, the researcher becomes just one more piece of furniture in the life of the group (something that's there but benign). This is a good outcome; as the researcher becomes more familiar to them, group members interact normally. Group members may, however, ask the researcher to describe what he or she is recording, how the person perceives the group, and/or what he or she is concluding about the group. Not everyone in the group may see themselves and their group as the researcher—the outsider—does. They may ask the researcher to revisit or revise conclusions. Researchers sometimes make hard decisions to report what they believe they see and experience, rather than appease the group members by painting a less than accurate picture of the group. When researchers enjoy relationships with the individuals or groups they are studying, it can be difficult to disappoint them by reporting less than flattering observations. Part of being a researcher, however, is to have integrity in what gets reported. Researchers need to report honestly and accurately what they have seen, heard, and experienced, even if it is unpopular with the group being studied.

Independent Funding

One way to avoid the pressure of censorship is to ensure that researchers have independent funding for the research (funding not provided by the group being studied).

Many social researchers are university faculty. In order to do long-term projects they need funding to help with travel costs, buying time off their university position, and so on. Sometimes groups being studied offer to defray the costs of research by providing compensation. While it seems generous for the group to offer research funding, it can be problematic. What happens when the group asks to see a researcher's preliminary report only to find that the objective researcher has accurately pinpointed both group strengths and weaknesses? What if the group then asks the researcher to modify his or her results? The financial pressure may impact what the final report/research describes, disqualifying it from being objective science.

When researchers vie for maximizing access to the group, they also need to obtain independent funding. Both of these allow the data to come unfiltered to the researcher, avoiding the pitfalls of organizational censorship, either through limiting the researcher's access to different areas of the group or through financial pressure to report what is more flattering to the group.

Risk

One last hurdle to consider before undertaking field research is to evaluate how much and what type of risks might be involved in doing the research—for the researcher and the subjects. We've mentioned that researchers need to consider thoughtfully whether they are going into the field overtly or covertly. For either option, are there risks involved in doing the research, from who or what is being studied, or what the researcher is learning? Recall the Humphreys's *Tearoom Trade* study, where Humphreys participated as a watchqueen, choosing to study the tearoom subculture as a covert researcher. In the course of his research he was picked up by the police and jailed. Since he was a covert researcher, he could not risk telling the police who he was and what he was doing for fear of either losing his contacts in the tearoom subculture (by outing himself) or by being pressed into informing on his subjects to the police, who may have taken his data and used it to shut down the tearooms. Another risk Humphreys took lay in the nature of the group he was studying. One day while acting as watchqueen, Humphreys and several men in the tearoom were set upon when a "gang of ruffians" came into the tearoom, locked the door, and physically assaulted the men participating in the tearoom activity (Humphreys 1975).

Just because a researcher takes the high road and goes into a study as an overt social researcher, some activities are not protected and still involve risks. An example of overt research taking an unexpected turn for the researcher is the case of Rik Scarce. He was a graduate student at Washington State University conducting field research on radical environmental movements. In the course of his research, Animal Liberation Front (ALF) activists raided a laboratory at Washington State University. Being known on campus as someone studying radical environmental movements, Scarce was approached by federal prosecutors who believed he had interviewed the people involved in the laboratory raid. When Scarce refused to answer some of the prosecutor's questions on the basis of both his First Amendment rights and the American Sociological Association's ethical guidelines (which stipulate that confidentiality must be maintained by researchers even if legal force is applied (ASA 1999: 3 cited by Lee-Treweek and

Linkogle 2000:18), he was jailed for 159 days. Scarce reports that his release came only after the judge in the case realized he would not betray the promises of confidentiality made to his research participants, effectively preventing him from revealing whether or not he had interviewed particular activists (Lee-Treweek and Linkogle 2000; Scarce 2005a, 2005b).

Unlike journalists who have precedence for protecting their sources, social scientists are not given the same consideration. As Scarce (2005a:B24) noted in an article advocating legal protection for social scientists, "The lack of a federal shield law prompts social scientists to limit their assurances of confidentiality to research participants for fear of going to jail. As a result, we practice self-censorship, deliberately restricting the topics we study. Absent a shield law, it takes a brave scholar to conduct in-depth interviews with, for example, polygamous fundamentalist Mormons or violent inner-city drug dealers." While researchers abide by their professional codes of ethics and maintain their integrity in the face of betraying confidential information on their subjects, doing so may result in risky and unintended consequences. In undertaking field research, researchers must entertain all possibilities of where the research may take them, including the possibility of their "going native."

Going Native. Part of the rights of passage for graduate students getting ready to do field research are the stories that abound about "going native" or becoming so involved in participating in field research that the researcher loses objective perspective. One popular urban myth that makes the graduate student rounds is about a graduate student doing field research in the local college town's prostitution scene. The story goes that she became so empathetic with the prostitutes she studied, that she began taking tricks with them on the street. One night, after not coming home, her roommate contacted the police, who subsequently found her dead—from a pimp who thought she was encroaching on his territory. Of course this was an urban myth. No such study and no such graduate student existed. The myth was designed to scare graduate students into thinking about the risks involved in doing field research. In the latter respect, however, the myth is useful. Is there a risk that researchers may become so involved in their research that they are unable to maintain the appropriate objectivity in order to do the study as a social scientist, and not as a group member? These issues related to risk—as an overt researcher, covert researcher, and by going native—are important to consider before entering into the field.

ENTERING THE FIELD

Once researchers have clearly defined and relevant research questions, have maximized access to the group, and have addressed the potential risks involved in the research, they are then able to enter into the field. As we noted previously, the best case scenario for a field researcher is to go into the field as an overt researcher—allowing those being observed to know they are being observed. The trick to doing this is to be aware of the Hawthorne effect—where people behave differently when they know they are being observed.

The Hawthorne Effect

In social science the *Hawthorne effect* refers to the idea that when being observed, people change their behavior. This observational effect was discovered over the course of several studies engineered by the management of the Hawthorne plant for the Western Electric Company in Chicago. Based on the principles of "scientific management" created by Frederick Taylor in 1911, a series of experiments were conceived by researchers to study the relationship between illumination and productivity, as well as the effects of rest time and work hours on productivity. In the illumination phase of the experiments, researchers were surprised by their results. One group of workers at the electric plant (the experimental group) were subjected to decreased lighting to see how illumination/lighting impacted productivity, as compared to a control group who received constant lighting conditions. While researchers expected productivity to decline in the experimental group, the study results showed that both the experimental and control groups increased their productivity (in fact, lighting levels for the experimental group reached the equivalent of working by moonlight before negatively impacting productivity). The common explanation for the surprising result was in the observation itself. The experimental group was pulled off of the regular line in order to control the amount of light to which they were exposed. Being singled out as a group for the experiment—in effect, by being aware of their being observed by the scientists—was credited with their increased productivity, regardless of the amount of light available.

While the Hawthorne effect has been a useful concept to use when illustrating the effects researchers may have on their subjects, the effect has come under fire from others who attribute the surprising results from the studies to a host of other uncontrolled variables (for example, workers may have been hoping for a chance to be transferred out of the regular manufacturing department, they may have been afraid of losing their jobs, they may have felt more agency in determining new work procedures, etc.; see Wickström and Bendix 2000). While the general aim of the original Hawthorne studies was to link work practices with productivity, in the social sciences the Hawthorne effect is useful when observing people. For specific designs (field research, for example), it is helpful to realize that being a social science researcher and observing others may cause unanticipated consequences, disrupting the normal interaction of those being observed.

If you have ever traveled to a culture that is significantly different than your culture of origin, you probably have studied some of the different cultural norms (for example, driving or walking on the left side of the street/sidewalk when in Britain). As prepared as you are, there are usually some surprises when you arrive in the new culture. Researchers who have done their homework and selected an appropriate topic and place to conduct their field work will find the same dynamic. In order to compensate for their entry into the field and the disruption that may cause, researchers expect to spend a significant amount of time in the field.

Neutralizing Observer Effects

A general rule for a field researcher is to be in the field for a significant amount of time (typically two years). Recognizing that the people you observe will react differently than usual to being observed, researchers should go into the field expecting to neutralize their effect. *Neutralizing observer effects* entails allowing the research subjects to gradually get used to having an observer among them. Part of neutralizing observer effects includes having researchers keep a low profile as they move into the field. Keeping a low profile helps researchers in two ways: it allows them to establish themselves as a fixture within the group while affording them the ability to observe the group's norms, interactions, and language. Within several months, those who are being observed will return to their normal interactions, having gotten used to having the researcher in their midst. Having used the initial entry time into the group to learn more about the group, the researcher is then armed with a more thorough understanding of the group's cultural norms and interactions, to begin the more formal observation.

TAKING DATA IN THE FIELD

Field research is unique in a variety of ways, including the fact that it is where you go, rather than what you do. Now that the researcher is ready to begin to collect observations, how does he or she do that? Two keys to collecting data in the field are qualitative interviewing and field notes. *Field notes* are the primary form the data take. While researchers conduct their field work, they are constantly taking in information. These observations should be recorded as soon as possible. In some types of field research, it is easy for researchers to carry around their field notes—with a laptop, recorder, smart phone, or simply pen and paper. In many cases, however, field researchers need to learn how to catalog information systematically by memory in order to be physically recorded at a later time.

Interviewing in the Field

Field notes are composed, much of the time, with the results of qualitative interviews. Qualitative interviews are unique in that they are significantly less structured than quantitative interviews. Often there is a general plan for the type of data a researcher is collecting, but without specific questions and/or without a specific time table for collecting the data. Some field research has a more structured interview process, for example, McKinney collaborated with a colleague to interview postbaccalaureate students undertaking year-long ministry internships to see how the internship experience impacted their vocational aspirations (did participating in the internship reinforce their desire to work in ministry, did it diminish their desire to work in ministry, etc.). For this field work, each intern answered a semistructured set of questions. McKinney and Drovdahl (2007) fashioned a set of questions that targeted three areas of vocational discernment to learn how internship experiences impacted interns' sense of themselves, the skills they brought to the internship/skills they learned from the internship, and their future vocations. For many field researchers, however, this level of structure is too narrow for the scope of their research. Much of the practice of qualitative interviewing relies on the researcher's dexterity in crafting unscripted moments to collect data.

In many cases, field researchers keep in mind that much of their interviewing appears as if they are simply interacting with a respondent in a casual conversation. Qualitative interviewing takes great skill and practice. Researchers need to be subtle in how they elicit information and flexible in directing their inquiries. Even for skillful interviewers it is often the case that someone being interviewed veers off topic. Field researchers can try to subtly direct the conversation to get back on course; however, if the researcher is too directive, the subject's recognition that he or she is talking to a "researcher" might kick in, inhibiting or truncating responses.

Field Notes

When doing these informal conversations, researchers are asking questions, listening to the answers, and interpreting what they hear. These data are then recorded as field notes. **Field notes** are the careful and systematic recording of what the researcher observes during field work. When recording observations, field researchers write down specific and detailed accounts of what they are observing. Not only do researchers record what they actually observed, they also record what they *think* they observed, or the interpretation of how they characterize what their observations mean in the context of the group.

Recording field notes requires skill and self-discipline. Observations should be recorded as soon as possible and as thoroughly as possible. Field researchers are inundated with data—trying to rely on memory and putting off recording observations can result in lost data. As the data are recorded, researchers are also interpreting what the data mean. Are there connections to larger concepts or between events? Researchers attach meaning to what may seem obvious but should be anticipated. For example, are women or men more likely to participate in an activity; are there stratification systems around gender, race, age, and so on.

By keeping careful field notes, the organizing and analysis of the data become more streamlined. Social scientists are interested in finding patterns across events. As they attend to interpreting their experience in the field and recording their notes, researchers coming out of the field have more focus. Remember, researchers should go into the field with a clear vision of why they are going; it will make organizing and analyzing findings much easier.

RELIABILITY AND VALIDITY

Whereas the data garnered through field research are unsurpassed in providing data on what people actually do in real life, there are some issues that may impact the data's reliability. The first issue is from deception. Deception within field research takes on two forms, deception through the researcher (when he or she chooses to be covert) and deception through the subjects.

Field researchers are usually the lone researcher in the field; thus no one else is equipped to evaluate what the researcher has actually observed in the field. By the reporting of intricate rituals and processes the researcher presents, we tend to assume the data are valid, but are they reliable—or true? There is pressure in the academic world to find new or unusual practices or elements of life, which can lead researchers to embellish their findings. We live in a "publish or perish" environment. Because we have no other authority with which to compare the field researcher's findings, how would we know if researchers are misrepresenting what they observed?

One controversial study of the Yanomami people in the Amazon region characterized them as one of the most violent cultures ever studied. Napoleon Chagnon's (1977) anthropological work on the Yanomami people found that a significant percentage of the adults in the group had a relative that had been murdered. When Chagnon would ask how many people each man had killed, the subjects responded with astoundingly high numbers. The numbers of killings the Yanomami reported, however, were disproportionately high. It turns out the Yanomami believe they can kill others with a curse. Therefore when people they did not like died, a Yanomami man might claim he had killed his "enemy" due to a curse (and many Yanomami might claim the killing of the same person). Thus in reporting, did Chagnon misrepresent the culture? Other researchers that later studied the Yanomami painted a very different picture of the culture (Peters 1998, Tierney 2002). Tierney (2002) reports that when Chagnon collected his data, he often misrepresented himself (for example, by lying or being unclear with how or why he was collecting data). Having multiple researchers study the same group gives a broader picture of the group. In doing field researcher, however, having several accounts of one group may not always be an option.

Another form of reliability bias that can occur is through the use of informants. While informants—people connected to the group who can provide insider information— can be very useful, they can also be misleading. Informants are often assumed to be well informed, thus their appellation. While they can give general information about the group, you need to ask the question of why they would be willing to give you insider information about the group, especially if the information is sensitive or secretive. Is the informant placed strategically within the group, with access to the information they claim to have, or is the informant willing to inform on the group because they are somewhat marginal to it? Informants are useful, but field researchers need to assess the position and access the informant may or may not have. We described a researcher earlier in the chapter who relied on an informant during the preliminary study of a religious group. He (and we suspect the informant) were then surprised to find that the group had rules governing research on the group. Clearly, the informant was not so integral to the group that he was aware of these rules. Information obtained from an informant should be used thoughtfully and judiciously.

Generalizing from field studies can also be problematic. Data obtained in the field are detailed and important, describing processes that no other research design can describe. These detailed data are a great strength of field research; it is also a weakness in that it makes the findings limited in generalizability. First, generalizing within the group should be done using **proportional facts**. Using descriptors like "the majority," "most," "few," and "typically" should be backed up by the data. Did the researcher take a census of the whole group in order to make a generalization that "most" of them did a certain task? While using proportional statistics is helpful for readers, researchers need to make sure that the statistic is reflected by a careful accounting of the data.

Invalid generalizations are a second generalizability issue. **Invalid generalizations** occur when trying to generalize to groups that were not studied. Let's say we wanted to do a study of Girl Scout troops and the strategies they used to mobilize their annual cookie drive. We could spend several years observing how local Girl Scout troops do this and then publish our findings. There would be a tendency subsequently to generalize our findings to all Girl Scout troops. That generalization would be invalid, however, since we can only generalize to the groups we actually observed. Field research does not involve any type of relevant population or probability sampling; thus we are unable to take our findings and apply or generalize them to other groups, no matter how similar they may seem. Our findings will inform other researchers who seek to study similar groups, but we (and they) cannot assume our findings describe any group other than the ones specifically studied in the course of the field research.

Field research is both exploratory and descriptive. We tend to assume that the great detail described within field research reflects what the researcher actually found. Since the concepts and patterns described by a field researcher are so rich in detail, we tend to assume that they are valid—that they are measuring what they say they are measuring. This rich, descriptive data provide a window into social life that is unsurpassed by other research designs. Because field researchers get "up close and personal" with people, there are some ethical considerations we need to highlight for this particular research design.

ETHICS

As a brand-new graduate student at the University of Chicago, sociologist Sudhir Venkatesh became involved with the gang subculture in the Robert Taylor Homes on Chicago's south side. For the next seven years, Venkatesh immersed himself in the gang culture that regulated much of the daily life within the public housing complex. With the publication of his third book, *Gang Leader for a Day*, Venkatesh (now a Columbia University sociology professor) received some backlash on his "ambivalent" stance regarding ethical concerns raised by his field research (Carr 2009; see also Charles 2009; Clampet-Lundquist 2009, Venkatesh 2008; Young 2009).

While his field research encompassed a variety of issues related to daily life in the Robert Taylor Homes housing projects, the majority of Venkatesh's account in *Gang Leader for a Day* deals with the relationship he developed with J.T., a leader with the Black Kings gang in the housing project. As Venkatesh (2008:29) reports, "It was pretty thrilling to have a gang boss calling me up to go hang out with him." The adrenaline rush of being afforded such unparalleled access to this culture is understandable. What comes into question, however, is Venkatesh's ability to maintain his objectivity while he participated within the gang culture (as denoted by the book's title, Venkatesh was given the opportunity to be a gang leader for a day).

As a participant-observer, Venkatesh readily admits that he became a "hustler," someone trying to get ahead by getting something from others. One of his critics notes how Venkatesh's hustling had consequences—not just for his ability to maintain objectivity, but also for his ability to maintain confidentiality for his research subjects (Clampet-Lundquist 2009). As part of his research, Venkatesh learned about some of the other "hustling" going on in the projects—off-the-books ways members of the community earned money. As part of the daily operations of the gang, J.T. and the Black Kings took a "tax" on all money-earning activities. Venkatesh consulted with J.T. and another member of the community about other participants' hustles, breaking the other participants' confidentiality. Venkatesh's "fact-checking" came with a cost to those who had not revealed to the gang that they were earning money—money not "taxed" by the gang (Charles 2009; Clampet-Lundquist 2009).

Apart from keeping scientific objectivity and subject confidentiality, a third concern pertaining to Venkatesh's research is his ability (or inability) to overcome his preconceptions about the culture he was engaging. Charles (2009:209) writes that Venkatesh does not discuss "the importance of his preconceived attitudes and beliefs and tendency to adhere to negative racial stereotypes." She writes that for someone immersed in this gang culture, Venkatesh should have understood his own biases. Consider, for example, Venkatesh's (2008:248) expectations regarding attending a gang party: "I had envisioned half-naked women sitting poolside and rubbing the bosses with sunscreen while everyone passed around marijuana joints and cold beer." Did Venkatesh's preconceived ideas impact his ability to see and interpret what was going on within this particular subculture?

Doing social research is complex. As Charles (2009:209) notes regarding Venkatesh's research, "Despite claims that we check our preconceptions at the door when we do our jobs, social science researchers are human. Like everyone else, our attitudes and beliefs are shaped . . . by our own experiences in the world." This statement is true for all social science research, but when dealing so closely with individuals as we do in field research, we must be more conscious of our impact on these individuals. Three issues that are particularly salient to ethical practices in field research are the issues of privacy, deception, and participant-observation.

In the process of doing field research, we learn things that violate others' **privacy**. One key ethical concern for researchers is to take care to protect the names, people, and places where their field research occurs. For Venkatesh's participants, their privacy was violated when he made their information ("hustling") available to other members of the community. Researchers must take great care to protect the people we study. Like any field researcher, Venkatesh was hustling—getting information from people in order to complete his own research agenda; this is what we do. There is usually very little gain for people participating in social research, while researchers earn advanced degrees and/or prestige by publishing our findings. Because we are sometimes hustlers, it is imperative that we offer as much privacy and protection as possible for our subjects.

One way to do this is to carefully disguise when, where, and who we study. For example, in a study (1997) analyzing intake interviews with homeless clients in a large city, the account included in the publication made every effort to disguise who these clients (and social workers) were. The description included in the publication noted, "This research was conducted at a human service agency (hereinafter, Homeless Assistance, or HA) that served homeless persons in a midsized city in the southern United

States (hereinafter, River City)" (Spencer and McKinney 1997:187). By generalizing to "HA" in "River City," researchers tried to ensure that no one would be able to discern exactly what organization was studied, who the clients and social workers were, and in what city the research took place. These tactics maximized the privacy of those involved in the research, protecting them.

Another issue, related to protecting the privacy of those being studied, is the question of observing in public versus private spaces. The general rule in ethics is to obtain permission to observe whenever possible. But what if you want to observe interactions in public spaces? Public space is by definition open space. Can you simply go there to observe interactions without asking permission from the people you are observing? Is it ethical to talk/watch/record people if they do not know they are being observed? This is an unresolved debate within disciplines that utilize field research. It seems nearly impossible to get permission before the fact in many cases of public space observation. But we also need to remember that ethically, being upfront about what we as researchers are doing is the best practice.

Deception also plays a part in the ethical concerns of field research. In general we expect that deception should be limited. Choosing to do field research as an overt researcher is the best choice from an ethics perspective (although even when doing research openly, there is no guarantee that the researcher will not encounter a number of other ethically ambiguous areas). There are, however, cases where being a covert researcher is acceptable. When Lofland and Stark began their research into the Unification Church, they made the decision to study the NRM as covert researchers. Having no previous literature to guide them in determining approximately how long the religious conversion process could be, the researchers chose to use deception. By taking their case to the group's leader, Reverend Kim, Lofland and Stark were able to gain access to the group with minimal disruption to the process of conversion they intended to study. Whenever deception is being employed in a study, it must be approved by an institutional review board (IRB). Being approved by an IRB allows the researcher to proceed with as many safeguards in place for participants as possible.

One very unique ethical concern for field researchers is their level of **participation**. When doing field research, it is common to enter into a group to experience their activities firsthand. While this allows researchers to get the detailed data and experience that we expect from this research design, it comes with costs. How might the researcher's participation impact the overall activity or interaction among the group? Does the researcher's participation endanger the integrity of the activity? Finally, can the researcher maintain objectivity when he or she spent so much time participating with the group? Note some of the critiques of *Gang Leader for a Day*. Based in his preconceived notions regarding the gang activity in a low-income housing project, and his excitement for living so closely with such a unique subculture, Venkatesh's report of this subculture is called into question—did his participation impact his ability to remain objective? Did his participation change the daily activities within the culture (it did for those who had not reported their "hustling" income to the gang). When doing field research, social scientists are charged with being aware of their own biases, keeping to their objectivity, and protecting their participants.

Exercise: Field Research in Action

Think about some of the interactions in which you are interested. How would you define what these interactions are? Where would you go to study these interactions? Imagine spending time studying the residential culture at your campus as a field researcher. Where would you go to watch interactions at the dorm, in the fraternity/sorority houses (individual rooms, common rooms, the laundry room)? What might you find? Would you expect differences in interactions due to gender norms, class status norms, and so on?

Create a study guide to outline what customs people within different residential situations may practice (e.g., would you find different strategies of bringing alcoholic beverages into a Greek house versus a residence hall room)? Are there patterns of interaction that occur in the campus dining hall? Spend some time in the places where you are able to observe some of these interactions. Are there any surprises you see that you did not anticipate? If you could do more research about these behaviors, how would you modify what or where you observe? What do you conclude about the patterns of behavior you observed?

18

CONTENT ANALYSIS

People who major in the social sciences are drawn to observing people, but people are not always the object of our study. **Content analysis** is the study of **cultural artifacts**, or the things that humans have created, rather than people themselves. Content analysis is what we call an "unobtrusive measure," or a way of studying social behavior without directly affecting it. We've already looked at another unobtrusive measure—aggregate-level data analysis. Like aggregated data, content analysis does not directly look at individuals and thus is not "reactive" (Marshall and Rossman 1995). Content analysis can take anything that people have created, including comic strips, magazine articles, movies, books, pottery, meeting minutes, sales receipts, phone books, Web pages, fairy tales, song lyrics, and paintings, in order to examine what these artifacts can tell us about human life and interaction.

As one research design, content analysis is valuable in allowing us to examine communication systematically: what does a set of artifacts tell us about how, to whom, why, and what is communicated? For example, one early content analysis documented that the post–Civil War lynching of black men was not the result of black men raping white women, a commonly circulated myth. But why was this myth a common understanding of black men and/or of lynching? Drawing on the accounts of lynching reported in white newspapers, Ida B. Wells Barnett undertook a content analysis of

Understanding and Applying Research Design, First Edition. Martin Lee Abbott and Jennifer McKinney.
© 2013 John Wiley & Sons, Inc. Published 2013 by John Wiley & Sons, Inc.

newspaper articles to see if lynching was indeed the product of black men raping white women. Wells Barnett found that of the 728 lynchings that occurred between 1881 and 1891, in only a third of the cases were the black men even accused of rape (Tolnay and Beck 1995). The primary charge against the black men who were lynched was insolence (not keeping to their place), rather than any kind of violence (Tolnay and Beck 1995). Based on the evidence, Wells Barnett challenged the rape/lynching connection to illustrate that the association served to legitimate the mob violence that resulted in both the lynching of black men and the raping of black women during the period of Reconstruction in the South (Feimster 2009).

Findings like Wells Barnett's are the unique craft of content analysis—documenting a gap between public perception and the baseline reality. Social science excels at the systematic examination of what is happening in the world in order to compare what is happening in public perceptions (e.g., see our discussion in Chapter 1 and Chapter 13 regarding the commonsense belief that religion in America has always been in decline versus what the data illustrate about religious participation in America). Likewise, there's a general cultural understanding that with the women's liberation movement of the 1970s, women have achieved parity with men in everything from the sexual revolution, to equal pay for equal work, to equal sexual objectification in the media. While the quantitative data consistently show that parity between the sexes in these areas has not yet happened (for example, women still lag behind men in earnings), how can content analysis help us to illustrate how or why a lack of parity between women and men has been achieved?

By examining cultural artifacts, content analysis can uniquely address the widespread belief that men are being equally sexually objectified within the media; that they, like women, are now seen only as sexual objects. In 2011 an article in the journal *Sexuality & Culture* analyzed the sexualization of men and women on the cover of *Rolling Stone* magazine. Taking a census of all magazine covers from 1967 to 2009, Hatton and Trautner (2011:256) found that both women and men have been increasingly sexualized, although women "continue to be more frequently sexualized than men." How can the researchers make this conclusion? Let's walk through the process of selecting artifacts, coding the patterns that appear in the artifacts, and then generalizing to what the data tell us about the artifacts, in this case how covers of *Rolling Stone* magazine sexualizes women and men.

DEFINING THE POPULATION

Hatton and Trautner's (2011) research question dealt with the sexual objectification of men and women in the media. The media are quite broad—should their data have included all magazines, newspapers, Web pages, broadcast and cable network shows, advertisements, films, or billboards? And within each of these media outlets, should all data be included—advertisements within magazines or between segments of television and cable shows, the trivia segments attached to the preliminaries at movie showings, all articles, music played on radio or Internet music sites, number of news commentators within a broadcast, the music used to introduce stories? For some research questions, the sheer amount and variety of pertinent artifacts can become quite overwhelming. Researchers need to think carefully about their question and narrow (in some cases) the range of pertinent artifacts or materials.

For Hatton and Trautner's question of how (and how much) women and men have been sexualized over time, they selected just one magazine to use as the basis for testing their research question. Is one magazine enough? To justify using one discrete set of materials, Hatton and Trautner made a case in their article that *Rolling Stone* is a long-standing and influential magazine. As a cultural icon, *Rolling Stone* is a well-known and popular magazine, focusing on music and music culture, but also pop culture in general (including politics, film, television, and current events). The magazine has also played a key role in addressing the culture for more than four decades since its inception in 1967. Hatton and Trautner's careful articulation of why they chose one magazine illustrates how to begin testing a research question through content analysis. Choosing *Rolling Stone* magazine is a good choice for reasons they explained—*Rolling Stone* reaches a lot of people and is an influential magazine. What appears on the cover of *Rolling Stone* is inevitably seen and consumed as icons of popular culture (there's even a song about having "made it" as an artist when you've appeared "on the cover of the *Rolling Stone*").

After selecting *Rolling Stone*, the authors next needed to specify what parts of the magazine would be used in the analysis—the whole magazine including articles, advertisements, photographs, or just the articles, just the advertisements, or just the photographs. Hatton and Trautner chose to focus on the magazine covers, since it is the cover that most clearly displays the icons of popular culture. Even this decision, however, needed to be refined. Between its first issue in 1967 through the end of 2009, *Rolling Stone* magazine published 1046 covers. Hatton and Trautner verified the magazine's covers by comparing them with two histories of the magazine, as well as a third-party Web site that has catalogued all *Rolling Stone* covers. Once all magazine covers were collected, the authors decided to exclude 115 of them from their analysis. Why would they exclude covers? The covers the researchers excluded did not feature humans— neither males nor females. Since the core of the analysis entailed documenting possible changes in the sexual objectification of women and men over time, magazine covers that contain no men or women should have been eliminated from the materials. As you can see, selecting the population of pertinent materials can be quite complex. Selecting materials to address a research question should be very carefully thought out and very clearly articulated when describing them within the context of an article or book (as Hatton and Trautner did, making a compelling case for the influence, and thus inclusion, of the one magazine for their study).

CENSUS OR SAMPLE?

While Hatton and Trautner chose to take a census of all *Rolling Stone* covers, they could have chosen to take a random sample of the pertinent artifacts. Depending on the research question and the scope of the analysis (and data), taking a random sample of materials can also be useful. For Hatton and Trautner, time was an important variable. Taking a census of their materials allowed them to clearly see change over time, making a census an important choice. Utilizing a random sampling technique can also be appropriate. McKinney's master's thesis used content analysis to look at the depictions of religious groups and people in the media—again, a choice that can be quite onerous. She also chose to focus on one influential media source, the *New York Times*. Refining this choice further, she included only the first section of the newspaper, the national report. Since the *New York Times* has been in publication for more than a hundred years, McKinney chose to do a two-year systematic random selection of newspapers in order to select the articles (not advertisements or other content) for her analysis.

Again, depending on the scope of the research question, taking a census or a sample of the pertinent materials can be appropriate. As researchers craft and consider their research questions, they also need to consider what materials (artifacts) will be appropriate measures of the phenomena. Once the population of materials (the type or category of materials) has been chosen, researchers consider which and how much of these materials should be included in the analysis. The census and probability sampling techniques discussed in Chapter 8 should help guide the researcher in defining how to select the pertinent materials. In some cases it may be possible for a researcher to select all of the pertinent materials, thus taking a census of the materials (e.g., Hatton and Trautner's [2011] selected all covers of *Rolling Stone* magazine). In some cases, however, it may not be feasible to include or observe all of the pertinent materials to study your research question. When including all materials is prohibitive, choosing a probability sampling method can help you to select a random sample of the pertinent materials.

Once the decisions have been made regarding the selection of the appropriate materials, the next step is to begin getting familiar with the materials. Social scientists are interested in finding patterns across materials. Since content analysis focuses on written, verbal, or visual data, the task at hand is to transform the raw data—magazine covers, newspaper articles, and so on—into a standardized form in order to begin to categorize the patterns emerging from the data. Coding materials helps classify them into larger conceptual frameworks. These frameworks, however, are subject to interpretation, so researchers need to be clear in defining what the frameworks involve. Like in field research, having a clear research question helps researchers to go back to the general hypothesis—what is the research question, what is the main focus of the research? Evaluate the materials; do they suggest content that falls into what the research question expects the data to show? How does the researcher know what the data illustrate?

CODING IN CONTENT ANALYSIS

The key to doing content analysis lies in the coding of the content. **Coding** is the breaking down and conceptualizing of the data in order to make sense of it in new ways (Strauss and Corbin 1990). Coding materials is part of the process of analyzing the data by examining the data, comparing the data, and interpreting what the data tell us. One of the primary ways researchers interact with cultural artifacts while doing content analysis is by using grounded theory. **Grounded theory** is a way of observing patterns within data by coding patterns into larger conceptual frameworks. Grounded theory allows the data speak for itself, since researchers develop theory based in what emerges from the reality of the data—thus the theory is derived from what the data illustrate (rather than imposing a theoretical perspective on the data, the data emerge within the context of analysis and inform to create a theory of why the data look the way they do) (Strauss and Corbin 1990).

In the context of coding, researchers should be asking questions of the data—what are the data conveying, how do the data convey it, do patterns differ from each other, how do patterns differ from each other? Coding in content analysis should be flexible. As the researcher becomes more familiar with the materials, patterns will emerge. When a researcher finds what appears to be a pattern, that pattern needs to be tested against the remaining data—does the pattern hold up, is it truly a unique pattern from the other data? Coding is a constantly changing process as researchers create names for patterns, only to realize later that the pattern they think they saw does not really describe a clear pattern across the data. So in the initial stages of coding, researchers should be flexible, recognizing that patterns that seem to emerge may disappear when compared to the overall set of artifacts.

For example, McKinney's (1995) general research question for her master's thesis was "How do the media report on religious groups?" Having heard contradictory expectations from religious groups regarding media coverage (conservative religious groups saying the media was biased against religious groups while progressive religious groups deemed religion to be missing from media reports altogether), McKinney wanted to see if there were patterns regarding the reporting of religious groups; a general enough question. As she began to code her materials, the most obvious decision was to code each *New York Times* article by the religious group mentioned in each article. Interestingly, no pattern by religious group initially appeared (for example, Catholics, Methodists, and Baptists had a variety of positive and negative articles; there was no consistent pattern in the news articles by religious organization). Having coded all of the articles by religious group and finding no pattern, McKinney had to go back to the materials and look for alternative patterns.[1]

Open Coding

As Strauss and Corbin (1990) write, one option for coding is "open coding," which generally involves three steps: labeling the phenomena, discovering categories, and naming categories. When looking at the data and trying to discern patterns, you want to label the patterns you see by asking,

a. What pattern do you see happening in the data?
b. What does the pattern represent?
c. How is one pattern similar to/different from other patterns?

---

[1] Eventually McKinney found that news stories on religious individuals and groups ran on a continuum of positive to negative images (a "Saints to Sinners" continuum). Rather than specific groups being treated in particular ways, larger configurations of groups were treated to similar news stories—that is, "conservative" religious groups were more likely to be treated negatively in the news stories, whereas more mainstream religious groups were treated more positively.

> Once patterns have been determined, can they be grouped into larger categories? When categorizing patterns,
>
> a. Group concepts together
> b. Pull together other groups of data
>
> By naming categories, the data represent what is being illustrated by the data, allowing a researcher to
>
> a. Develop the patterns in the data more fully
> b. Clearly express what the data mean.

As McKinney coded her data on how religious groups were represented in the media, she found patterns in the data that indicated that religious groups were very helpful. She also found patterns in the data that produced images of religious groups that were downright dangerous. For the positive imagery she created one category of data called, "The Good Samaritan" frame (*frames* are ways of making sense of patterns that consistently employ terms and images to depict specific phenomena [Entman 1991; Goffman 1974]). In this frame the images of religious groups and individuals were depicted as highly positive. For example, they were seen as helping their communities and others in need by distributing food during natural disasters, collecting donations, leading fundraising campaigns, and championing the disadvantaged. The pattern was captured with this quote: "With a shoestring budget, a single assistant, and dozens of volunteers, [a good Samaritan] has created a job for himself as a role model, problem solver, and merchant of hope" (McKinney 1995:14).

The Good Samaritan pattern was distinctly different from the data McKinney categorized as "The Ministry of Hate" frame. Within the Ministry of Hate frame, religious individuals and groups were characterized as forcing their views on others, taking over political parties and local political boards. These groups and individuals were portrayed as being controlling, divisive, fervent, and extreme. They were further described as hate groups or fear-mongering groups. By labeling the phenomena (for example, positive or negative images) and discovering categories (grouping the positive images or negative images to see patterns), and then naming categories (Good Samaritan, Ministry of Hate), the data take on more developed and meaningful implications.

CODING *ROLLING STONE*

Hatton and Trautner analyze *Rolling Stone* covers to see if men and women have become more sexualized over time. How did they measure what is sexualized? Are there differences in sexualization? For example, are some images slightly sexualized while others are highly sexualized? Much like creating categories for variables in survey research, content analysis requires skill in thinking through how to conceptualize measures. Acknowledging that sexualization runs on a continuum, Hatton and Trautner created a scale (0 to 23 points) of 11 separate variables to measure sexualization. These measures included coding for clothing/nudity (images scored 0 points for wearing nonrevealing clothing and 5 points for being completely naked), touch, poses, text, head versus body shots, and so on.

Coding can be done by a lone researcher, whose task is then to make sure the codes are both valid and reliable. Do two pieces of your materials seem to say the same thing only in different ways? Why? What impressions are you getting from the materials? Content analysis takes a significant familiarity with the materials being analyzing to tease out the nuances of the patterns in the data.

RELIABILITY AND VALIDITY

Since researchers impose codes and categories on materials, in many ways we cannot fully know if they are reliable or not. This is the task of the researcher; to be diligent in analysis in order to fully and clearly articulate the patterns, substantiating them with the data. One check on reliability is through the type of content a researcher is focusing on—manifest and latent content. **Manifest content** is the explicit content of the materials—why are they being presented (for example, why does *Rolling Stone* magazine exist?).

There is a secondary type of content to code, the **latent content**, which is the implicit meaning of the materials. For example, Hatton and Trautner's findings from their content analysis of *Rolling Stone* magazine was that both men's and women's images were increasingly sexualized. The gender difference between women and men, however, is striking. While men's sexualization increased from 11 percent to 17 percent between 1967 and 2009, women's sexualization increased from 44 percent to 83 percent. This paints a picture of women being primarily sexual while men are only peripherally so. This latent content paints a very particular vision of women versus men. Another interesting latent message that comes through the content analysis is the absence of women (which makes their sexualization even more significant). Hatton and Trautner note that of the more than a thousand magazine covers they analyzed, there were 726 images of men and 280 images of women. That discrepancy is striking, illustrating that in a magazine that focuses on popular culture, women are significantly less visible (and when women are visible, according to these data, they are portrayed as highly sexualized).

One of the hallmarks of qualitative analysis is the rich detail used to describe the processes being studied. We tend to assume that the patterns reported are valid—that they measure what they are purported to measure. There are, however, three biases that can impact the validity of the results of a content analysis: deposit bias, survival bias, and invalid generalizations. **Deposit bias** (sometimes called selection bias) occurs when only a portion of the appropriate artifacts are included to be coded in the analysis. Say, for example, that we were interested in learning about the daily lives of sixteenth-century European peasants. What materials might we use to measure their daily lives? From sixteenth-century Europe, we might rely on accounts from landlords on the number of peasants and the types of jobs they had. We might also look to the tools that peasants used in their daily lives. Church records may give us insight into how often peasants attended church, confession, and so on. There may also be diary accounts of life during the sixteenth century that would help us craft a database of materials to look at the life of peasants. There is a definite bias to these materials. If we are interested in the daily lives of sixteenth-century European peasants, except for the tools that they used perhaps, none of these materials can speak directly for the peasants. Sixteenth-century European peasants were likely illiterate and did not leave recorded accounts of their lives. This lack of material creates the deposit bias—only a portion of the pertinent artifacts are included (e.g., the tools are included but nothing else directly related to peasant life). Anything gleaned from materials other than what comes directly from describing daily peasant life will not accurately depict peasant life, creating a deposit or selection bias.

A second bias that can impact the validity of a content analysis is survival bias. In **survival bias**, only some portion of the pertinent materials have been retained (or exist) in order to do analysis. In looking at the messages inherent in silent film, a researcher could employ a random sampling technique to create a database of silent film. In early film there was no "decency code" early on. Whereas many people would expect "old" movies to be less sexualized, without any kind of a rating code filmmakers in the early days of film could be quite racy. Content analysis is particularly adept at analyzing how communication has changed over time. In order to research how sexualized images were during the silent film period, a researcher could try to take either a census or sample of silent films. Unfortunately, the analysis of silent film would lead to a survival bias. It is estimated that less than 10 percent of all American silent films ever made still exist today. Even if a researcher was able to analyze all extant silent film, whatever the researcher gleaned would not be able to be generalized to all silent film—there are simply not enough silent films that have survived to be able to do an analysis that can be generalized to silent film.

The third bias that can impact the validity of a content analysis is the invalid generalization. **Invalid generalizations** occur when researchers (or readers) try to generalize their findings to anything other than the materials they analyzed. For example, Hatton and Trautner analyzed the covers of the *Rolling Stone*. Any generalizations they make can then only be made about the covers of the magazine (they cannot even generalize to the magazine as a whole regarding the sexualization of women; had they analyzed the articles within the magazine, they would then only be able to generalize their findings to the content of the articles). Sometimes we are tempted to generalize the findings of a content analysis to people. Since content analyses do not use people as the unit of analysis, we cannot generalize to people; we can only generalize to the materials specifically used in the analysis. Therefore, based on Hatton and Trautner's analysis, we can only conclude that women *on the covers* of *Rolling Stone* magazine are more sexualized than men *on the covers* of *Rolling Stone* magazine.

Content analysis allows researchers to utilize cultural artifacts to study what these artifacts convey about human life. Some of the advantages to doing content analysis as a research design include the relatively low costs of using materials in terms of both time and money. Content analyses do not require having a trained staff (like face-to-face interviews), or any special lab space and/or equipment (like in experiments). Because content analysis is an unobtrusive measure, there is no danger to people. The only drawbacks to doing content analysis is that the research is limited to recorded materials; if there are no existing materials, there can be no analysis.

Unobtrusive Measures

Data come in a variety of expressions. We have discussed content analysis as a way of analyzing information and providing insight into social phenomena using cultural artifacts (like magazine covers). We noted additional sources of information that could be coded and analyzed like comic strips, magazine articles, movies, books, pottery, meeting minutes, sales receipts, phone books, Web pages, fairy tales, song lyrics, and paintings.

There is another class of objects we might consider as well, those that exist in physical residues of social interaction. For example, if we are interested in the extent to which college students cram for exams, we might collect the empty energy drink containers discarded into resident hall trash bins during midterm and final exam weeks compared to the first week of the term.

Admittedly, operationalizing variables by using these residues causes questions about the adequacy of reliability and validity, but gathering information from such sources can provide fresh ways of considering social phenomena. In many ways, using such information as a reflection of social meaning is not that dissimilar from the way archeologists codify and study some behaviors and attitudes of civilizations long past.

There are some classic and recent sources to explore if you are interested in these methods. Webb et al. (1999) provide the most comprehensive analysis, but the more recent discussions of garbology (see, e.g., Rathje and Murphy 2001) are also quite informative, although more limited in scope.

Exercise: Content Analysis of Popular Magazines

Based on the content analyses we've discussed in this chapter, what you would expect to find in conducting your own content analysis on a popular magazine? Choose a popular magazine (if you do not have one handy, visit your local bookstore, grocery store, or library to find a print version of a magazine). Create a general hypothesis of the types of content you would expect to find in the magazine you've selected. Think about the magazine as a case study in content analysis. Follow the processes that we've discussed to analyze the magazine, looking at manifest and latent content, by using these guidelines:

1. Hypothesize what you expect to find in the magazine (e.g., what type of articles do you anticipate finding, what type of products do you expect to be advertised, who do you expect to see represented—what gender(s), age ranges, social classes, races/ethnicities do you expect to see?).

2. Decide what you would like to look at: ads, content (articles), or both ads and content.

3. Begin by getting a sense of how the magazine is organized, by counting the total number of pages of the magazine. Count the total number of pages for ads and articles. What percentage of the magazine are ads versus articles? What does this tell you about what the magazine is trying to accomplish (for example, is it more content or more advertisement)?

4. Choose a strategy for how to select the ads and/or content you are going to analyze. Should you take a census (i.e., all of the ads and/or content in the magazine), or should you take a random sample (what kind of random sample should you use; how will you choose an appropriate sample size for the materials)?

5. Begin to familiarize yourself with the data (your materials). Determine an analytic strategy to identify categories or themes in the materials (do you begin with a general idea of what themes you expect to find, or do you start with the materials, creating themes that emerge from them?).

6. Based on the data, what do you conclude about the content (ads and/or articles) in your magazine (that is, what are the prevalent manifest and latent themes you've found)?

PART IV
STATISTICS AND DATA MANAGEMENT

STATISTICAL PROCEDURES UNIT A: WRITING THE STATISTICAL RESEARCH SUMMARY

When you summarize the results of your research, remember that these summaries are designed to be written for a research audience. The settings might include a professional association at which you are highlighting your findings (as in a poster session), a brief blurb for a book, the findings section of a team research report, or a proposal to company representatives as part of a consulting project.

Create the summary keeping the following points in mind, and use established writing guidelines (e.g., Chicago, American Psychological Association, Modern Language Association). Include figures and tables *only* when they are necessary in explaining or describing the findings. Unnecessary visual summaries can be distracting and redundant.

The following guidelines are organized by the steps you need to take to complete the statistical analysis. After you have stated the hypotheses, address each of the following as ingredients for your decision whether to reject the null hypotheses and how the findings relate to the overall research problem.

1. State the general *research question* or theory, and then list the hypotheses and study variables.

Understanding and Applying Research Design, First Edition. Martin Lee Abbott and Jennifer McKinney.
© 2013 John Wiley & Sons, Inc. Published 2013 by John Wiley & Sons, Inc.

2. Describe the *purpose of the statistical test(s)* you will use to address whether to reject the null hypothesis, accept the alternate hypotheses, and so on. The following are two examples of using different statistical procedures:

> A one-sample *t* test was conducted on PWPVS scores of postal workers in a large suburban postal terminal to evaluate whether their mean score was significantly different from a score of 45, the assumed population mean of the PWPVS for U.S. postal workers.
>
> A multiple linear regression analysis was conducted to examine whether worker satisfaction is predicted by type of job, satisfaction with coworkers, perceived social support, and satisfaction with supervision. Results of this analysis will show the proportion of variance in worker satisfaction contributed by the combination of predictor variables, and it will measure the unique contribution of each predictor to the variance in worker satisfaction.

3. Report the *descriptive statistics* (mean, standard deviation, standard error, etc.).

4. Report and briefly discuss the *assumptions* of the test and whether your data met these assumptions. Cite the evidence for meeting the assumptions (e.g., whether Levene's test results indicate violations of equal variance, whether skewness and kurtosis figures are beyond acceptable limits and indicate different statistical tests). Each statistical procedure includes a separate set of assumptions, but most all (parametric tests) focus on whether outcome variable groups are normally distributed, group variances are equal, and interval data are used (for the outcome variable particularly, although specific tests can differ).

5. Report the results of the *omnibus* statistical test including test ratio, df, and sig. levels. The omnibus test is the overall test that indicates a statistically significant finding among all components of the data. An example of this is the ANOVA test in which an overall significance test establishes whether all paired group differences among a factor, taken together, indicate nonchance differences in the dependent variable. The following examples rel*ate to* ANOVA and multiple regression respectively:

   - The omnibus F ratio (14.55) was significant ($p = 0.02$) indicating that we can reject the null hypothesis of equal group differences on the dependent variable.
   - The model was significant (F = 99.07, $p < 0.001$) indicating an R^2 of 0.128. The resulting regression equation was $Y_{pred} = 1.17 + 0.14X_1 + 0.11X_2$

6. Report on any *post hoc or individual predictor* tests that are part of the statistical procedure. In the case of ANOVA, this includes tests like the Tukey's HSD, Scheffe, LSD, Bonferroni, and so on, in which the researcher must determine if pairwise differences among independent variable groups are significantly different. For multiple regression, the researcher should note whether each predictor variable is significant as determined by individual *t* tests. In chi-square tests, follow-up tests would include the results of a two variable analysis within each condition of a separate control variable.

7. Report and discuss the *effect size*. This is a crucial part of every statistical analysis and must be reported to indicate the impact of the test variable(s) on

the dependent variable. Each statistical test has a separate method for determining effect size, so consult a statistical text (e.g., see Abbott 2011) for the appropriate procedures.

For example, in *t* tests, you might report, "According to the criteria for Cohen's d, the effect size of 0.367 is considered small to medium. This means . . ."

For regression results, you might conclude, "Both predictors explain 27 percent of the variance in the outcome variable ($R^2 = 0.272$), and the squared part correlation for predictor one (part $r^2 = 0.15$) indicates that it uniquely explains 15 percent of the variance in the outcome variable.

For chi square, you could note that "Cramer's V (0.120) results for the first panel of the multivariate test indicates small differences between the two test variables in the first category of the control variable."

8. *Interpret* the statistical results. All of the foregoing steps apply to the *statistical* findings. The researcher must now translate these technical findings into the language of the study so that the audience can understand how they apply to their problem of interest. Many people do not understand how technical findings like "the pairwise differences among factor A indicate significant differences between low and high levels" translates into real words!

In a regression study, for example, we might conclude that predictor one is a significant predictor of the outcome variable and results in a 5 percent change in the outcome measure with every 10 percent change in the predictor (when other predictors are not allowed to affect the results). Stated simply, "When pay (predictor one) increases by 10 percent, there is a 5 percent increase in worker job satisfaction (among workers with similar levels of job tenure (predictor two)."

Elaborate briefly what conclusions you might draw regarding the research question. The greatest danger in this step is to over-conclude. That is, researchers (even seasoned researchers) make the mistake of exceeding the boundaries of what the data actually report. Here is an example of over-concluding by a student reporting significant results in a test linking a new instructional procedure in a district of 10 schools to increased math achievement in those schools: "Therefore this procedure should be promoted in all schools and teachers not using them should be retrained."

In this case, the student was stretching the findings just a bit! The problem of the ecological fallacy (see earlier sections in this book) is a particular problem in the interpretation section.

Here are some brief examples (from Abbott 2011) of interpretation sections for three statistical procedures:

ANOVA: "Taken together, the data analyses indicate that groups of FR have an impact on schools' reading achievement results. In particular, as the percentages of students qualified for FR increase in the schools, the percentage of students meeting the reading achievement assessment declines . . . [The] F test (24.22) is significant ($p < 0.05$) and the effect size substantial ($\eta^2 = 0.47$), indicating that 47 percent of the schools' reading achievement scores are affected by grouping on the FR variable (Low, Medium, and High). A Tukey HSD analysis indicated that each of the three FR groups' reading achievement percentages were significantly different from the other groups."

Chi square: "As you can see from the figure, when the respondents have been victims of a crime, they are much more likely to have a high fear of crime. Over 54 percent of victims have a fear of crime compared to 29.4 percent of those not victimized. This difference in fear persists even when crime is on the decline as it was in our study."

Multiple regression: "The findings indicate that the two predictors, eskills and eaccess, are significant predictors of eimpact. Both predictors explain about 27 percent of the variance (i.e., $R^2 = 0.275$) in eimpact, a finding considered large. . . . Each predictor is a significant predictor . . . but eskills is the more powerful of the two in explaining the variance of eimpact. According to the sample data, it is important for elementary teachers to have access to technology for making an impact on the classroom, but is it much more important that they perceive that they have technology skills. Having technology skills explains almost ten times more variance in eimpact than having access to technology."

STATISTICAL PROCEDURES UNIT B: THE NATURE OF INFERENTIAL STATISTICS[1]

Descriptive statistics examine the ways a set of scores or observations can be characterized. Is the set of scores normally distributed? What is the most characteristic indicator of central tendency? Are there extreme scores? What information about the nature of the scores can be obtained from examining histograms and other graphs?

Thus far, we have examined raw score distributions made up of individual raw score values. For example, a set of 40 schools each had percentages of students qualified for FR, or percentages of students passing math and reading achievement scores. We tried to understand the nature of these variables; whether they were normally distributed and how the percentages were distributed compared to the standard normal distribution.

In the real world of research and statistics, practitioners almost always deal with sample values, since they very rarely have access to population information. Thus, for example, we might want to understand whether the job satisfaction ratings of a sample of 100 software engineers is characteristic of *all* software engineers, not just our 100. Since it is practically impossible to get job satisfaction ratings for all software engineers, we must measure the extent to which sample values approximate or estimate the overall population values. We are using the picture at hand (of our sample group) to

---

[1] Parts of this section are adapted from M. L. Abbott, *Understanding Educational Statistics Using Microsoft Excel® and SPSS®* (Wiley, 2011), by permission of the publisher.

Understanding and Applying Research Design, First Edition. Martin Lee Abbott and Jennifer McKinney.
© 2013 John Wiley & Sons, Inc. Published 2013 by John Wiley & Sons, Inc.

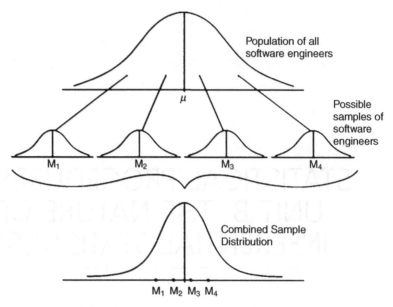

Figure SPUB.1. Creating a sampling distribution.

get a better picture of the overall (population) group that we have no way of picturing.

Inferential statistics are methods to help us make decisions about how real-world data indicate whether dynamics at the sample level are likely to be related to dynamics at the population level. That is to say, we need to start thinking about our *sample distribution* of software engineers as being one possible sample group of 100 taken from the overall population compared to a number of other such samples (of 100) that could be taken. We can then study our sample and see how the statistical information might compare to the average set of information derived from all the other possible sample groups. Is our sample mean, for example, likely to be close to the average mean of all the other sample groups?

Take a look at Figure SPUB.1. This figure shows that if you were to take repeated samples from the population of software engineers (we have only shown four samples, but in theory, you would take many samples), you could take just the mean values from each of the samples and make up a separate "combined" distribution of samples. If we were to then take our study sample of 100 software engineers, we could see how our study mean value compares to all the other sample means. If the study mean was close to the mean of the set of sample means, we could say that our sample mean is representative of the population mean (Figure SPUB.1).

What we just described is actually a basic principle of inferential statistics known as the ***central limit theorem***. Statisticians who have gone before us discovered several things about taking repeated samples from distributions. In the central limit theorem, they found that means of repeated samples taken from a population will form a standard normal distribution (assuming a large sample size) even if the population was not normally distributed. The combined sampling distribution that results will have important properties to researchers conducting inferential studies.

The combined sampling distribution would have the following properties:

- Most of the sample means would lay close to the overall population mean with some spreading out into the tails. Thus the sample means would be normally distributed.
- The mean of the combined sampling distribution of means would be equal to the population mean. That is, if you added up all the sample means and divided by the number of sample means, the resulting average of sample means would equal the population mean.
- The standard deviation of the sampling distribution will be smaller than the standard deviation of the population, since we are only using the individual mean scores from each sample distribution to represent the entire set of sample scores. Using only the mean of each sample group results in lopping off most of the variability of individual scores around their contributing group means with the result that the sampling distribution will have a smaller spread.

As you can see in Figure SPUB.1, the four hypothetical samples are taken from the population, and their individual means make up a separate distribution (the combined sampling distribution). You can see that the individual means, which represent their sample distributions (M_1, M_2, M_3, M_4), lay close to the population mean in the new distribution. Figure SPUB.1 shows this process using only four samples for illustration. In real-life studies, the researcher does not have to take repeated samples from the population. Because of the long history we have of statistical measurement, we can accept the fact that the theorem is valid and we can use it in our study.

Here is how we could use the central limit theorem in an actual study.

1. Take a sample of software engineers (using appropriate sampling procedures).
2. Measure the characteristics (job satisfaction) of the sample distribution (mean, standard deviation, etc.).
3. Use specialized statistical procedures that will allow us to compare the sample mean to the (unmeasured, unknowable) mean and standard deviation (job satisfaction) of the population and make a decision as to whether our sample is likely to be related to that population.

Here are some possible results:

1. The sample job satisfaction mean is judged to be close to the mean of all the sample job satisfaction means, and therefore we conclude that our sample of software engineers is no more satisfied or dissatisfied than any of the other software engineers in the population.
2. The sample mean is judged to be far away from the mean of all the sample means, opening up some possibilities for us to consider:

a. We might consider that our software engineers are more satisfied (or less satisfied) than all the other possible samples because they underwent some training to make them more content with their jobs. If this is the case, we would conclude that the training changed the software engineers to such an extent that they no longer are similar to all possible software engineers.

b. Alternatively, we might conclude that the study software engineer sample was different from the population because it was a weird sample, or we just happened to have them take the job satisfaction inventory on a day when the company announced a series of potential future layoffs.

If you have taken a class in statistics, you will recognize the previous example as a (single sample) t test. In terms of our possible outcomes, The first outcome would indicate that a job satisfaction contentment training program was not effective. It was therefore *not* judged to be statistically significant. In this case, that means the sample mean was not far enough away from the population mean to conclude they were likely not the same sorts of software engineers.

The second outcome (for either reason) would indicate a **statistically significant** difference between the sample mean and the population mean if the sample mean value was far enough away from the population mean. In this case, the sample mean would be considered an unlikely value to have just happened by chance (statisticians usually consider values beyond the middle 95 percent of the standard normal curve to be significant, thus making a conclusion like "significant beyond the 0.05 level").

If we concluded that the sample was significantly different from the population mean because it reflected some policy or training that we did, we would have confidence in the power of our training. If we concluded that the sample was different because it was a weird sample, then we might say that this finding was really in error (statisticians would call this an *alpha error* or assuming a finding was significant when it really shouldn't have been).

In summary, inferential statistics is a process of comparing a sample value (or set of sample values) to an unknown population value (or values) based on time-tested statistical principles. We have therefore seen that a researcher's sample values can be used to make statistical decisions (i.e., statistically significant) or conclusions about the likelihood of differences being the result of some action on the part of the researcher (policy or training) or on the basis of error.

We just used the example of a single sample t test. Since more complex research designs use more than one sample group (i.e., experimental group versus control group in an experiment), there are other statistical tests that use different formulas to make the same kinds of conclusions. Each statistical test (if they are parametric) compares a sample value or values to population values using a sampling distribution.

For example, if we used two sample groups of software engineers (giving one the training and the other not), we could see if the difference between our sample groups was similar to the difference between all possible sets of sample group differences). This would be an independent samples t test. If we had used three such samples (giving one group no training, one group slight training, and one group intensive training), we would use an ANOVA (analysis of variance) to decide whether there we significant differences between the sample group means (relative to the sampling distribution of variances).

In all these cases, we are comparing sample values to population values, not examining individual scores within a sample. We no longer think of our sample values individually, but as *one set that could be derived from a population along with many more such sets*; our sample set of values are now seen as simply one possible set of values alongside many other possible sample sets. That is the difference between inferential and descriptive statistics. We therefore change the nature of our research question:

1. Are our sample values normally distributed? (Descriptive statistics)
2. Do our sample values likely reflect the known (or unknown) population values from which our sample supposedly came? (Inferential statistics)

PROBABILITY

Human actions are rarely, if ever, determined, but they are fairly predictable. One has only to consider the many ways in which the things we (think) we choose to do are really those things that are expected. Marketing specialists have made billions of dollars on this principle by targeting baby boomers. Sociologically, when people repeat actions in society, they create patterns (or ruts) that persist. Think of the many ritualized behaviors you enact every day. (For example, one of us has a very rigid schedule in the morning that starts with making coffee; changing this would be catastrophic!) The result is that human behavior can be characterized by predictability.

When we speak of predictable actions, we note that when actions are repeated, they form "ruts" which come to typify behavior. Thus, presented with a similar situation, the individual most likely follows the behavior that creates the rut. For example, when someone arises in the morning, they may typically make coffee and read the newspaper. They are equally free to do other behaviors (weed the garden, drive their car onto a lawn, etc.), but they will most likely make coffee. Over time, they have repeated this behavior, and it becomes almost second nature. Strictly speaking, **Probability** is the field of mathematics that studies the likelihood of certain events happening out of the total number of possible events. Thus, what are the chances that a poker player will draw a five of hearts if they need it to complete their straight flush? All the available cards have an equal chance of being chosen (unless the dealer is a crook), so how likely is the player to get their card? Stated in *formula language*, **empirical probability** is simply the number of occurrences of a specific event divided by the total number of possible occurrences. So, whether a player gets the five of hearts presents a probability of one card divided by the remaining cards in the deck of cards. This is different from the "predictable" language of probability we noted earlier. Formally, probability deals with *independent* possibilities (the likelihood of a five of hearts out of the remaining cards), whereas using predictable choices represents dependent possibilities since each action has been affected by previous actions.

We have already seen that behaviors, attitudes, and beliefs have a great deal of variability. Why is there such variability? Why do people not always believe the same thing and act the same way? In descriptive statistics, we learn the ways to understand the extent of the variability and whether the resultant distribution of behaviors and beliefs conforms to a normal distribution. But that does not explain the *why*.

Human actions and beliefs have many causes. We cannot understand all of them. The fact that variance exists may be due to our inability to understand the full range of forces acting on the individual at any particular moment. But it may also exist because we cannot fully explicate individual choice or action.

Probability involves the realm of expectation. By observing and measuring actions and behaviors over time, we develop expectations that can help us better predict outcomes. If we observe that a group of workers produce a certain sick time ratio (for example 15 percent of total days worked), we might predict the same outcome on future occasions if there are no changes in working conditions or company policy. Our expectation, being based in observation, will help us predict more accurately. This still does not explain *why* the workers take the same amount of sick time, but it does point out an area for investigation. We may never discover all the reasons for sick time rates, but the study of probability gets us closer to a more comprehensive understanding.

PROBABILITY, THE NORMAL CURVE, AND *P* VALUES

If you think about the normal curve, you will realize that human actions can take a number of different courses. Most responses tend to be clustered together, but there will be some responses that fall in different directions away from the main cluster. Therefore, we can think of the normal curve as a visual representation of the fact that we do not have certainty, but probability in matters of such things as attitudes and buying behavior, test scores, and aptitudes.

In inferential statistics, statisticians think of the normal curve in terms of probability. Since approximately 68 percent of the area (or a proportion of 0.68) of the normal curve lies between one standard deviation, positive or negative, for example, we can think of any given case having a 0.68 probability of falling between one standard deviation (SD) on either side of the mean.

Knowing what we do of the distribution of area in the standard normal curve, we can observe that possible scores beyond 2 SDs (in either a positive or negative direction) are in areas of the distribution where there are very few cases. Thus randomly selecting a score from a distribution in these small areas would have a much smaller probability than randomly selecting a score nearer the mean of the distribution. In a normally distributed variable, only 5 percent of cases or so will fall outside the $\pm$ 2 SD area. Therefore, with 100 cases, selecting a case randomly from this area would represent a probability of 0.05 (since $5/100 = 0.05$).

Returning to our example from the earlier section, we can see how the job satisfaction of our software engineer sample might be considered either likely or not likely depending on how the sample value compared to the sampling distribution mean. Consider Figure SPUB.2 that shows one possibility (the significantly different possibility).

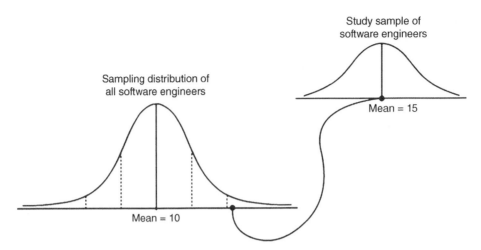

Figure SPUB.2. The significance of the study group.

As you can see in the figure, the combined sampling distribution shows a job satisfaction mean of 10 (out of 20 total), and you can see where the SDs fall by the dotted lines. The figure shows a hypothetical result of our study group obtaining a job satisfaction average of 15. When we use the appropriate statistical procedure (*t* test) to translate the sample mean values into the combined sampling distribution, we see that it falls beyond the cutoff for +2 SDs.

If this was an actual finding, we would conclude that the job satisfaction level of the sample was statistically significantly higher than the population, since the study sample mean fell that far above the population mean. Since 2 SDs (technically ± 1.96 SDs) on a standard normal comparison curve excludes about 5 percent of the area of the curve (and therefore results in a probability of happening by chance of less than 5 percent), the researcher would call this statistically significant. This 5 percent exclusion area is typically the standard by which statistical findings are considered significant. The researcher might therefore conclude that the significant finding had a probability level less than 5 percent, or $p < 0.05$.

The p values that are reported in statistical studies therefore correspond to the probability of obtaining a given finding by chance. Each statistical procedure has nuances for how this is expressed, but essentially it is shorthand for how statisticians decide whether or not a given finding is excessive, relative to the sampling distribution information.

POPULATIONS (PARAMETERS) AND SAMPLES (STATISTICS)

Parameters *refer to measures of entire populations and population distributions.* This is distinguished from *statistics, which refer to measures of sample data taken from populations.* We need to distinguish these measures, since inferential statistics is quite specific about the measures available for analysis. As you progress in using research designs and appropriate statistical applications, you will find that statisticians and researchers distinguish population values from sample values in their symbols; population parameters are typically represented by Greek symbols.

THE HYPOTHESIS TEST

Hypotheses tests are the formal logical processes established to make a scientific decision. The decisions that result from a hypothesis test are typically couched in terms of the probability levels of the outcome and the extent to which you can say whether or not a finding is statistically significant. Ultimately, the process is used to support or refute a hypothesis so that a researcher can make a statistical decision about a specific research problem and possibly inform a theory.

There is a generally agreed upon set of steps that form the general procedure for a hypothesis test (with some variations for each statistical procedure). Here are the steps with our (contrived) results applied so you can see how it works:

1. The *null hypothesis*: Researchers begin by considering a statement that can be measured and then verified or refuted. They begin with the assumption that there will be no difference between the mean of the study sample and the mean of the population. The object of the research process is to see if this is an accurate assumption or if our sample—population difference violates that assumption by being either too large or too small. The null hypothesis in our example is that the job satisfaction level of the study sample of software engineers is not different from the job satisfaction level of all software engineers in the population. This is known as the null hypothesis because the statement posits no difference: zero, null, nada.

2. The *alternative (or research) hypothesis*: This statement is created in order to present the finding that would refute the null hypothesis; thus the *alternative* finding. In our study, we proposed that a sample of software engineers might show higher (or possibly lower!) job satisfaction than the population if we would train them to be happier. Technically, our alternative hypothesis allows findings to be *not equal to* and thus either higher or lower than the population values.

3. *The critical value of exclusion*: The 5 percent exclusion area. Recall that we need to have a benchmark to help us decide whether our actual calculated results are considered typical or atypical. Actually, we are using this benchmark to help us decide which hypothesis (null or alternative) is more accurate. As we discussed before, for this particular situation researchers use a 5 percent benchmark. That is, if the calculated/transformed study sample mean of job satisfaction falls into the 5 percent exclusion area of a standard normal distribution (represented by the values in the sampling distribution tails), then it would be considered atypical. In probability terms, this would represent a probability of

occurrence of ($p < 0.05$). Stated differently, the distance of our sample mean from the population mean would be considered not likely to occur just by chance; some reason other than chance (our training program) would create a finding this extreme.

4. *The calculated value*: This value is the calculated result of the appropriate statistical test. In our example, it would be the calculated single sample T ratio. If we had used two groups instead of one (men and women, for example), this would be the calculated independent samples *t* test, and so on with other statistical procedures

5. *Statistical decision: Possibly reject null hypothesis*: This step asks us to compare what we calculated (step 4) to the benchmark (step 3) in order to see which hypothesis (null or alternative) is more likely. In our study, the job satisfaction ratings of our sample group of software engineers was much higher than the population (i.e., the sample mean was in the 5 percent exclusion region of probability). Thus we would reject the null hypothesis. That is, we would conclude that the difference between the study sample and the population was too big to be the result of chance. The training may be responsible for moving the sample group mean far away from the population mean.

STATISTICAL SIGNIFICANCE

In probability terms, any finding of $p < 0.05$ is considered statistically significant. Researchers and statisticians have a specific definition for statistical significance: it refers to the likelihood that a finding we observe in a sample is too far away from the population parameter (in this case the population mean) *by chance alone* to belong to the same population.

PRACTICAL SIGNIFICANCE: EFFECT SIZE

Recall our discussion of effect size in former sections. Researchers and statisticians have relied extensively on statistical significance in the past to help make statistical decisions. You can see how this language (i.e., using p values) permeates much of the research literature; it is even widespread among practitioners and those not familiar with statistical procedures.

The emphasis in statistics and research now is on **practical significance**, or *effect size*, which refers to the impact of a finding, regardless of its statistical p value. The two issues are related to be sure. However, effect size addresses the issue of the extent to which a difference or treatment in a test of difference results in extreme test values. That is, how much impact does a research variable have to achieve in order to result in a change in a sample value?

Consider our hypothetical finding. We might reject the null hypothesis and conclude a statistically significant result. But this finding only has to do with the probability of whether the finding is a chance finding. *The effect size consideration is a completely different issue*. It does not concern itself with probability, but rather how far away from the population mean has our sample mean ended up from the population mean? In practical terms, this would be like saying, "The *p* values are fine, but I am astounded to find out that my software engineers are 40 percent more satisfied [or whatever, I just made up the number] than the typical software engineer. My training program must really be working well (or something else I do not see might be causing the job satisfaction to be higher)."

DATA MANAGEMENT UNIT A: USE AND FUNCTIONS OF SPSS[1]

MANAGEMENT FUNCTIONS

In this section, we cover the essential functions that will allow you to get started right away with your analyses. Before a research procedure is created, it is important to understand how to manage the data file.

Reading and Importing Data

Data can be entered directly into the spreadsheet or it can be read by the SPSS program from different file formats. The most common format for data to be imported to SPSS is through such data programs as Microsoft Excel or simply an ASCII file where data are entered and separated by tabs. Some large data files, like General Social Survey (GSS), have already been converted into SPSS file format and are available simply by downloading the data from the Web site.

Using the drop-down menu command "File-Open-Data" (initiated in the upper left of the main menu ribbon in either Data View or Variable View) creates a screen that enables the user to specify the type of data to be imported (e.g., Excel, Text, etc.). The

---

[1] Some of the material in this section is adapted from Abbott (2011) with permission of the publisher.

Understanding and Applying Research Design, First Edition. Martin Lee Abbott and Jennifer McKinney.
© 2013 John Wiley & Sons, Inc. Published 2013 by John Wiley & Sons, Inc.

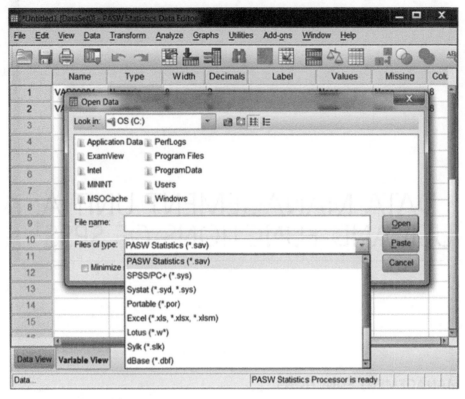

Figure DMUA.1. The SPSS screens showing import data choices.

user is then guided through an import wizard that will translate the data to the SPSS spreadsheet format.

Figure DMUA.1 shows the screens that allow you to select among a number of "Files of Type" when you want to import data to SPSS. These menus resulted from choosing "File" in the main menu and then "Open Data." The small drop-down menu allows you to choose a number of different file types to import data. As you can see in Figure DMUA.1, there are many common types including SPSS files, Excel files, and a number of others including text files (that you can see if you use the navigation bar within the small Files of Type menu).

Sort

It is often quite important to view a variable organized by size or other consideration. You can run a statistical procedure, but it is a good idea to check the position of the data in the database to make sure the data are treated as you would expect. In order to create this organization, you can "sort" the data entries of a variable in SPSS.

Figure DMUA.2 shows part of the 2010 GSS database—a sample ($N = 10$) of cases and two variables: health of respondent ("health") and sex of respondent ("sex"). As

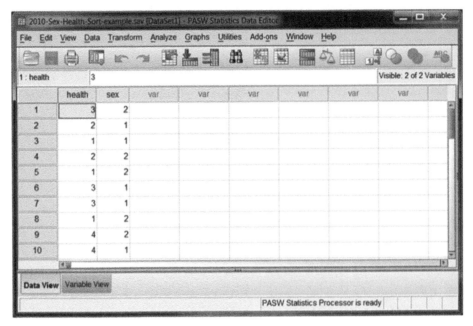

Figure DMUA.2. The SPSS screen showing the unsorted "Sort-example" data.

you can see, these two variables consist of a series of numbers that represents the coding of the data. For sex, the codes are Male = 1 and Female =2.

Figure DMUA.3 shows the Variable View of the spreadsheet that allows you to examine the codes for health. If you click on the cell for "health" under its Values column, you will see a separate menu box that lists the values of each of the numbers in the database. This box shows that 1 = Excellent, 2 = Good, and so forth.

Sorting the data in SPSS is straightforward. The user can simply select "Data" from the main menu bar, and then "Sort." This results in the screens shown in Figures DMUA.4a and DMUA.4b in which we have selected sex as the sorting variable. When you select "Data–Sort Cases" as shown in Figure DMUA.4a, you will then be shown the separate menu box shown in Figure DMUA.4b in which you can select "sex" by simply highlighting it and moving it to the Sort by: window using the arrow key.

If we choose sex, as shown in Figure DMUA.4b, we can specify a sort that is either "Ascending" (alphabetical order beginning with "A" if the variable is a string variable or starting with the lowest value if it is a numerical value) or "Descending." Selecting "sex–Ascending" results in the screen shown in Figure DMUA.5. There are many reasons to perform a sort, among which is that you can inspect the values of a related study variable on values of the sorted variable. Thus, for example, you can see in Figure DMUA.5 that among this example set of cases, males' (sex = 1) health is generally a bit poorer than females (sex = 2). The data show that the males' health values appear to be a bit higher (and therefore poorer) than those of the females.

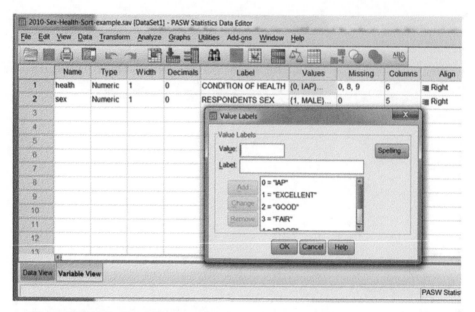

Figure DMUA.3. The SPSS screen showing the Value Labels for the health variable.

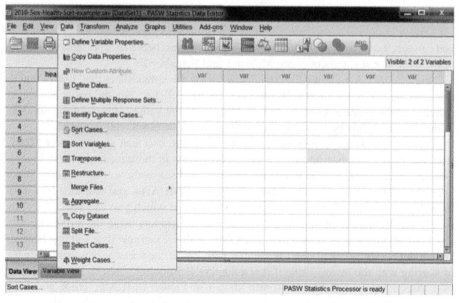

Figure DMUA.4a. SPSS data screen showing the "Sort Cases" function.

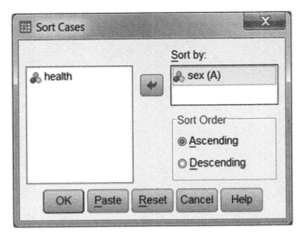

Figure DMUA.4b. SPSS data screen showing "Sort Cases" specifying sex as the sorting variable.

Figure DMUA.5. The SPSS screen showing variables sorted by sex.

SPSS allows multiple sorts. Figure DMUA.6 shows the specification window for a multiple sort first by sex and then by health. Users can generate this result simply by listing multiple variables in the "Sort Cases–Sort by" window. Note that the each of the variables can independently be sorted in ascending fashion (signified by an "A" next to the variable name) or descending (signified by a "D"). Figure DMUA.6 also indicates the nature of the variable (string or numeric) by the small symbols next to

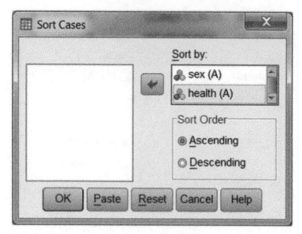

Figure DMUA.6. The SPSS Sort Cases window showing a sort by multiple variables.

Figure DMUA.7. The SPSS screen showing the database results of sorting by multiple variables.

each variable. Sorting (in either ascending or descending order) will therefore appropriately arrange the variables according to their type.

Figure DMUA.7 shows the result of this multiple sort. As you can see, the health values are nested within each category of sex. This makes visual inspection of the data somewhat easier. Compare this screen with the one shown in Figure DMUA.5. Sorting by multiple variables allows you to see the patterns a bit more clearly.

ADDITIONAL MANAGEMENT FUNCTIONS

SPSS is very versatile with handling large data sets. There are several useful functions that perform specific operations to make the analyses and subsequent interpretation of data easier. We do not cover all of these, but the following sections highlight some important operations.

Split File

A useful command for students and researchers is "Split File," which allows the user to arrange output specifically for the different values of a variable. Using our sort example from Figure DMUA.5, we could use the "Split File" command to create two separate files according to the sex variable and then call for separate statistical analyses on each of the related sets of cases of the other study variable (in this case, health).

By choosing the "Data" dropdown menu, we can select "Split File" from a range of choices that enable us to perform operations on our existing data. Figure DMUA.8 shows the submenu for "Data" with "Split File" near the bottom.

When we choose "Split File," we can then select which variable to use to create the separate data files. This is the Organize output by groups button shown in Figure DMUA.9. As you can see, if you choose this button, you can specify "sex" by clicking on it in the left column and moving it to the "Groups Based on:" box by clicking the arrow button.

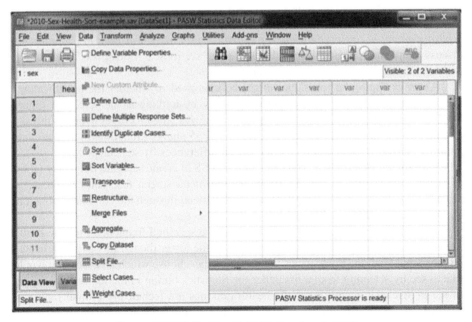

Figure DMUA.8. The "Split File" option in SPSS.

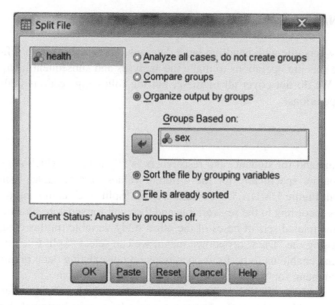

Figure DMUA.9. Steps for creating separate output using "Split File."

As Figure DMUA.9 shows, we selected the option "Organize output by groups" and then clicked on the variable "sex" in the database. By these choices, we are issuing the command to create two separate analyses for whatever statistical procedure we call for next, since there are two values for the sex variable ("1" and "2"). *When we perform a split file procedure in SPSS, it does not change the database; rather, it simply creates separate output according to whatever statistical procedure you want to examine.* We discuss many such statistical procedures in other sections. For now, it is important to understand that SPSS has this useful function.

As an example, the researcher might be interested in whether the health ratings differ by sex. We inspected the data in this way by looking at the screen in Figure DMUA.7. However, researchers cannot trust their eyes when it comes to analyzing and finding patterns in data. We can use some of the SPSS analysis functions to help provide a more objective way of examining the health differences by sex category.

With the file split by sex, you can call for SPSS to create separate frequency tables of health ratings. Figure DMUA.10 shows the menu screen that you can create to provide these tables. You obtain the Frequencies menu through the command ribbon in the original spreadsheet window.

As you can see in Figure DMUA.10, we have specified frequencies for the health variable. Recall that SPSS has already split the file, even though this is not indicated on the menu. With the Frequencies window, you can choose which analyses you would like to see by selecting from a list under the Statistics button in the upper-right corner of the Frequencies menu. Since the health rankings are technically ordinal rankings, we will simply use the default analysis which is a frequency table (note the checked box, "Display frequency tables," in the lower left of the box.

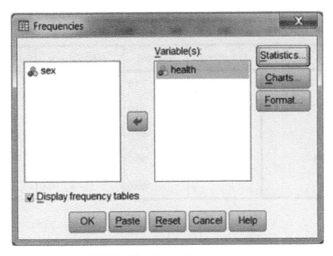

Figure DMUA.10. The SPSS Frequencies menu.

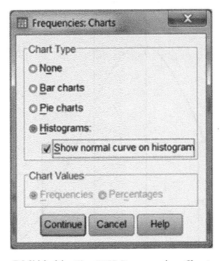

Figure DMUA.11. The SPSS Frequencies: Charts menu.

If you want a graphic display as well as a table of percentages, we can select the Charts button just below the Statistics button. Figure DMUA.11 shows the menu that appears when you choose the Charts button. Note that we have called for a histogram that includes a superimposed normal curve line.

When we make these selections, SPSS returns separate frequency analyses for males and females, since we have split the file by sex. As you can see in Figure DMUA.12a (the first part of the output, for Male respondents), most (three of five) of the cases show Fair or Poor health ratings. The frequency table (first panel of Figure DMUA.12a) is the numerical equivalent of the histogram. As you can see, two of the respondents indicated "Fair" health.

RESPONDENTS SEX = MALE
CONDITION OF HEALTH

		Frequency	Percent	Valid Percent	Cumulative Percent
Valid	EXCELLENT	1	20.0	20.0	20.0
	GOOD	1	20.0	20.0	40.0
	FAIR	2	40.0	40.0	80.0
	POOR	1	20.0	20.0	100.0
	Total	5	100.0	100.0	

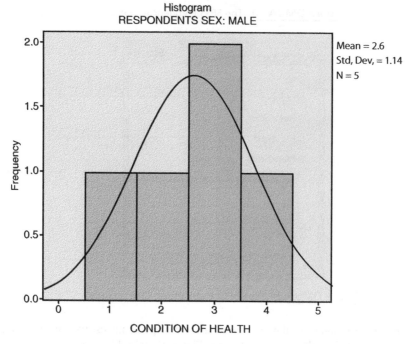

Figure DMUA.12a. The SPSS split file results for health ratings (Male).

Figure DMUA.12b shows the second part of the split file frequency results. As you can see from the histogram, the cases tend to indicate better health ratings than the male findings. Both histogram and frequency table show more cases among the lower (i.e., better) health ratings. Two of the cases reported excellent health.

Please note that when you use this procedure, it is necessary to "reverse" the Split File steps you used after you have created the desired output. Otherwise, you will

RESPONDENTS SEX = FEMALE
CONDITION OF HEALTH

		Frequency	Percent	Valid Percent	Cumulative Percent
Valid	EXCELLENT	2	40.0	40.0	40.0
	GOOD	1	20.0	20.0	60.0
	FAIR	1	20.0	20.0	80.0
	POOR	1	20.0	20.0	100.0
	Total	5	100.0	100.0	

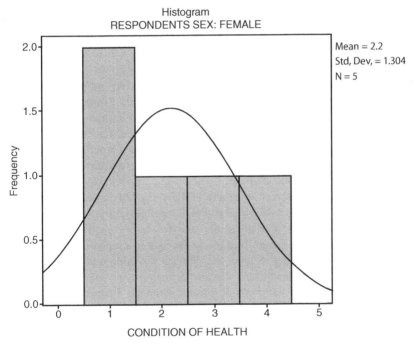

Histogram
RESPONDENTS SEX: FEMALE

Mean = 2.2
Std, Dev, = 1.304
N = 5

Figure DMUA.12b. The SPSS split file results for health ratings (Female).

continue to get "split results" with every subsequent statistical analysis you specify. SPSS will continue to provide split file analyses until you "turn it off" by selecting the first option, "Analyze all cases, do not create groups" at the top of the option list in the Split File submenu. You can see this option near the top of the submenu in Figure DMUA.9.

Compute (Creating Indices)

One of the more useful SPSS management operations is the Compute function, which allows the user to create new variables from existing variables. For this example, we use two variables from the 2010 GSS database that measure socioeconomic status (ses) to compute a new variable.

Socioeconomic status (ses) is a central concept in social science. It is a broad measure of one's social standing in that it measures education, income, and occupational prestige. Researchers have devised a way to combine these measures to yield one index that serves as a general indicator of social standing.

The GSS, as we noted earlier, provides a ses measure for all respondents. The sei variable is an index measure of the respondent's ses, and pasei is the respondent's father's ses rating. By using SPSS to divide the respondent's ses index by the father's ses index, we can create a new variable (generational ses mobility or "gensei") that indicates the extent of social mobility that has occurred from father to respondent. If the resulting number is greater than 1.0, then the respondent has eclipsed the ses of the father.[2]

We can compute gensei using the main menus in SPSS. At the main menu, select the "Transform" and then "Compute Variable" option. This will result in a dialogue box like the one shown in Figure DMUA.13. In this example, we can create a new variable ("gensei") by dividing the current "sei" variable by the "pasei" variable. The first step is to name the new variable by entering it into the "Target Variable:" the window at the upper left of the screen. Then, you can create a formula in the "Numeric Expression:" window. As Figure DMUA.13 shows, we clicked on "sei" from the list of variables and placed it in the window by clicking on the arrow button. Then, we entered a "/" mark using the keypad below the window. Last, we placed "pasei" in the window to complete the formula: sei/pasei.

As you can see from the screen in Figure DMUA.13, you can use the keypad in the center of the dialogue box for entering arithmetic operators, or you can simply type in the information in the "Numeric Expression:" window at the top. You will also note that there are several "Function group:" options at the right in a separate window. These are operations grouped according to type. Scrolling to the bottom allows the user to specify "statistical functions" like means, standard deviations, and so on. You can select whichever operation you need and enter it into the Numeric Expression window by clicking on the up arrow next to the Function Group window.

ANALYSIS FUNCTIONS

Over the course of our study in this book, we will have practice at conducting statistical procedures with SPSS. All of these are accessible through the opening "Analyze" drop-down menu as shown in Figure DMUA.14. The screen in Figure DMUA.14 shows the

---

[2] In this example, we are not considering the age, gender, or sex difference of respondent vis-à-vis the respondent's father. The example illustrates only the process of computing a new index from existing variables.

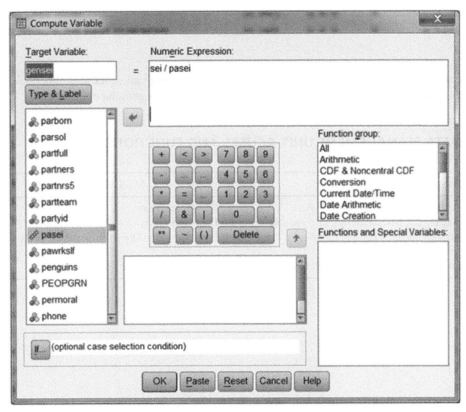

Figure DMUA.13. SPSS screen showing the "Compute" function.

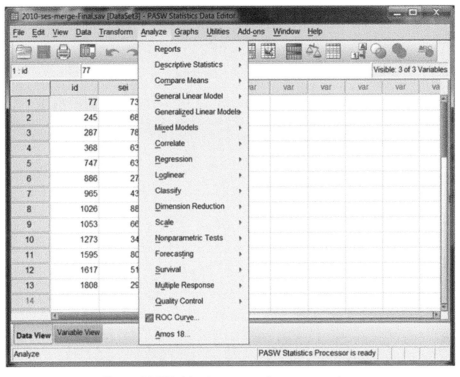

Figure DMUA.14. The SPSS Analyze menu options.

contents of the Analyze menu. We will not be able to cover all of these in this book, but you will have the opportunity to explore several of the submenu choices.

DATA MANAGEMENT UNIT A: USES AND FUNCTIONS

Use and Function	SPSS Menus
Read and Import data	File–Open Data
Sort the database according to a variable	Data–Sort Cases
Arrange output according to different categories of a variable	Data–Split File–Organize output by Groups
Compute a new variable or index	Transform–Compute Variable

DATA MANAGEMENT UNIT B: USING SPSS TO RECODE FOR T TEST

USING SPSS TO RECODE QUESTIONNAIRE ITEMS

In order to use these as variables in the example, we will make certain assumptions and methods to prepare the data for the t test. We use SPSS to "Transform" the data from both variables. The SPSS frequency reports for these variables are shown in Figures DMUB.1 and DMUB.2.

As you can see in both Figure DMUB.1 and DMUB.2, we placed a solid line just under the first "Total" row. This separates the data we will use in the example from other data categories. Under "Missing" are categories like "IAP" (Inapplicable) that refers to whether the data are used in different waves of the survey, "Don't Know," and "No Answer." We are excluding these categories in our example of how to use the independent t test.

Recoding "health1"

Using SPSS, we can recode the values of a variable by choosing the "Transform-Recode" options. There are actually two recode options: "Recode into Same Variables"

Understanding and Applying Research Design, First Edition. Martin Lee Abbott and Jennifer McKinney.
© 2013 John Wiley & Sons, Inc. Published 2013 by John Wiley & Sons, Inc.

RS HEALTH IN GENERAL

		Frequency	Percent	Valid Percent	Cumulative Percent
Valid	1 Excellent	282	13.8	24.3	24.3
	2 Very good	355	17.4	30.6	54.8
	3 Good	345	16.9	29.7	84.5
	4 Fair	162	7.9	13.9	98.5
	5 Poor	18	.9	1.5	100.0
	Total	1162	56.8	100.0	
Missing	0 IAP	857	41.9		
	8 DONT KNOW	1	.0		
	9 NO ANSWER	24	1.2		
	Total	882	43.2		
Total		2044	100.0		

Figure DMUB.1. The frequency report for "health1."

HOW FAIR IS WHAT R EARN ON THE JOB

		Frequency	Percent	Valid Percent	Cumulative Percent
Valid	1 Much less than you deserve	165	8.1	14.6	14.6
	2 Somewhat less than you deserve	350	17.1	30.9	45.5
	3 About as much as you deserve	534	26.1	47.2	92.7
	4 Somewhat more than you deserve	62	3.0	5.5	98.2
	5 Much more than you deserve	20	1.0	1.8	100.0
	Total	1131	55.3	100.0	
Missing	0 IAP	857	41.9		
	8 DONT KNOW	25	1.2		
	9 NO ANSWER	31	1.5		
	Total	913	44.7		
Total		2044	100.0		

Figure DMUB.2. The frequency report for "fairearn."

and "Recode into Different Variables." We use the latter, since that allows us to create a "new" variable from the original variable and in so doing leave the original data intact. Figure DMUB.3 shows the SPSS screen that results when you choose this type of recode.

You can see in Figure DMUB.1 that the values for health1 are from "Excellent" through "Poor" where each category is given a numerical referent (1 through 5 corresponding to each category). In this example, we are using perceived health as the

Figure DMUB.3. Using the SPSS "Recode" feature for "health1."

outcome measure and therefore are considering these numbers (1 to 5) as an interval measure.

If we select the "Recode into Different Variables" choice in Figure DMUB.3, a screen appears that allows you to select which variable to recode (Figure DMUB.4). As you can see, we have chosen "health1" from the list on the left side of the screen and placed it in the middle window using the arrow key. We also entered a name for our new variable, since we are going to create a new variable (that we named "GenHealth") from the original variable ("health1"). We do this by entering the new name in the "Name" window on the top right side of the screen and then "enter" it in the middle window by hitting the Change button. This places both original and new variables together so that the researcher knows what the recode involves: "health1→GenHealth."

Once we have named our new variable, we can choose the Old and New Values button near the bottom middle of the screen. This creates a new screen shown in Figure DMUB.5.

Figure DMUB.5 shows how we can change the values of the original variable into new values in the new variable. We do this to demonstrate how to use this feature of SPSS, since you may use it repeatedly in research applications. In this case, we notice that the original values in health1 are coded with "1" referring to "Excellent" and the other values ending with "5" referring to "Poor." It is fine to leave the values as they are, but in many statistical applications, it is helpful to have the larger values indicate more positive conditions. Thus we would need to reverse the values by assigning a "5"

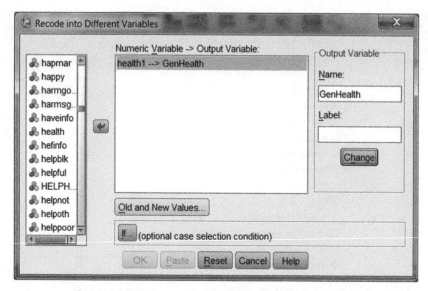

Figure DMUB.4. Recoding health1 to GenHealth in SPSS.

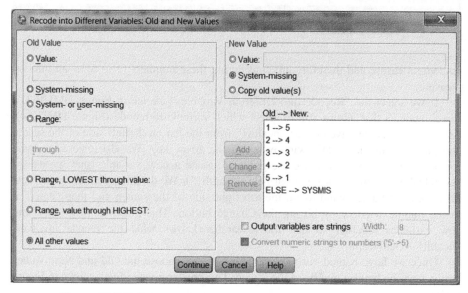

Figure DMUB.5. Recoding values from health1 to GenHealth in SPSS.

to "Excellent" through "1" for "Poor." Figure DMUB.5 shows this in the "Old→New" window. As you can see, the left half of the screen is devoted to the original variable, and the right half is devoted to the new variable. Thus, to change a value from original to new variable, you would enter a value on the left side in the "Value" box at the top and a new target value in the "Value" box at the top of the right side. Then, choosing

GenHealth

		Frequency	Percent	Valid Percent	Cumulative Percent
Valid	1.00 Poor	18	.9	1.5	1.5
	2.00 Fair	162	7.9	13.9	15.5
	3.00 Good	345	16.9	29.7	45.2
	4.00 Very Good	355	17.4	30.6	75.7
	5.00 Excellent	282	13.8	24.3	100.0
	Total	1162	56.8	100.0	
Missing	System	882	43.2		
Total		2044	100.0		

Figure DMUB.6. The recoded values of health1 as part of GenHealth.

the Add button in the middle of the right side would place the changed values in a separate summary box. Thus, as you can see, the top change appears as "1→5," which indicates that we are changing "Excellent" from an original value of 1 to a new value of 5. We will not change the original data by doing this, since we are creating a new variable from the original variable. Continuing in this manner, we can recode all the remaining values as shown in the summary box. You will note that we chose "All other values" (bottom of the left side) and recoded them to "System missing" for the new variable (shown on the top right side) so that our study could only include the values 1 to 5.

Performing this recode results in a new variable ("GenHealth") with the values "reversed" as shown in Figure DMUB.6. Compare these values with those shown in the original ("health1") frequencies reported in Figure DMUB.1.

Recoding "fairearn"

We can use the same SPSS process (Recoding into Different Variables) to change the values of the outcome variable in our study, "fairearn." In this case, however, we will not reverse the values from the original but rather create categories in a new variable that combine categories from the original. You can refer to the preceding figures to see the different screens, but the following are descriptions of the steps:

- Recode the "fairearn" variable to a new variable: "deserveearn"
- Categorize the new variable so that it regroups the original values into two new values (1 and 2)

The SPSS screen shown in Figure DMUB.7 includes the way we performed the recoding for this variable. The resulting new variable "deserveearn" has only two categories that contain the different values of the original variable.

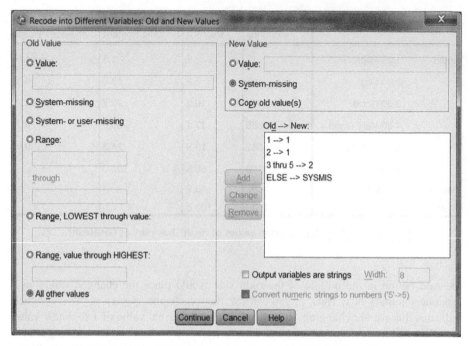

<u>Figure DMUB.7.</u> Using SPSS to recode the values of "fairearn" to "deserveearn."

deserveearn

		Frequency	Percent	Valid Percent	Cumulative Percent
Valid	1.00 Earnings are Less than Deserved	515	25.2	45.5	45.5
	2.00 Earnings are As Much or More than Deserved	616	30.1	54.5	100.0
	Total	1131	55.3	100.0	
Missing	System	913	44.7		
Total		2044	100.0		

<u>Figure DMUB.8.</u> The values and categories for the new variable "deserveearn."

Figure DMUB.8 shows the frequencies of the new variable that you can compare with the values of the original variable shown in Figure DMUB.2. By combining these categories, we have "dichotomized" the original variable into two categories so that we can use this as the predictor variable in the *t* test. As with health1, dichotomizing variables is not favored by all researchers, since some may argue that combining categories distorts the data and creates outcomes that are an artifact of the (re) categorization process. This argument has merit, but this method is commonly used (especially as a way to use the *t* test), especially in those circumstances where the number of respondents in the value categories is vastly uneven. It is also helpful in interpreting broad differences in the predictor variable, as it is used in this example. We are mainly interested in the primary differences between workers who believe earnings are not up to what they deserve and those who believe the earnings are at least what they deserve. In this example, we are not interested in the gradations of opinion regarding whether earnings are deserved. As you can see in Figure DMUB.8, categorizing the "deservee-arn" data in this way creates somewhat equivalent groups (by number).

DATA MANAGEMENT UNIT B: USES AND FUNCTIONS

Use and Function	SPSS Menus
Change the values of an existing variable by creating a new variable and specifying changes from the old to the new values	Transform–Recode into Different Variables
Change the values of an existing variable	Transform–Recode into Same Variables

DATA MANAGEMENT UNIT C: DESCRIPTIVE STATISTICS[1]

Statistical procedures are best used to discover patterns in the data that are not directly observable. Bringing light to these patterns allows researchers to understand and engage in problem solving. This section describes descriptive statistics for each of the major levels of data: nominal, ordinal, interval, and ratio. Since statistical procedures are designed for specific kinds of data, we believe it is important to discuss each procedure separately, according to its level. SPSS can be used with all of these levels of data, as you will see.

DESCRIPTIVE AND INFERENTIAL STATISTICS

Statistics, like other courses of study, is multifaceted. It includes both descriptive and inferential processes. **Descriptive statistics** are methods to summarize and boil down the essence of a set of information so that it can be understood more readily and from different vantage points. We live in a world that is bombarded with data; descriptive statistical techniques are ways of making sense of it. Using these straightforward methods reveals *numerical and visual patterns* in data that are not immediately

[1] Some of the material in this section is adapted from Abbott (2010), with permission of the publisher.

Understanding and Applying Research Design, First Edition. Martin Lee Abbott and Jennifer McKinney.
© 2013 John Wiley & Sons, Inc. Published 2013 by John Wiley & Sons, Inc.

apparent. Stated differently, these methods allow us to see the world as it is, not necessarily by common sense.

Inferential statistics are a different matter altogether. These methods allow you to make predictions about unknown values on the basis of small sets of sample values. In real life, we are presented with situations that cannot provide us with certainty. Would a method for teaching mathematics improve the scores of all students who take the course? Can we predict what a student might score on a standardized test? *Inferential statistics allow us to infer or make an observation about an unknown value from values that are known.* Obviously, we cannot do this with absolute certainty; we do not live in a totally predictable world. But we can make inferences within certain bounds of probability. Statistical procedures allow us to get closer to certainty than we could get without them.

DESCRIPTIVE STATISTICS

Descriptive statistics include graphical and numerical procedures to assist the researcher to understand and see patterns in data. Typically, a researcher gathers data, which, unexamined, exists as a series of numbers with no discernible relationship. By using descriptive statistical techniques, the researcher can present the data in such a way that whatever patterns exist can be assessed numerically and visually.

DESCRIPTIVE PROCEDURES FOR *NOMINAL* AND *ORDINAL* DATA

In the chi square chapter where we discussed the contingency table, we noted that nominal data are best described by noting the percentages of cases that fall within each category of a particular variable. For example, Figure 7.1 shows what percentage of the General Social Survey (GSS) respondents could be categorized as having high or low work autonomy based on their response to the item dealing with freedom to make decisions at work. Since this is a nominal (categorical) variable, we can describe the data simply by noting the percentages of cases that fall in either category.

You will also notice in Figure 7.1 that the second study variable, health condition, can also be described with percentages, even though this variable is an *ordinal variable*. The categories of health condition are distinctly different, but they have the quality of "more than–less than" that characterizes ordinal data.

When we discussed contingency table analyses, we noted that you can describe the relationship between the study variables using measures of association. Chi-square procedures typically use contingency coefficient, phi, or Cramer's V to describe the relationship between nominal level variables. (If our analysis had used two ordinal variables, we could have used Kendall's Tau-b to describe the association.)

There are ways to describe nominal and ordinal data *visually* (graphically) as well as numerically. The primary way to do this is through the use of the **histogram**, which is a graph that displays the number of cases that fall within each category of a study

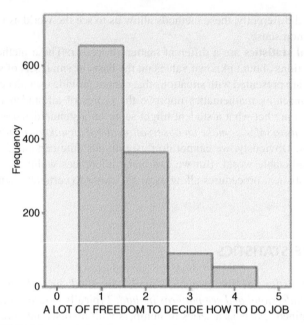

Figure DMUC.1. The SPSS histogram for the work autonomy study variable.

variable. Figure DMUC.1 shows a histogram describing the work autonomy variable
that we used in the chi-square analysis for Figure 7.1.

As you can see, Figure DMUC.1 shows the "frequency" on the vertical (y) axis
and separate columns for each category of the GSS item on the horizontal (x) axis. The
numbers for the horizontal axis represent coded values for the categories created by
GSS. Thus the question posed to respondents was, "[I have] a lot of freedom to decide
how to do [my] job."

Figure DMUC.2 shows the main category codes for this GSS item ("wkfreedm").
You can view "value labels" for study variables by looking at the variable view of
the main data screen and clicking on a specific variable's "Values" column. In Figure
DMUC.2, we have selected the values column for the "wkfreedm" variable. Each variable in the database is given a brief name (shown in the "Name" column to the far left),
and many are assigned labels corresponding to the wording of the survey item (shown
under the "Label" column for each variable). By clicking anywhere in these cells, you
can see the information supplied for each variable. The "Value Labels" box that appears
for the wkfreedm variable is shown in Figure DMUC.2. As one example, the category
of "1" means "Very true." It is these label codes that are reproduced in the horizontal
axis of the histogram, as shown in Figure DMUC.1.[2] One reason for this is that study

---

[2] When we used this variable in the chi-square study (with health condition), we combined the wkfreedm
categories into "high" or "low" by assigning categories 1 and 2 to "high" and categories 3 and 4 to "low."

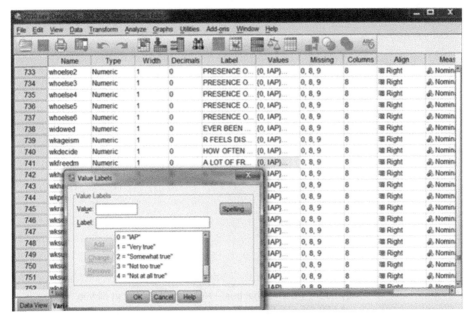

Figure DMUC.2. The SPSS Variable View window showing value labels for "wkfreedm."

variables are often interval data and the category codes are continuous (i.e., the different values have equal distances).

When they are viewed as interval data, the columns of data are joined together with no distances between the columns to reflect the continuous nature of the data. The value of each column is placed in the middle of the column. (Sometimes, "grouped" data are represented in which each side of each column represents the class limits of each category's values.)

You can see the difference in SPSS graphs by looking at the same variable (wkfreedm) in a "bar graph," shown in Figure DMUC.3. Both graphs look very similar, with most of the respondents reporting stronger agreement with the statement (i.e., taller columns on the left sides of the graphs). But the bar graph makes no assumptions about the levels of measurement of the variable represented. As you can see, the bars are separated, and the variable labels are reported as they appear in the original database (if the person compiling the data included them!). The columns of "counts" of data are shown as separate categories, in this case befitting the level of measurement of this variable. The researcher can decide whether to treat the data as ordinal or nominal in the study.[3]

---

[3] In the chi-square study, we combined ordinal categories and "transformed" this to a nominal variable with two categories. Technically, we could still treat it as ordinal, but "categorizing" variables in this way treats them as nominal variables. This process had minimal impact on the association measures in the chi-square analysis.

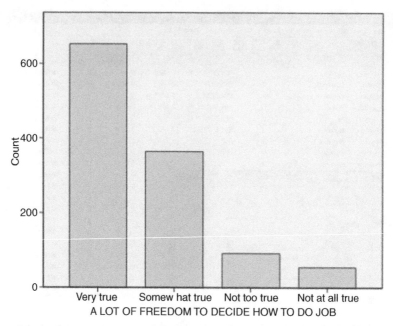

Figure DMUC.3. The SPSS bar graph for the work autonomy study variable.

Using SPSS Graphs for Nominal and Ordinal Data

The process for using SPSS to generate graphs like those in Figures DMUC.1 and DMUC.3 is accomplished through the use of the main menu ribbon. Figure DMUC.4 shows how to obtain either the bar graph or the histogram. As you can see, we selected "Graphs" on the main menu ribbon and then "Legacy Dialogs" in the submenu. SPSS graphics allows you to use a variety of means to customize figures, but the "Legacy Dialogs" represents the more common type of figures used by researchers. You can see in Figure DMUC.4 that the bar graph is the topmost choice in the list.

If we choose "Histogram," as shown in Figure DMUC.4, we are presented with a separate specification window for the histogram. Figure DMUC.4a shows this window in which we have moved wkfreedm from the list of variables on the left to the "variable:" window on the right using the arrow button. You can explore the additional specification features for this graph, but simply choosing OK will create the histogram shown in Figure DMUC.1.

Using SPSS Frequencies Command for Nominal- and Ordinal-Level Data

As we mentioned earlier, data can be described numerically as well as visually. We have already used one SPSS procedure for describing nominal (and ordinal) data, that of crosstabs. In the chi-square chapter, we used the example of work autonomy and health condition to explain how to use the crosstabs procedure and how to interpret the results. In the preceding section we discussed visual (graphical) methods of describing

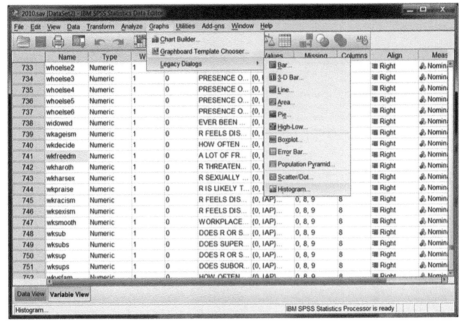

Figure DMUC.4. The SPSS Legacy Dialogs specification window for wkfreedm.

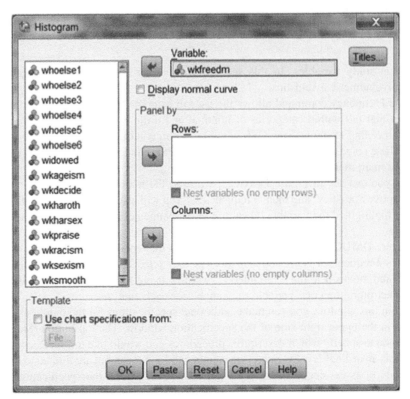

Figure DMUC.4a. The SPSS Histogram specification window for wkfreedm.

Figure DMUC.5. The SPSS Frequencies window for obtaining descriptive statistics.

individual study variables. In this section, we discuss how to use SPSS to obtain descriptive numerical statistics.

The Frequency command allows the user to examine the number and percentage of cases that fall within categories of nominal and ordinal variables. We can use one of the preceding examples (the GSS variable, health condition) to demonstrate how to obtain these results. Figure DMUC.5 shows the SPSS window for Frequencies derived from the main menu ribbon.

As you can see in Figure DMUC.5, we used the main Analyze menu to specify "Descriptive Statistics" and then "Frequencies." You will also notice in Figure DMUC.6 that we highlighted the variable ("health") in the Name column that we will use in the example.

Figure DMUC.6 shows the Frequencies menu that appears when the "Descriptive Statistics–Frequencies" selection is made. The user selects the variable from the list on the left and moves it to the "Variable(s):" window using the arrow key. Note that you may select more than one variable to examine for each analysis.

From this window, you can make additional specifications for the output using the buttons in the upper right side of the specification window. The top button (Statistics) allows you to specify which descriptive procedures you would like to run. Recognizing that this is an ordinal-level variable, we cannot call for a mean but are limited to median and mode to assess central tendency. Figure DMUC.7 shows the specification menu that appears when you select Statistics. As you can see, we selected only "Median" and

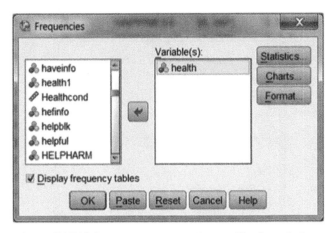

Figure DMUC.6. The SPSS Frequencies specification window.

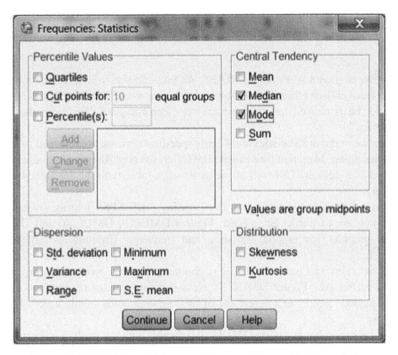

Figure DMUC.7. The SPSS Frequencies: Statistics specification window.

"Mode" in the "Central Tendency" panel of choices. The default for Frequencies is to show number and percentage of cases in categories of the variable, which are measures of dispersion.

Before finalizing the procedure, note that we can choose the Charts button (see Figure DMUC.6) to specify visual (graphic) representations. This additional

Figure DMUC.8. The SPSS Frequencies: Charts specification window.

specification is shown in Figure DMUC.8. As you can see, we called for a bar chart but could have chosen a number of other graphic measures. Note also that the chart we choose can be portrayed in "Frequencies" or "Percentages." We chose the default: "Frequencies."

When we make the Statistics and Charts specifications, we are returned to the main Frequencies menu. Note that (see Figure DMUC.6) the box "Display frequency tables" is checked as a default. This will allow us to see a table with the frequency of cases by category.

When we finalize the procedure (by pressing OK), SPSS returns several output tables. Three are of particular interest. Figures DMUC.9, DMUC.10 and DMUC.11 show the graphic (bar graph), statistics, and frequency (frequencies table) output, respectively.

The bar chart in Figure DMUC.9 is similar to that for work autonomy that we discussed earlier (see Figure DMUC.3). Note that most of the respondents (i.e., the mode) indicated "good" as a descriptor of their general health. Only a very few indicated "poor."

The statistics table in Figure DMUC.10 indicates both the median and mode information based on the GSS codes for the categories (1 = Excellent, 2 = Good, etc.). As you can see, the median and mode are both 2, which indicates that the category coded with a 2 (Good) obtained the most respondent selections (Mode) and that this category was in the middle of the overall distribution of respondent selections (Median).

The frequency table in Figure DMUC.11 contains a great deal of information. As you can see, there are five columns of information for each variable category. The first column lists the categories for the variable, including the missing response categories, "IAP" ("inapplicable to a set of respondents), "DK" (don't know), and "NA"

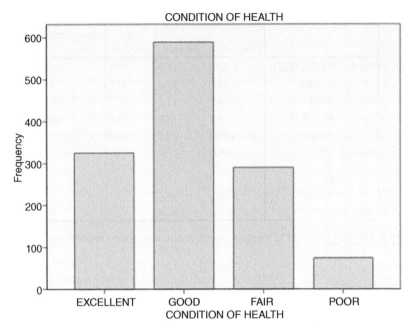

Figure DMUC.9. The SPSS bar chart obtained from Frequencies: Charts.

Statistics

CONDITION OF

HEALTH

N	Valid	1278
	Missing	766
Median		2.00
Mode		2

Figure DMUC.10. The SPSS statistics table obtained from Frequencies.

(nonapplicable). The second column ("Frequency") shows the number of respondents who made selections for each category of health (or were classified according to the IAP category, for example). The third column ("Percent") shows the *percentage of the total number of responses* of the cases in a particular category. The fourth column ("Valid Percent") shows the percentage of respondents in the primary categories. The last column ("Cumulative Percent") shows the accumulating percentage from the first to the last primary category of responses. Thus 325 respondents indicated "Excellent" health, which represented 15.9 percent of all response categories including "missing" (325/2044), but which represented 25.4 percent of the responses in the four nonmissing categories (325/1278).

CONDITION OF HEALTH

		Frequency	Percent	Valid Percent	Cumulative Percent
Valid	EXCELLENT	325	15.9	25.4	25.4
	GOOD	589	28.8	46.1	71.5
	FAIR	290	14.2	22.7	94.2
	POOR	74	3.6	5.8	100.0
	Total	1278	62.5	100.0	
Missing	IAP	763	37.3		
	DK	2	.1		
	NA	1	.0		
	Total	766	37.5		
Total		2044	100.0		

Figure DMUC.11. The SPSS frequencies table obtained from Frequencies: Statistics.

Researchers should carefully decide which percentages to report. In most cases, you can report only the valid percentages, but it is important to point out the number of missing cases as well.

DESCRIPTIVE PROCEDURES FOR INTERVAL DATA

Many social science research variables (interval level) are **normally distributed** in that they conform to the bell curve where data pile up in the middle and tail off in both directions. Normal distributions are important in research because many research techniques (especially those designed for interval data) require variables to be normally distributed. Researchers typically examine four dimensions of a distribution of data to determine whether it is normal: central tendency, variability, skewness, and kurtosis.

Numerical Procedures: Central Tendency

Simply looking at a set of numbers is not the best way to understand the patterns that may exist. The numbers are typically in no particular order, so the researcher probably cannot discern any meaningful pattern. Are there procedures we can use to understand these patterns *numerically*?

Central tendency measures suggest that a group of scores can be understood more comprehensively by using a series of numerical procedures. As these measures suggest, we can understand a lot about a set of data just by observing whether or not most of the scores cluster or build up around a typical score. That is, do the scores have a *tendency* to approach the middle from both ends? There will be scores spreading out around this central point, but it is helpful to describe the central point in different ways and for different purposes. The primary question the researcher asks here is, "Can we identify a 'typical' score that represents most of the scores in the distribution?" The following are the most commonly used central tendency measures.

Mean. The **mean** is the arithmetic average of a set of scores. To calculate a mean, the researcher needs at least interval data because you need to be able to add, subtract, multiply, and divide numbers to calculate it. If you have less than interval data, it would not make sense to use these arithmetic operations, since you could not assume the intervals between data points are equal. (For example, you could not get a meaningful "average sex," since sex is nominal level.)

Calculating the mean value uses one of the most basic formulas in statistics, the average:

$$\frac{\sum X}{N}$$

This formula uses the "Σ" symbol, which means "sum of." Therefore, the average, or mean value, can be calculated by adding up a set of numbers, or "summing" them, and then dividing by how many numbers there are in the set by the number of data observations (N). To take an example, the data values in Table DMUC.1 represent actual values from a school database from Washington state.[4] These are a sample of schools with fourth grade ($N = 40$) percentage of students qualified to receive free or reduced lunches. This variable is important to social researchers, since it represents one of the only ways to gauge the family income level of the students. Since the values are percentages of all students in the school, the data are interval level.

Using the values in Table DMUC.1, we can calculate the mean by summing the 40 numbers to get 1979.41. If we divide this number by 40, the amount of numbers in the set, we get 49.49.

$$\frac{\sum X}{N} = \frac{1979.41}{40} = 49.49$$

TABLE DMUC.1. School Data for Central Tendency

40.0	62.7	49.0	12.2
62.1	73.3	52.0	25.8
58.3	86.3	37.3	31.1
67.7	83.6	28.1	31.8
43.6	49.6	78.8	37.1
29.7	100.0	38.3	86.0
49.7	46.4	67.1	5.7
44.5	37.4	37.0	58.2
53.8	41.2	85.0	29.6
53.6	43.1	45.7	17.3

[4] The data are used courtesy of the Office of the Superintendent of Public Instruction, Olympia, Washington.

What does the mean of 49.49 indicate? If you inspect the data in Table DMUC.1 you will see that 100 percent of the students in one school qualified for free or reduced lunch while 5.7 percent of the students at another qualified. That is quite a difference! What is the *typical* percentage of students who qualified for free or reduced lunch? That is, if you had to report one score that most typified all the scores, which would it be? This is the mean, or average value. It expresses a central value (toward the middle) that characterizes all the values.

Median. Another measure of central tendency is the **median**, or middle score among a set of scores. This isn't a calculation like the mean, but rather it identifies the score that lies directly in the middle of the set of scores when they are arranged large to small (or small to large). In our set of scores, the median is 46.05. If you were to rank-order the set of scores by listing them small to large, you would find that the direct middle of the set of scores is between the twentieth (45.7) and twenty-first (46.4) numbers in the list. In order to identify the direct middle score, you would have to average these two numbers to get 46.05 ((45.7 + 46.4)/2). An equal number of scores in the group of scores are above and below 46.05.

The median is important because sometimes the arithmetic average is not the most typical score in a set of scores. For example, if I am trying to find the typical housing value in a given neighborhood, I might end up with a lot of houses valued at a few hundred thousand and five or six houses valued in the millions. If you added all these values up and divided by the number of houses, the resulting average would not really characterize the typical house because the influence of the million-dollar homes would present an inordinately high value.

To take another example, the values in Table DMUC.2 are similar to those in Table DMUC.1 with the exception of seven values. In order to illustrate the effects of "extreme scores," we replaced each percentage over 70 with a score of 100.0. If you calculate an average on the adjusted values in Table DMUC.2, the resulting value is 52.16.

TABLE DMUC.2. Adjusted Free or Reduced Lunch Percentages

40.0	62.7	49.0	12.2
62.1	100.0	52.0	25.8
58.3	100.0	37.3	31.1
67.7	100.0	28.1	31.8
43.6	49.6	100.0	37.1
29.7	100.0	38.3	100.0
49.7	46.4	67.1	5.7
44.5	37.4	37.0	58.2
53.8	41.2	100.0	29.6
53.6	43.1	45.7	17.3

Changing six of the original values resulted in the mean changing from 49.49 to 52.16. But what happens to the median when we make this change? Nothing. The median remains 46.05, since it represents the middle of the group of scores, not their average value. In this case, which is the more *typical* score? The mean value registers the influence of these large scores, thereby "pulling" the average away from the center of the group. The median stays at the center.

This small example shows that only a few extreme scores can exert quite an influence on the mean value. It also shows that the median value in this circumstance might be the more typical score of all the scores, since it stays nearer the center of the group. Researchers should be alert to the presence of extreme scores, since they oftentimes strongly affect the measure of central tendency. This is especially true any time the values reflect money such as housing values, household income, and so on.

Mode. The **mode** is the most frequently occurring score in a set of scores. This is the most basic of the measures of central tendency, since it can be used with virtually any set of data. Referring to Table DMUC.1, you will see that there are no values exactly the same. This is often the case when we use "continuous" data (like the percentages of free or reduced lunches by school). When there are equivalent values in the database, the mode is a typical score or category, since data most often mass up around a central point. In this case, it makes sense that the mode, at the greatest point of accumulation in the set, represents the most prevalent score.

One data pattern that social scientists need to look for is the **bimodal** distribution of data. This situation occurs when the data have two (or more) clusters of data rather than massing up around the middle of the distribution. You can detect a bimodal distribution numerically by observing several values of the same number. But in larger databases, it is more difficult to do this. In this case, it is easier to use visual means of describing the data. We discuss these further in a later section, but consider the example in Figure DMUC.12. The data in Figure DMUC.12 are the GSS 2010 respondents' socioeconomic indexes (sei). As you can see from the graphic (histogram), there are two more or less distinct clusters of data rather than just one. There is a cluster near the index score of 30 and another cluster around the index score of 65.

In this situation, what is the most appropriate measure of central tendency? The data are interval, so we could calculate a mean. The mean for these adjusted scores is 48.99. However, would this mean value truly be the most characteristic, or typical, score in the set of scores? No, because the scores in the set of data (in Figure DMUC.12) no longer cluster around a central point; they cluster around two central points. Therefore, it would be misleading to report a mean of 48.99, even though it is technically a correct calculation. The discerning researcher would report that there are two clusters of data, indicating a bimodal distribution. In the case of the data reported in Figure DMUC.12, it appears that "most" of the GSS respondents indicated sei indexes quite divergent from one another.

Central Tendency and Levels of Data. The mean is used with interval (or ratio) data, since it is a mathematical calculation that requires equal intervals. The median and mode can be used with interval as well as "lower levels" of data (i.e.,

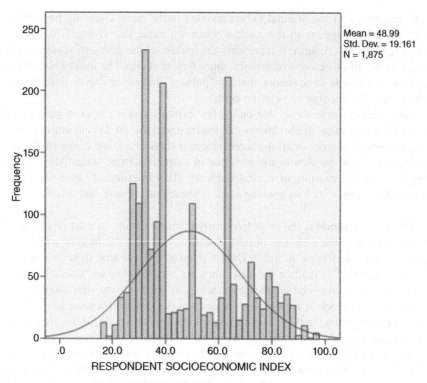

Figure DMUC.12. The sei data from GSS 2010.

ordinal and nominal), whereas a mean cannot. Using either median or mode with interval data does not require a mathematical calculation; it simply involves rank-ordering the values and finding the middle score or the most frequently occurring score, respectively. The mean cannot be used with ordinal or nominal data, since we cannot use mathematical calculations involving addition, subtraction, multiplication, and division on these data, as we discussed earlier.

The median is a better indicator of central tendency than the mean with "skewed" or imbalanced distributions. We have more to say shortly about skewed sets of scores, but for now, we should recognize that a set of scores can contain extreme scores that might result in the mean being unfairly influenced and therefore not being the most representative measure of central tendency. Even when the data are interval (as, for example, when the data are dealing with monetary value, or income), the mean is not always the best choice of central tendency despite the fact that it can use arithmetic calculations.

The mode, in contrast, is helpful in describing when a set of scores fall into more than one distinct cluster (bimodal distribution). Consider Figure DMUC.12 that shows an example with interval data. The mode is primarily used for central tendency with nominal data.

Numerical Procedures: "Balance" and Variability

We continue to explore descriptive statistics in this chapter. This time, we examine the extent to which scores spread out from the mean of a distribution of values. It is important to understand the characteristic score or value of a distribution, as we saw with central tendency, but it is also critical to understand the extent of the *scatter, variability, or dispersion* of scores away from the center. How far away do scores fall, do the scores fall equally to the left and right of the mean, and to what extent do the scores bunch up in the middle relative to their spread? The answers to these and similar questions will help us to complete our description of the distribution of values.

Skewness. *Skewness* is a term that describes whether, or to what extent, a set of values is not perfectly balanced but rather *trails off* to the left or right of center. We will not discuss how to calculate skew, but it is easy to show. If you look at Figure DMUC.1, you can see that the number of cases trail off to the right side of the histogram. The number of cases do not bunch up in the middle of the histogram as they would if the (interval) data were normally distributed (i.e., in the bell curve shape). This is an example of a "positive skew," since the values trail off to the right (generally in the direction of greater values of the variable). The data can also trail off to the left of center in which case it would represent a "negative skew."

You can see the skew more easily if you superimpose the normal curve onto the histogram as we have done in Figure DMUC.13. This figure is from the GSS database in which respondents were asked about their number of hours of e-mail per week. (Note that you can superimpose the normal curve on the figure by checking the "Display normal curve" box in the histogram specification window as shown in Figure DMUC.4a.) Clearly, the data are not normally distributed! There are many respondents who indicate a great many hours of e-mailing per week. So much so, that the researcher might question the nature of the item and the respondents. (For example, does this include one's occupational use of e-mail? Only personal use? etc.)

Kurtosis. **Kurtosis** is another way to help describe a distribution of values. This measure indicates how peaked or flat the distribution of values appears. Distributions where all the values cluster tightly around the mean might show a very high point in the distribution, since all the scores are pushing together and therefore upward. This is known as a *leptokurtic* distribution. Distributions with the opposite dynamic, those with few scores massing around the mean, are called *platykurtic* and appear flat. "Perfectly" balanced distributions show the characteristic bell curve pattern, being neither too peaked nor too flat. The distribution of responses in Figure DMUC.13 clearly indicates a leptokurtic distribution, since an inordinate number of respondents indicated only a few hours of e-mailing per week.

Range. One simple way to measure variability is to use the **range**, the numerical difference between the highest and lowest scores in the distribution. This represents a helpful global measure of the spread of the scores. But remember it is a global measure and will not provide extensive information. Nevertheless, the range provides

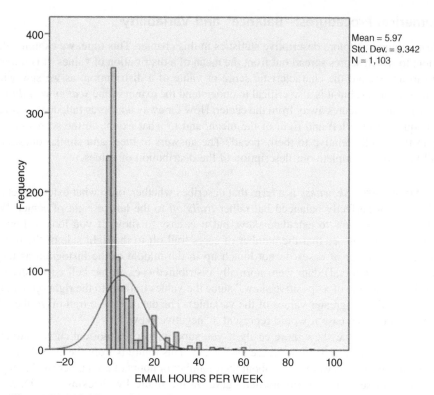

Figure DMUC.13. The SPSS wkfreedm histogram with superimposed normal curve.

a convenient shorthand measure of dispersion and can provide helpful benchmarks for assessing whether or not a distribution is generally distributed normally.

Percentile. The **percentile** or percentile rank is the point in a distribution of scores below which a given percentage of scores fall. This is an indication of *rank,* since it establishes a score that is above the percentage of a set of scores. For example, a student scoring in the 82nd percentile on a math achievement test would score above 82 percent of the other students who took the test.

Therefore, percentiles describe where a certain score is in relation to the others in the distribution. The usefulness of percentiles for educators is clear, since most schools report percentile results for achievement tests. The general public also sees these measures reported in newspaper and Web site reports of school and district progress.

Education researchers use a variety of measures based on percentiles to help describe how scores relate to other scores and to show rankings within the total set of scores, including quartiles (measures that divide the total set of scores into four equal groups), deciles (measures that break a frequency distribution into 10 equal groups), and the interquartile range that represents the middle half of a frequency distribution (since they represent the difference between the first and third quartiles).

Standard Deviation and Variance. The standard deviation (SD) and variance (VAR) are both measures of the dispersion of scores in a distribution. That is, *these measures provide a view of the nature and extent of the scatter of scores around the mean.* So, along with the mean, skewness, and kurtosis, they provide a way of describing the distribution of a set of scores. With these measures, the researcher can decide whether a distribution of scores is normally distributed.

The **variance** (VAR) is by definition the square of the SD. Conceptually, the VAR is a *global measure of the spread of scores,* since it represents an average squared deviation. If you summed the squared distances between each score and the mean of a distribution of scores (i.e., if you squared and summed the deviation amounts), you would have a global measure of the total amount of variation among all the scores. If you divided this number by the number of scores, the result would be the VAR, or the *average squared distance of the cases from the mean.*

The SD is the square root of the VAR. If you were to take the square root of the average squared distances from the mean, the resulting figure is the ***standard deviation***. That is, it represents a *standard* amount of distance between the mean and each score in the distribution (not the average *squared* distance, which is the VAR). We refer to this as *standard,* since we created a standardized unit by dividing it by the number of scores, yielding a value that has known properties to statisticians and researchers. We know that, if a distribution is perfectly normally distributed, the distribution will contain about six SD units, three on each side of the mean.

Both the SD and the VAR provide an idea of the extent of the spread of scores in a distribution. If the SD is small, the scores will be more alike and have little spread. If is large, the scores will vary greatly and spread out more extensively. Thus, if a distribution of test scores has a SD of 2, it conceptually indicates that *typically* the scores were within 2 points of the mean. In such a case, the overall distribution would probably appear to be quite scrunched together, in comparison to a distribution of test scores with a SD of 5.

Sample SD and Population SD. We have more to say about this difference in other chapters when we discuss inferential statistics. For now, it is important to point out that computing SD for a sample of values, as we did with the ratio data, will yield a different value depending on whether we understand the distribution of data to represent a complete set of scores or merely a sample of a population.

Remember that inferential statistics differs from descriptive statistics primarily in the fact that, with inferential statistics, we are using sample values to make inferences or decisions about the populations from which the samples are thought to come. In descriptive statistics, we make no such attributions; rather, we simply measure the distribution of values at hand and treat all the values we have as the complete set of information (i.e., its own population). When we get to considerations of inferential statistics, you will find that, *in order to make attributions about populations based on sample values, we typically must adjust the sample values, since we are making guesses about what the populations look like.* To make better estimates of population values, we adjust the sample values.

SPSS has no way of distinguishing inferential or descriptive computations of SD. Therefore, they present the inferential SD as the default value. We will show how to determine the differences and examine the resulting values using SPSS.

OBTAINING DESCRIPTIVE (NUMERICAL) STATISTICS FROM SPSS

Obtaining descriptive statistics from SPSS® for interval and ratio data is straightforward. The primary consideration for the user is which statistical information is needed. The procedures and output available depend on a number of factors including the level of data for the variable(s) requested. In the preceding section, we discussed using the SPSS Frequencies procedure for nominal and ordinal levels of data. This procedure can also be used with interval data, but it is more convenient to use the "Analyze–Descriptive Statistics–Descriptives" command from the main Analyze menu.

As you can see from Figure DMUC.14, we are showing the example of creating descriptive statistics for the percentage of (seventh grade) students in Washington schools who met the standard for the math assessment in 2010. This is an aggregate measure in that the scores represent the collected percentage of students by school. These data are used widely in evaluation research, but remember that they are not individual data, which restricts the nature and extent of the conclusions that can be made using them in research.

Despite the nature of the data, the descriptive statistics outcomes can be derived using the same procedure. We are requesting descriptive statistics for an interval-level variable that we can use to determine, among other outcomes, the extent to which the distribution of data approximates a normal distribution.

Figure DMUC.14. The SPSS Descriptive Statistics–Descriptives menu.

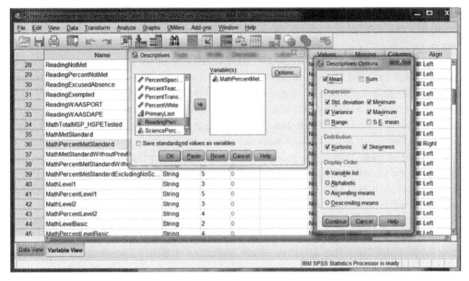

Figure DMUC.15. The SPSS descriptive statistics specification menus.

Descriptive Statistics

	N	Minimum	Maximum	Mean	Std. Deviation	Variance	Skewness		Kurtosis	
	Statistic	Statistic	Statistic	Statistic	Statistic	Statistic	Statistic	Std. Error	Statistic	Std. Error
MathPercentMetStandard	484	0	100	52.64	18.173	330.271	-.043	.111	-.398	.222
Valid N (listwise)	484									

Figure DMUC.16. The SPSS descriptive statistics output.

When we initiate this choice in SPSS, we are presented with several choices for descriptive statistics outcomes. As you can see in Figure DMUC.15, we specify a variable ("Math Percent Met Standard") in the "Variable(s):" window of the "Descriptives" submenu, and then we can choose the Options button to choose the output measures we want. In the example in Figure DMUC.15, we chose the Mean, SD, VAR, minimum and maximum values, skewness, and kurtosis.

By making these choices, SPSS returns the output table shown in Figure DMUC.16. As you can see, we now have "descriptors" for 484 schools (with seventh grades) in Washington. We can see the four main descriptors of distributions: central tendency (mean = 52.64), dispersion (SD = 18.173, and VAR = 330.271), skewness (−0.043), and kurtosis (−0.398).

Interpreting skewness and kurtosis is a bit of an art, but there are some guidelines that may be helpful. Skewness (the balance or imbalance of a set of interval data) is best interpreted by dividing the skewness statistic (−0.043) by its standard error (0.111). If the resulting number is less than 2 or 3, the distribution is probably "balanced" or does not appear lopsided. You can use the sign (positive or negative) of the skewness value to indicate which way the skew tends (i.e., negative to the left and positive to the right), but the *magnitude* of the result indicates whether or not the skewness is excessive.

This guideline is greatly affected by the overall size of the data set, however. Typically, the Std. Error of Skewness reported by SPSS will be *smaller* with larger numbers of values in the distribution. So large data sets (200 to 400) might have very small Std. Error of Skewness numbers and result in the overall skewness result being very *large* (since dividing by a smaller number yields a larger result). Smaller data sets will typically have large Std. Error of Skewness numbers with resulting small skewness results.

In light of these issues, the researcher needs to consider the *size of the distribution* as well as the *visual evidence* to make a decision about skewness. In our example the skewness result is (−0.387) (derived by −.043/.111), which, according to our guideline, represents a balanced distribution.

The kurtosis finding is interpreted in the same way as the skewness finding. In our example, the kurtosis number is −1.79 (derived from −0.398/0.222) which is within the acceptable guideline. The negative result indicates a flatter distribution; a positive number indicates a more peaked distribution.

Using SPSS Explore

There are other ways to obtain descriptive statistic outcomes from SPSS, but one we note here is the "Explore" procedure. If you look at Figure DMUC.14, you will see that the "Explore" procedure is obtained through the "Analyze–Descriptive Statistics" menu (the Explore choice is located just below "Descriptives"). When you make this choice, SPSS returns the menu shown in Figure DMUC.17.

As you can see, we specified Math Percent Met Standard in the "Dependent List:" window by using the arrow key. At this point, we have several additional choices to

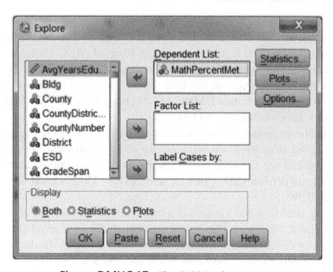

Figure DMUC.17. The SPSS Explore menu.

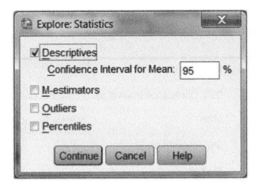

Figure DMUC.18. The SPSS "Explore: Statistics" menu.

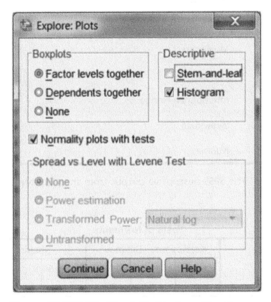

Figure DMUC.19. The SPSS "Explore: Plots" menu.

help in our specification. We will only examine some of these, since several relate to inferential processes, a discussion for later sections.

The first additional specification results from the Statistics button in the upper right corner of the Explore menu. By choosing this button, we will be able to call for descriptive analyses, as shown in Figure DMUC.18.

The Explore menu (shown in Figure DMUC.17) also includes a choice for "Plots" through the button in the upper-right corner. By choosing this option, you will see the following submenu ("Explore: Plots") in which you can choose a range of outputs to help assess the nature of the variable distribution. Figure DMUC.19 shows these options.

Descriptives

			Statistic	Std. Error
MathPercentMetStandard	Mean		52.64	.826
	95% Confidence Interval for Mean	Lower Bound	51.01	
		Upper Bound	54.26	
	5% Trimmed Mean		52.69	
	Median		52.50	
	Variance		330.271	
	Std. Deviation		18.173	
	Minimum		0	
	Maximum		100	
	Range		100	
	Interquartile Range		26	
	Skewness		−.043	.111
	Kurtosis		−.398	.222

Figure DMUC.20. The SPSS descriptive output from the "Explore: Statistics" menu.

Tests of Normality

	Kolmogorov -Smirnov[a]			Shapiro-Wilk		
	Statistic	df	Sig.	Statistic	df	Sig.
MathPercentMetStandard	.024	484	.200*	.997	484	.394

a. Lilliefors Significance Correction

*. This is a lower bound of the true significance.

Figure DMUC.21. The SPSS descriptive output from the Explore: Plots menu.

In the following section, we discuss visual assessments of descriptive statistics. For now, we focus on the numerical output, which are available by checking the box "Normality plots with tests" in the middle of the Explore: Plots menu.

Figure DMUC.20 and DMUC.21, respectively, show the numerical output that results from the choices made in Figures DMUC.18 and DMUC.19.

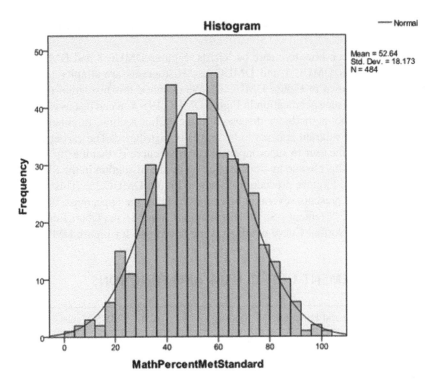

Figure DMUC.22. The SPSS histogram derived through the Explore: Plots menu.

The output in Figure DMUC.20 largely reproduces the output obtained from the Descriptive Statistics–Statistics procedure (see Figure DMUC.16). There are additional values (e.g., median, range, and others), but the SD, VAR, skewness, and kurtosis figures are available.

The output in Figure DMUC.21 is much different in nature in that they represent *inferential tests of the normality of the distribution of the test variable*. Since we will discuss inferential statistics in different sections, we point out here that the Kolmogorov-Smirnov and Shapiro-Wilk tests help researchers to assess whether the test variable (Math Percent Met Standard) is considered normally distributed (i.e., balanced, not bimodal, not lopsided, etc.). If the "Sig." value for both tests is greater than 0.05, then the distribution is considered within normal bounds. In this example, both Sig. values exceed 0.05 (0.200 and 0.394, respectively), so we can consider the Math Percent Met Standard variable to be normally distributed. These statistical tests are sensitive to the conditions of different variables (e.g., sample size), so interpret them cautiously.

As we have mentioned, it is always good to look at the visual output of statistical analyses in order to gain a better descriptive picture of a variable or variables in a study. We turn to this now with respect to interval-level variables.

OBTAINING DESCRIPTIVE (VISUAL) STATISTICS FROM SPSS

We have already seen how to create bar charts (Figures DMUC.8 and DMUC.9) and histograms (Figures DMUC.4 and DMUC.4a). Histograms are simpler to see with interval data, as shown in Figure DMUC.22. We obtained this histogram through the Explore procedure (see specification in Figure DMUC.19). You can just as easily create the histogram by the methods we discussed in the earlier section. In either case, you will note that the histogram appears to be normally distributed. The earlier histogram procedure allowed the user to superimpose the normal curve so that the "fit" is clearer. You can still make this change by double clicking on the histogram in the SPSS output that results from the Explore procedure (shown in Figure DMUC.22). This results in a "Chart Editor" that presents several choices for changing the appearance of the histogram. If you select, "Elements–Show Distribution Curve" in the Chart Editor screen, you can select the Normal Curve overlay (as we have done for Figure DMUC.22).

DATA MANAGEMENT UNIT C: USES AND FUNCTIONS

Use and Function	SPSS Menus
Create a histogram showing the categories of a variable in bars	Graphs–Legacy Dialogs–Histograms (there are alternative ways of doing this, but this is the simplest)
Create descriptive statistics by showing the frequency and percent of the variable categories (values)	Analyze–Descriptive Statistics–Frequencies
Create descriptive statistics for interval-level variables	Analyze–Descriptive Statistics–Descriptives
Create descriptive statistics for an interval variable according to the categories of a predictor variable (including separate tests of normality)	Analyze–Descriptive Statistics–Explore

STATISTICAL PROCEDURES
UNIT C: z SCORES[1]

In other sections, we have discussed how to describe distributions of (interval level) raw scores graphically and in terms of central tendency, variability, skewness, and kurtosis. Using what we learned from examining these descriptive statistics, we can confirm whether our data are normally distributed or if we must use different procedures.

THE NATURE OF THE NORMAL CURVE

The normal distribution is very common in social science research, so we need to deepen our understanding of some of the properties of the normal *curve*. We call the normal distribution a curve, since the histogram forms a curve when the top midpoints of the bars are joined together. Technically, this is called a **frequency polygon**. If you look at Figure DMUC.22, you will see the SPSS histogram for the school-based math achievement variable. In the figure, SPSS overlaid the normal curve on top of the histogram so you can see the extent to which the data approximate a normal distribution. As you can see from that figure, if you were to connect the top midpoints of the

---

[1] Some of the material in this section is adapted from Abbott (2011), with permission of the publisher.

Understanding and Applying Research Design, First Edition. Martin Lee Abbott and Jennifer McKinney.
© 2013 John Wiley & Sons, Inc. Published 2013 by John Wiley & Sons, Inc.

histogram bars with a line, it would not be the smooth line you see, but rather a more jagged line. However, as a database increases its size, the histogram approximates the smooth normal curve in variables that are normally distributed; the jagged line becomes filled in as more cases are added.

When we speak of the normal distribution and how our sample data set is normally distributed, we actually speak about our data *approximating* a normal distribution. We refer to the *perfect* normal distribution as an ideal so that we have a model distribution for comparison to our *actual* data. Thus the normal curve is a kind of perfect ruler with known features and dimensions. In fact, we can mathematically chart the perfect normal curve and derive a picture of how the areas under the curve are distributed. Because of these features, we refer to the perfect normal distribution as a **standard normal distribution**.

The perfect normal curve, shown in Figure SPUC.1, has *known proportions of the total area between the mean and given standard deviation units*. A standard normal curve (also known as a *z distribution*) has a mean of 0 and a standard deviation of 1.0. This is always a bit puzzling until you consider how the mean and standard deviation are calculated. Since a perfect distribution has equal numbers of scores lying to the left and to the right of the mean, calculating the mean is akin to adding positive and negative values resulting in 0. Dividing 0 by *N,* of whatever size, will always equal 0. Therefore the mean of a perfect standard normal distribution is equal to 0.

The standard normal distribution has a standard deviation equal to 1 unit. This is simply an easy way to designate the known areas under the curve. Figure SPUC.1 shows that there are six standard deviation units that capture almost all the cases under the perfect normal curve area. This is how the standard normal curve is arranged mathematically. So, for example, 13.59 percent of the area of the curve lies between the first

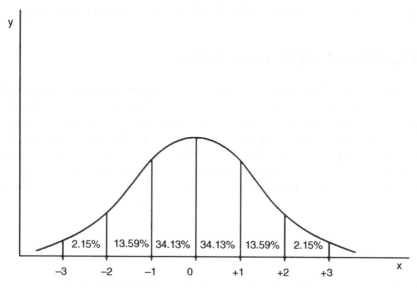

Figure SPUC.1. The normal curve with known properties.

(+1) and second (+2) standard deviation on the right side of the mean. Since the curve is symmetrical, there is also 13.59 percent of the area of the curve between the first (−1) and second (−2) standard deviation on the left side of the curve, and so on.

Remember that this is an ideal distribution. As such, we can compare actual, or raw, data distributions to it as a way of understanding our own raw data better. Also, we can use it to compare two sets of raw score data, since we have a perfect measuring stick that relates to both sets of imperfect data.

There are other features of the standard distribution we should notice.

- The scores cluster in the middle, and "thin out" toward either end.
- It is a balanced or symmetrical distribution, with equal numbers of scores on either side of the middle.
- The mean, median, and mode all fall on the same point.
- The curve is "asymptotic" to the x-axis. This means that it gets closer and closer to the x-axis but never touches it, since, in theory, there may be a case very far from the other scores—off the chart, so to speak. There has to be room under the curve for these kinds of possibilities.

THE STANDARD NORMAL SCORE: *z* SCORE

When we refer to the standard normal deviation, we speak of the *z* score, which is a very important measure in statistics. *A **z** score is a score expressed in standard deviation units.* Thus a score of 0.67 would be a score that is two-thirds of one standard deviation to the right of the mean. If a score was expressed as a *z* score of (−3.5), we would recognize immediately that this score would have an inordinately low value, relative to the other values, since it would fall 3.5 standard deviations below the mean where there is only an extremely small percentage of the entire curve represented.

Because it has standardized meaning, the *z* score allows us to understand where each score resides compared to the entire set of scores in the distribution. It also allows us to compare one individual's performance on two different sets of (normally distributed) scores. It is important to note that *z* scores are expressed not just in whole numbers but as decimal values, as we used in the preceding example. Thus a *z* score of −1.96 would indicate that this score is slightly less than two standard deviations below the mean on a standard normal curve as shown in Figure SPUC.2.

CALCULATING *z* SCORES

The *z* scores are important because they provide a perfect standard of measurement that we can use for comparison to raw score distributions that may not be perfectly normally distributed. Raw score distributions like math achievement of our study schools in the descriptive statistics section are near normally distributed but still not perfectly so. Look again at figure DMUC.22 to see the difference between the raw score of the histogram and the perfect normal curve that is superimposed on the histogram.

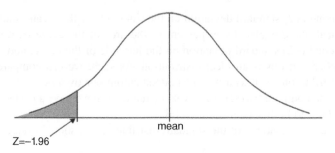

Figure SPUC.2. The location of $z = (-)\,1.96$.

Because raw score distributions are not always perfectly distributed, we must perform descriptive analyses to see if they are within normal boundaries. Thus, by looking at the skewness and kurtosis of a distribution, we can see if the raw score distribution is balanced and close to a normal shape; if the mean, median, and mode are on the same point (or close), that is another indication that the data approximate a normal distribution. Finally, we can use the visual evidence of the histogram to help us understand the shape of the distribution.

The formula for transforming raw score values (x) to standard normal scores (z) is as follows, where X is the raw score, M is the mean, and SD is the (population) SD:

$$Z = \frac{X - M}{SD}$$

We can see the value of z scores by discussing an example using the U.S. Census data.[2] In this example we compare the death rates from different causes by creating z scores. The following are the data for deaths in the states from diabetes (per 100,000 of the population in 2007) and the death rate for motor vehicle accidents (per 100,000 of the population in 2007). The death rates show different distribution data, so it is not possible to compare them directly. Table SPUC.1 shows the different descriptive measures for both variables from the SPSS Descriptive Statistics menu.

As you can see, the means and SDs of the variables are quite different. Using the z score formula, we can create comparable measures. Arizona has a diabetes death rate of 17.40 and a 17.60 motor vehicles death rate. These appear to be almost the same. But which represents the higher death rate in Arizona?

$$\text{Diabetes: } Z = \frac{X - M}{SD} \quad Z = \frac{17.40 - 23.253}{4.455} \quad Z = -1.314$$

$$\text{Motor Vehicles: } Z = \frac{X - M}{SD} \quad Z = \frac{17.60 - 15.824}{5.707} \quad Z = 0.311$$

As you can see, *the z scores for the different death rates are very different even though the raw scores look very similar.* The death rate for diabetes is quite low (almost

[2] Source: U.S. Census Bureau, The 2011 Statistical Abstract, Births, Deaths, Marriages & Divorces: Deaths, 121.

TABLE SPUC.1. Descriptive Statistics for Deaths from Diabetes and Motor Vehicle Accidents, 2007

Descriptive Statistics

	N	Minimum	Maximum	Mean	Std. Deviation	Skewness		Kurtosis	
	Statistic	Statistic	Statistic	Statistic	Statistic	Statistic	Std. Error	Statistic	Std. Error
number of deaths from diabetes per 100,000 population 2007 (SA 2011)	51	12.90	35.50	23.2529	4.45526	.367	.333	.684	.656
number of deaths from motor vehicle accidents per 100,000 population 2007 (SA 2011)	51	6.70	31.60	15.8235	5.70721	.638	.333	-.011	.656
Valid N (listwise)	51								

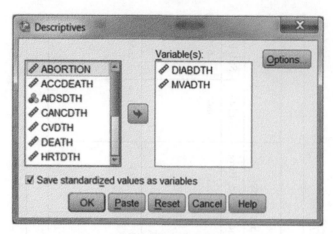

Figure SPUC.3. The SPSS specification window for creating *z* scores.

1 and 1/3 SD below the mean!), whereas the death rate for motor vehicle accidents is above the mean (about 1/3 SD above the mean).

USING SPSS TO CREATE *z* SCORES

You could use the *z* score formula to calculate *z* scores for every raw score in a data set, but it would be time consuming and tedious. Using SPSS to create *z* scores is very easy using the "Descriptives–Descriptives" selection from the Analyze menu. Figure SPUC.3 shows the specification window in SPSS for creating *z* scores. You will recall that the Descriptives menu is obtained through the main menu, Analyze–Descriptive Statistics. Choosing this option results in the Descriptives menu in which we have called for descriptive statistics for the two variables in our database (death rates from diabetes and from motor vehicles).

Recall from our earlier discussion that we can use the Options button to further specify mean, skewness, and a variety of other statistical procedures. However, I can create *z* scores by simply checking the box at the lower-left corner of the Descriptives menu window. It is shown in Figure SPUC.3 as the box "Save standardized values as variables." When I check this box, new variables are added to the data set consisting of the *z* score values corresponding to the raw score values of the variables selected. Figure SPUC.4 shows part of the database in which two variables have been added: "ZDIABDTH" (*z* scores for variable DIABDTH) and "ZMVADTH" (*z* scores for variable MVADTH). As you can see from Figure SPUC.4, we have highlighted the data fields for Arizona to show how SPSS creates the *z* scores from the raw scores.

The *z* scores are very helpful for direct comparisons on specified variables, but the researcher can also identify extreme (raw score) values by looking for *z* scores that exceed ±2.00. If you recall, *z* scores between −2.00 and +2.00 capture about 95 percent

Figure SPUC.4. The SPSS database showing z score variables added.

Figure SPUC.5. The SPSS database showing sorted z scores for diabetes deaths by state.

of the area of the standard normal curve. Scores outside these boundaries may be suspect as outliers or extreme for one reason or the other (e.g., entered incorrectly into the database, special case, etc.). Researchers may want to examine the z scores by sorting the variables (using the procedures we discussed in an earlier section).

As you can see from Figure SPUC.5, we sorted the z scores for diabetes death rates (we only show the extreme cases). West Virginia represents an extreme score if we use

2.5 as a midpoint boundary for outliers. The others we show in Figure SPUC.5 exceed the 2.0 boundary in both directions (Nevada is in the negative direction; the other states represent positive values).

STATISTICAL PROCEDURES UNIT C: USES AND FUNCTIONS

Use and Function	SPSS Menus
Create z scores from raw scores	Analyze–Descriptive Statistics–Descriptives–(Check box entitled, "Save standardized values as variables")

GLOSSARY

Aggregate data A research design where data are comprised of units that are made up of many individual units (e.g., a state (aggregate unit) is comprised of many people [individual units]).

Aggregates Units that are composed of many individuals, such as cities, states, nations, schools, clubs, or churches, each composed of hundreds, thousands, or millions of people.

Alternative hypothesis The research assumption that is stated in contrast to the null hypothesis.

Anchor questions Survey items that distort the range and preferred selection of response categories, biasing the question.

ANOVA The analysis of variance test that assesses the extent to which the variance between group means and the grand mean of a distribution is large relative to the variance of individual scores of different sample groups. This test is typically used with three or more sample groups.

Antecedent variable A variable that is hypothesized to precede the other variables in an equation.

Areal bias A bias resulting when certain geographic areas are not selected as part of a sample, underrepresenting particular population characteristics.

Areal units Cases (units) composed of geographically defined boundaries—counties, states, nations, and so on.

Attrition When subjects drop out of an experiment.

Balanced categories Survey item categories that have parallel options (for example if "agree" was one category response, "disagree" should also be included).

Beneficence The principle that researchers have an obligation to secure the well-being of research participants.

Beta The *standardized* regression coefficient. Sometimes called *beta weight* and *beta coefficient.* Generally, the larger the value of Beta, the more impact a particular predictor variable has on the outcome variable while the other predictors are held constant.

Understanding and Applying Research Design, First Edition. Martin Lee Abbott and Jennifer McKinney.
© 2013 John Wiley & Sons, Inc. Published 2013 by John Wiley & Sons, Inc.

Between-group designs Those studies in which the researcher seeks to ascertain whether *entire groups* demonstrate unequal outcome measures.

Biased language Language used in survey items that serves to direct respondents to answer in particular ways.

Bimodal A set of data with two modes.

Case The smallest unit (person, situation, object, etc.) upon which data are obtained.

Categorical variable A variable that has a limited number of category options, often seven categories or less, and is measured at a nominal or ordinal level of measurement.

Census An official count of an entire population (all units in the set) and recording of certain information about each unit.

Central limit theorem The statistical notion that means of repeated samples taken from a population will form a standard normal distribution (assuming a large sample size) even if the population was not normally distributed.

Central tendency One of the ways to describe a set of data is to measure the way in which scores bunch up around a mean, or central point of a distribution. Mean, median, and mode are typical measures of central tendency.

Chi square A statistical procedure used with any level of data to assess whether the differences among categories of variables are statistically significant.

Chi-square test of independence A statistical test to assess whether the categories of two variables are linked with one another (i.e., are dependent on one another) or if they are not connected. Are the categories correlated with one another?

Closed-ended questions Survey items that have clear, discrete category options.

Cluster sample A probability sampling technique that randomly selects clusters (aggregate units), subsequently selecting the individuals within the clusters.

Coding The breaking down and conceptualizing of the data in order to make sense of it in new ways.

Coefficient of determination The squared value of r, Pearson's correlation coefficient, which represents the proportion of variance in the outcome variable accounted for by the predictor variable of a study.

Concepts Abstract ideas formed by combining the characteristics of a construct, which can then be linked through theory.

Confederate A person who is part of the experimental design but appears to be another subject.

Confidence interval An estimated range of values based on sample statistics that should contain the actual population value a certain proportion of the time.

Confidence level The probability that the parameter (the true value of a population characteristic) falls in the confidence interval.

Confirming evidence The observations we use, based on our own experiences, to find one person or situation that fits the given findings.

Conformity bias   The tendency for respondents to answer survey items in socially acceptable ways.

Constant   A measure that does not change or vary.

Content analysis   The systematic study of cultural artifacts including textual, visual, or aural data.

Contingency coefficient   An effect size measure for chi-square analyses. This measure is essentially a correlation with nominal data and is used to show the strength of association among study variables.

Contingency table   Presentation of data in rows and columns to show how data in rows are contingent on or connected to the data in column cells. Also called cross-tabs or cross-tabulation analysis.

Continuous variable   A variable that has an unlimited number of category options and is measured at an interval or ratio level of measurement.

Control variable   Used to take into account the effects of the variable on another set of variables (used for testing spuriousness).

Control/Comparison group   A group that provides the experimenter the opportunity to see how much of the eventual (outcome) change is due to the experimental treatment and how much is due to the nature of the experimental subjects.

Convenience "sample"   A nonprobability "sample" that selects the most convenient or available participants. Data from a convenience sample cannot be generalized to any population.

Correlation   A statistical process that measures how changes in two variables are related to one another.

Covert observation   The process of observing people as an "undercover" researcher; research participants are not aware that they are the object of study.

Cramer's V   An effect size measure for the chi-square test of independence with tables that exceed the 2×2 arrangement (i.e., that have more than two rows and/ or categories).

Criteria for causation   For one variable to cause changes in another, there should be a time order stipulated (when the independent variable comes first), a correlation between variables, and a nonspurious relationship after controlling for other variables.

Critical value of exclusion   The values on a reference distribution that exclude a certain percentage of the distribution and thereby establish limits of the probability of likelihood of a certain finding.

Cross-sectional studies   Surveys that take a snapshot of a respondent group at one particular time and place.

Crosstabs   The SPSS procedure that creates a contingency table and provides chi-square analyses.

Cultural artifacts   The materials that humans have created that tell us about human life.

Curvilinear correlation   The relationship between two variables in which the values of the two variables change in the same direction up to a point at which the changes in variables reverse direction.

Data distributions   The patterned shape of a set of data. Much of statistics for research uses the so-called normal distribution, which is a probability distribution consisting of known proportions of the area between the mean and the continuum of values that make it up.

Debriefing   A full description of a study provided to subjects after the study, correcting any misconceptions and addressing any anxiety or concerns participants may have encountered during the research.

Deductive   A form of reasoning that derives conclusions beginning from abstract concepts and working toward concrete evidence in order to test assumptions.

Dependent samples   Groups that have some structural linkage that affects the choice of group membership. Examples are using the same subjects (pre and post) for an experiment or comparing the results of a test using matched groups. Tests for dependent samples are also known as repeated measures, within subjects, and paired samples tests.

Dependent variable   A variable that is thought to be influenced by another variable, it is often referred to as the *effect* variable and depends on changes in another variable (the independent variable).

Deposit bias   Occurs when only a portion of the appropriate artifacts are included in the analysis.

Descriptive statistics   The branch of statistics that focuses on measuring and describing data by numerical and visual means such that a researcher can discover patterns that might exist in data not immediately apparent. In these processes, the researcher does not attempt to use a set of data to refer to populations from which the data may have been derived, but rather to gather insights on the data that exist at hand.

Dichotomous variable   A variable that has two answer categories or attributes.

Directionality   How variables are related to each other through a "positive" directionality, where the variables vary in the same direction (as one increases, so does the other) or through a "negative" directionality, where the variables vary in opposite directions from each other (as one increases, the other decreases).

Double-barreled questions   Survey items that are unclear because they imply measurement of two or more ideas.

Dummy variables   In regression, categorical predictor variables can be used or created such that the categories are expressed in separate subvariables with values of 1 and 0 (usually with values of 1 representing membership in a group and 0 representing nonmembership); in this way, the categories can be directly compared in their effect on the outcome variable, as in the case of comparison groups or with treatment groups in experimental designs.

Ecological fallacy   A mistaken interpretation of data in which conclusions about one level of data (individuals, for example) are made by analyzing different levels of data (e.g., groups).

Ecological validity   The extent to which laboratory conditions (setting, experimenters, procedures, etc.) are similar to those encountered in the natural settings, and therefore the extent to which the conclusions can be generalized to the world outside of the lab.

Effect size   Refers to the magnitude of a finding, not whether or not it is a chance or nonchance finding. Also referred to as "practical significance."

Effect size of correlation   This is an indication of the strength of a correlation referring to the amount of variation in the dependent variable accounted for in the predictor variable.

Empirical   Observable using the physical senses (sight, sound, taste, touch, and smell).

Empirical generalization   The summary statement based on empirical observations.

Empirical probability   The number of occurrences of a specific event divided by the total number of possible occurrences.

Eta squared   The proportion of variance in the outcome measure explained by the grouping on the independent variable.

Ethics   The standards of conduct for a given profession or group.

Exhaustive   Variable categories (attributes) that contain all possible responses to describe the variable.

Experimental group   A group that is exposed to the independent variable (or stimulus) to see how the variable impacts the dependent variable compared to the control group.

Experiments   A research design that involves taking an action and observing the effects of that action. The action is the independent variable (often termed "stimulus"), and the result or effect is the dependent variable.

Extreme scores   Outlying scores in a distribution that may result in a distortion of values of the total set of scores.

Face-to-face interview   One type of survey conducted in person between an interviewer and a respondent, where the interviewer reads the survey items to the respondent from an interview schedule.

Fidelity   The concept capturing the extent to which researchers and research settings apply the treatment in a consistent, standardized way.

Field notes   The careful and systematic recording of what the researcher observes during field work.

Field research   A qualitative research design where researchers go into the field, or a natural setting, to observe people, collecting very detailed information about individuals, groups, or interactions.

Frequency distributions   Tabular representations of data that show the frequency of values in groupings of data.

Frequency polygon   A graph that is formed by joining the midpoints of the bars of a histogram by a line.

F test   The statistical test comparing the between and within variances in an ANOVA test. The calculated ratio of between to within variance is compared to the exclusion values of the F distribution.

Generalize   To extrapolate the results of a representative sample to an entire population.

Grounded theory   A way of observing patterns within data by coding patterns into larger conceptual frameworks.

Hindsight bias   Assuming research findings as true after they have been given (sometimes referred to as the "I knew it all along" effect).

Histograms   Graphical representations of frequency distributions. Typically these are in the form of bar graphs in which bars represent groups of numerical values.

History effects   An effect that may impact subjects through events that happen simultaneously with the independent variable.

Hypothesis   A hypothesis states relationships between variables and explains *how* the variables are related (through directionality).

Hypothesis test   The formal process of assessing whether or not a test statistic is judged to be similar to the population elements from which it was drawn or statistically different from those elements.

Inaccurate observation   The semiconscious, casual, or inaccurate taking in of information that impairs our ability to observe the world as it is.

Independent samples   For statistical testing purposes, samples are independent if choosing the elements of one group has no connection to choosing elements of the other(s).

Independent *t* test   The statistical procedure that assesses the probability of whether two samples, chosen independently, are likely to be similar or sufficiently different from one on the basis of chance.

Independent variable   A variable that is assumed to influence another variable, it is often referred to as the *causal* variable and causes changes to a second variable (the dependent variable).

Indicators   Observable measures of a concept (another term for "variables").

Inductive   A form of reasoning that induces conclusions beginning from concrete observations and working toward the more abstract concepts.

Inferential statistics   Statistical methods that assist with making predictions about unknown (population) values on the basis of small sets of sample values.

Informant   A person who has insider information about a group.

Informed consent The opportunity for subjects to base their voluntary participation on a clear understanding of what a proposed study will do, any possible risks associated with the study, how their data will be handled (for example, how will researchers keep their data confidential), and if there are resources available for any adverse effects of the study.

Instrument effects Effects that occur when the tests used to measure an independent or dependent variable are flawed or biased, thus not measuring the impact of the independent variable on the dependent variable.

Interval data/measure Data with the qualities of an ordinal scale, but with the assumption of equal distances between the values and without a meaningful zero point. Typically, researchers use standardized test values (e.g., IQ scores) as an example of an interval scale in which the difference between IQ scores of 100 and 101 is the same as the distance between IQ scores of 89 and 90. This assumption allows the researcher to use mathematical properties (i.e., adding, subtracting, multiplying, and dividing) in their statistical procedures. The lack of a meaningful zero (e.g., what does an IQ of "0" mean?) is typically exemplified by the "0" on a Fahrenheit scale not referring to the "absence of heat" but rather simply to a point in the overall temperature scale.

Intervening variable A variable that is hypothesized to come between an independent and dependent variable an equation.

Interview schedule The survey instrument used in the face-to-face or telephone interview.

Invalid generalization Occurs when people try to generalize the findings of content analysis to anything other than the materials analyzed.

Justice The principle that the benefits and risks of the research are fairly distributed among groups, individuals, or societies.

Kurtosis The measurement of a distribution of scores that determines the extent to which the distribution is peaked or flat. Data distributions that have excessive kurtosis values may be overly peaked ("leptokurtic") or overly flat ("platykurtic").

Latent content The implicit meaning of materials in content analysis.

Levene's test The statistical test assessing the assumption of homogeneity of variance.

Linear relationship The relationship among study variables that forms a straight line if plotted on a graph. Violations of linearity might take the form of curvilinear relationships in which the graphed line is not straight but curved.

Line of best fit The line used in scatterplots to represent the direction and scatter of two variables plotted together. Also called the regression line (see Chapter 10).

Longitudinal studies Studies done over long periods of time (generally five years or longer) using the same respondents.

Manifest content The explicit meaning of materials in content analysis.

Mann-Whitney U test A nonparametric test of the difference between two independent samples of data that are ordinal or that are interval but do not meet the
assumptions of the independent t test.

Matching samples Samples that have been intentionally created to be equivalent on
one or several characteristics. Such a process affects the independence assumption
of the groups. Matched groups are therefore considered dependent samples.

Maturation effects The effects of a period of subject maturation, or growing into or
out of particular phase that impact the dependent variable.

Mean Average value in a set of data.

Median Middle-most score in a set of data.

Misclassification When answer categories are not exhaustive or mutually exclusive,
people may select from the categories given, classifying them into a category that
does not represent their true response.

Mode The most commonly occurring value in a set of data.

Multiple linear regression A statistical procedure assessing the influence on an
outcome variable of more than one predictor variable.

Mutually exclusive Variable categories that, by choosing one, automatically excludes
being able to choose another.

Negative correlation The relationship between two variables in which changes in
the values of one variable are related to changes in value of a second variable in
a different direction. Also called *inverse correlation.*

Nominal data/measure Data that exist as mutually exclusive categories (e.g., home
schooling, public schooling). These data can also refer to "categorical data,"
"dichotomous variables" (when there are two naturally occurring groups like male/
female), or "dichotomized variables" (when two categories are created from a
range of values or categories of other kinds of data like rich/poor from a list of
respondents' incomes in dollar values).

Nonparametric These are statistical procedures that do not make reference to theoretical distributions. Typically, statistical procedures using nominal and ordinal
data are considered nonparametric.

Nonprobability "samples" Collections or groups of people, referred to as "samples,"
that are not drawn from a relevant population.

Nonresponse bias A bias that occurs when a significant number of respondents do
not respond to a study.

Null hypothesis The assumption in an hypothesis test that there is no difference
between the study population yielding a particular sample statistic and the population from which the sample supposedly came.

Observation The process of taking into account empirical facts—observing using the
five senses of taste, touch, smell, seeing, and hearing.

Omnibus finding The combined effect of all predictor variables on the outcome
variable.

Omnibus test   In regression, the omnibus test reveals the combined effect of all the predictor variables on the dependent variable. Also known as the *joint effects*.

Online forums   Electronically hosted focus groups.

Online surveys   One form of a self-administered questionnaire that is delivered to respondents electronically, rather than through a hard copy questionnaire.

Open-ended questions   Survey items that allow respondents to answer with their own responses, rather than with predetermined categories.

Operationalize   The process of transforming concepts into variables that are empirically measurable.

Ordinal data/measure   Data that exist in categories that are ranked or related to one another by a "more than/less than" relationship like "strongly agree, agree, disagree, strongly disagree."

Original items   Survey items of measures that have not been previously tested.

Outlier   An extreme case that distorts the true relationship between variables, either by creating a correlation that should not exist or suppressing a correlation that should.

Overgeneralization   Generalizing a wider understanding and knowledge of phenomena based on little evidence.

Overt observation   The process of observing openly as a researcher, where research participants are aware that they are the object of study.

Panel studies   Studies using surveys that rely on the same respondents to answer multiple surveys over short periods of time (usually less than five years).

Paradox   The giving of potential benefits and denying of potential benefits.

Parameters   Characteristics or measures of entire populations.

Parsimonious   The process of explaining the most phenomena with the least amount of theoretical assumptions.

Participation   When doing field research it is common to enter into a group to experience their activities firsthand, often referred to as participant-observation.

Pearson's correlation coefficient   Named after Karl Pearson, this procedure is symbolized by r and used to measure the relationship between two interval-level variables.

Percentile   The point in a distribution of scores below which a given percentage of scores fall.

Phi coefficient   An effect size measure based on chi square and typically used in studies with 2×2 tables. When phi is squared, it expresses the proportion of variance in one variable explained by the other.

Population   All units or elements constituting a set or universe.

Positive correlation   The relationship between two variables in which changes in the values of one variable are related to changes in value of a second variable in the same direction.

Post hoc analyses   Individual comparison tests conducted among sample group values subsequent to a significant ANOVA finding.

Posttest   The measure of the dependent variable in an experiment.

Practical evaluation methods   These procedures indicate how much explanation of an outcome variable a correlation provides from the predictor variable. Also called *practical significance* and *effect size*.

Practical significance   The meaningfulness of a certain finding; usually established by considering the impact of the relationship between variables. Also known as *effect size*.

Pre-experimental design   An experimental design that does not have all of the features of a true experimental design (i.e., independent/dependent variables, pre-/posttests, experimental/control groups, or randomization).

Pretest   The measure of subjects on the dependent variable prior to the exposure of the experimental group to the independent variable. Also the procedure used to test the reliability and validity of new (original) survey items.

Privacy   A key ethical concern for researchers, who must take care to protect the names, people, and places where their field research occurs.

Probability   The field of mathematics that studies the likelihood of certain events happening out of the total number of possible events.

Probability proportional to size   A probability sampling technique that uses a modified cluster sample to collect data from proportionally defined units.

Probability sampling   Sampling techniques that ensure that every unit in a population has an equal probability to be selected for a sample, resulting in a *representative* sample of the population.

Proportional facts   The use of descriptors like "the majority," "most," "few," "typically," in order to assert the distribution of a variable.

Qualitative   Research designs that rely on quality of description, rather than quantity, relying on rich, descriptive detail in the reporting of human processes.

Quantitative   Research designs that rely primarily on describing or measuring phenomena in quantity, generally thought of as numerical quantity.

Quasi-experimental design   Experimental designs that do not have all the elements necessary to be considered a true experimental design. Typically, these designs do not use full randomization (usually because the conditions of real-world research prevent this) and therefore do not yield clear-cut outcomes that can be attributed only to the treatment variable.

Questionnaire   One type of survey that is usually self-administered by the respondent, who reads and responds to the survey instrument itself.

Quota "sample"   A nonprobability "sample" that collects data using a matrix and describes certain demographic characteristics of a target population.

R^2 The effect size indicator in regression; expresses the proportion of the variance in the outcome variable contributed by the predictor variables (also called the *multiple coefficient of determination*).

Random assignment   The process of assigning research subjects randomly: each subject has an equal chance of being assigned to the experimental group or the control group.

Range   The numerical difference between the highest and lowest values in a distribution.

Rates   Measures calculated using a common base for each unit so that variables are comparable across units.

Ratio data/measure   Interval data with the assumption of a meaningful zero point constitute ratio data (e.g., the amount of money people have in their wallets at any given time). The zero point allows the researcher to make comparisons between values as "twice than" or "half of," since the zero provides a common benchmark from which to ground the comparisons.

Relevant base   The common base in a rate that takes the most accurate population for rate calculation.

Reliability   The extent to which a given measuring instrument produces the same result each time it is used; a measure of consistency.

Representativeness   The ability for data from a probability sample to be generalized to the target population as a whole.

Research design   The modes of observation that allow scientists to collect observations in systematic and structured ways.

Respect for persons   The principle that people should be treated as "autonomous agents" who are allowed to consider for themselves the potential harms and benefits of a situation, analyze how risks and benefits relate to their personal values, and then to take action based on their analysis.

Response rates   The rate of response from those who have been sampled (or from a census).

Response set   The tendency for respondents to fall into a pattern of answering closed-ended questions in particular ways.

Restricted range   A problem in correlation studies in which the entire set of scores are not used in an association, but rather a selected group of the scores is used that that do not represent the total variability.

Sample   A random subset of a population.

Sampling   The process of selecting observations from a population.

Sampling fraction   Part of the two-step process in taking a systematic random sample where the total population is divided by the sample size. The resulting fraction is the nth number, added to the random start of the probability sample in order to select observations.

Sampling frame   The list of all units in a set (the population), from which probability samples are taken.

Scales of measurement   The descriptive category encompassing different classes of data. Nominal, ordinal, interval, and ratio data differ according to the information contained in their scales of values. Also known as *levels of measurement.*

Scattergram   A graph that shows the relationship between study variables when they are plotted together. Also called scatter diagram, scatterplot, or scatter graph.

Scientific method   The process of research whereby a problem is identified, a hypothesis is generated, and that hypothesis is systematically tested through empirical observation.

Selective availability bias   A bias that results from certain groups within a population who are unlikely to respond, compromising the randomness of a probability sample.

Selective observation   Observing only the evidence that supports a preconceived pattern that then serves to obscure the true pattern.

Simple random sample   A probability sampling technique where all of the units in a set (a population) have an equal chance to be chosen for the sample.

Single sample *t* test   A statistical procedure that assesses whether a sample mean is likely to come from a population about which we do not know the mean or standard deviation.

Skewness   A measurement of a data distribution that determines the extent to which it is "imbalanced" or "leaning" away from a standard bell shape (in the case of a normal distribution).

Snowball "sample"   A nonprobability sample that begins with one participant and then "snowballs" into more participants by participant referral of other participants to the researcher.

Social location   Where we fit into the social order and how that impacts the way in which we view the social world (e.g., how our social class, gender, our education, and/or political perspective, etc., impact how we interpret information).

Social units   Cases (units) derived from social boundaries—churches, clubs, sports teams, and so on.

Spearman's rho   A nonparametric correlation procedure that uses ranked (ordinal) data.

Spuriousness   A condition in which an assumed relationship between two variables is explained by another variable not in the analysis.

Standard deviation (SD)   The SD represents a *standard* amount of distance between the mean and each score in the distribution. It is the square root of the variance (VAR).

Standard items   Survey items that have been reliably and validly tested in previous surveys.

Standard normal distribution   A normal distribution that is perfectly shaped such that the percentages of the area under the curve are distributed in known and standard amounts around the mean. The mean has a value of 0 and the SD = 1. Also known as the *z distribution.*

Statistic   The observed value of a characteristic within a (sample) population.

Statistical significance   Refers to whether the measure of a variable can be said to be meaningfully greater or lesser than what would be expected by chance alone.

Stratified random sample   A probability sampling technique where some strata (characteristic) within a population is taken into account in order to sample that population.

Survey instrument   The survey itself is referred to as the "survey instrument" and is composed of written questions with response categories to be administered through face-to-face or telephone interviews, self-administered questionnaires, or online questionnaires or forums.

Surveys   Research design that uses a series of written and verbal prompts/items to quantify the personal opinions, beliefs, and ideas from a group of respondents.

Survival bias   Occurs when only some portion of the pertinent materials to be analyzed have been retained (or exist).

Systematic random sample   A probability sampling technique that uses a two-step process to select a random start within the sampling frame and then calculates a sampling fraction to determine which cases following the random start will be selected for the sample.

Telephone interview   One type of survey conducted over the telephone between an interviewer and a respondent, where the interviewer reads the questions from an interview schedule to the respondent.

Testing effects   When a pretest and/or posttest influence the measure of the dependent variable, rather than the independent variable.

Theory   A coherent group of assumptions that can be used to explain and predict phenomena.

Time order   Inferring that one variable (an independent variable) precedes another variable (a dependent variable).

Trend study   Two or more cross-sectional surveys with independent samples from the same population, with some questions in common, used to trace population trends.

Triangulation   The process of using multiple methods or data to derive conclusions.

Tukey's HSD test   One of the tests used to assess paired group differences when there is a significant omnibus test.

Unbalanced categories   Survey item response categories that do not have matching "degrees" of responses (e.g., "Agree," "Disagree somewhat," "strongly disagree").

Units of analysis   The things that a hypothesis directs us to observe; cases.

Unobtrusive measures   Measures that are nonreactive, meaning they that have no direct impact on people.

Unstandardized coefficients   The slope and intercept values expressed in the same units as the raw scores (i.e., not in standardized units).

Validity   Measuring the accuracy of a measure (variable); making sure the measure (variable) measures what we say it is supposed to measure.

Variable   A characteristic or aspect of something that varies.

Variance (VAR)   The variance is the average squared distance of the scores in a distribution from the mean. It is the squared SD.

Variance between   The variation of group means around the grand mean in an ANOVA analysis.

Variance total   The total variation (between and within) in an ANOVA analysis resulting from all individual scores varying around the grand mean.

Variance within   The variation of scores around their own group means in an ANOVA analysis. Also known as *error.*

Variation   The amount (or degree) of change found within a variable.

Weighted sample data procedures   Adjustments can be made to individual respondent data when these represent more than one case in the population.

Within subjects designs   Studies in which the researcher seeks to ascertain whether *subjects* in a group or matched groups change over time.

z score   A raw score expressed in standard deviation units. Also known as a *standard score* when viewed as scores of a standard normal distribution.

BIBLIOGRAPHY

Abbott, Martin Lee. 2010. *The Program Evaluation Prism*. Hoboken, NJ: John Wiley & Sons.

Abbott, Martin Lee. 2011. *Understanding Educational Statistics Using Microsoft Excel and SPSS*. Hoboken, NJ: John Wiley & Sons.

American Sociological Association (ASA). 1999. *Code of Ethics and Policies and Procedures of the ASA Committee on Professional Ethics*. New York: American Sociological Association.

Association of Religion Data Archives (ARDA). 2011. "General Social Survey 2010 Cross-section and Panel Combined." Retrieved November 7, 2011, from http://www.thearda.com/Archive/Files/Descriptions/GSS10PAN.asp.

Babbie, Earl. 2007. *Survey Research Methods*. Belmont, CA: Wadsworth.

Barker, Eileen. 1984. *The Making of a Moonie: Choice or Brainwashing?* Oxford: Blackwell.

Berger, Peter. 1967. *The Sacred Canopy: Elements of a Sociological Theory of Religion*. New York: Anchor Books/ Doubleday.

Berger, Peter. 1968. "A Bleak Outlook Is Seen for Religion." *New York Times*, April 5, p. 3.

Binder, Amy. 1993. "Constructing Racial Rhetoric: Media Depictions of Harm in Heavy Metal and Rap Music." *American Sociological Review* 58:753–767.

Bogod, David. 2004. "The Nazi Hypothermia Experiments: Forbidden Data?" *Anaesthesia* 59:1155–1159.

Bohrnstedt, George W., and David Knoke. 1982. *Statistics for Social Data Analysis*. Itasca, IL: F.E. Peacock.

Bolt, Martin. 1996. *Instructor's Manual to Accompany Myers' Social Psychology*. New York: McGraw-Hill.

Brandt, Allen M. 1985. "Racism and Research: The Case of the Tuskegee Syphilis Study," pp. 331–346 in Judith Walzer Leavitt and Ronald Numbers, eds., *Sickness and Health in America*. Madison: University of Wisconsin Press.

Brown, Eryn. 2011. "Why Time Is Spent Proving the Obvious." *The Seattle Times*, June 3, p. A5.

Campbell, Donald T., and Julian Stanley. 1963. *Experimental and Quasi-Experimental Designs for Research*. Belmont, CA: Wadsworth.

Carey, Benedict. 2011. "Psychological Missteps Teach Lessons in Treatment of Trauma." *The Seattle Times*, July 29, p. A3.

Carr, Patrick J. 2009. "Forum on *Gang Leader for a Day*." *Sociological Forum* 24(1):199–201.

Understanding and Applying Research Design, First Edition. Martin Lee Abbott and Jennifer McKinney.
© 2013 John Wiley & Sons, Inc. Published 2013 by John Wiley & Sons, Inc.

Chagnon, Napoleon. 1977. *Yanomamo: The Fierce People*. New York: Thomson Learning.

Charles, Camille Zubrinsky. 2009. " 'Gang Leader for a Day': A Rogue Sociologist Takes to the Streets." *Sociological Forum* 24(1):205–209.

Chaves, Mark. 1993. "Intraorganizational Power and Internal Secularization in Protestant Denominations." *American Journal of Sociology* 99:1–48.

Chaves, Mark. 1994. "Secularization as Declining Religious Authority." *Social Forces* 72:749–774.

Chaves, Mark. 1999. *How Do We Worship? A Report from the National Congregations Study*. Washington, DC: Alban Institute Press.

Chaves, Mark, Mary Ellen Konieczny, Kraig Beyerlein, and Emily Barman. 1999. "The National Congregations Study: Background, Methods, and Selected Results." *Journal for the Scientific Study of Religion* 38(4):458–476.

Clampet-Lundquist, Susan. 2009. " 'Gang Leader for a Day': A Rogue Sociologist Takes to the Streets." *Sociological Forum* 24(1): 202–204.

Cohen, Jacob. 1988. *Statistical Power Analysis for the Behavioral Sciences*. New York: Routledge.

Corbett, Michael, and Lynne Roberts. 2002. *A MicroCase Workbook for Social Research*. Belmont, CA: Wadsworth/Thompson Learning.

Dillman, Don A. 1999. *Mail and Internet Surveys: The Tailored Design Method*. Hoboken, NJ: John Wiley.

The Economist. 2010. "Census Day: Stand Up and Be Counted." March 31. Retrieved February 12, 2012, from www.economist.com/node/15819188/print.

Entman, Robert. 1991. "Framing US Coverage of International News: Contrasts in Narratives of the KAL and Iran Air Incidents." *Journal of Communication* 41:6–27.

Feimster, Crystal N. 2009. *Southern Horrors: Women and the Politics of Rape and Lynching*. Cambridge, MA: Harvard University Press.

Finke, Roger, and Rodney Stark. 2005. *The Churching of America 1776–2000: Winners and Losers in the Religious Economy*. New Brunswick, NJ: Rutgers University Press.

General Social Survey. 2010. "Appendix A: Sampling Design and Weighting." Retrieved February 12, 2012, from http://www3.norc.org/GSSWebsite/Documentation/.

General Social Survey (GSS). Smith, Tom W, Peter Marsden, Michael Hout, and Jibum Kim. *General social surveys, 1972–2010* [machine-readable data file] /Principal Investigator, Tom W. Smith; Co-Principal Investigator, Peter V. Marsden; Co-Principal Investigator, Michael Hout; Sponsored by National Science Foundation. NORC ed. Chicago: National Opinion Research Center (producer); Storrs, CT: The Roper Center for Public Opinion Research, University of Connecticut (distributor), 2011. (http://www3.norc.org/GSSWebsite/).

Goffman, Erving. 1974. *Frame Analysis*. New York: Harper Colophon Books.

Hadaway, C. Kirk, and Penny Long Marler. 2005. "How Many Americans Attend Worship Each Week? An Alternative Approach to Measurement." *Journal for the Scientific Study of Religion* 44(3):307–323.

Hatton, Erin, and Mary Nell Trautner. 2011. "Equal Opportunity Objectification? The Sexualization of Men and Women on the Cover of *Rolling Stone*." *Sexuality & Culture* 15:256–278.

Horowitz, Irving Louis, and Lee Rainwater. 1975. "Sociological Snoopers and Journalistic Moralizers: Part II," pp. 181–190 in Laud Humphreys, *Tearoom Trade: Impersonal Sex in Public Places*. Chicago: Aldine.

Hout, Michael, and Claude S. Fischer. 2002. "Explaining the Rise of Americans with No Religious Preference: Generations and Politics." *American Sociological Review* 67:165–190.

Huff, Darrell. 1982. *How to Lie with Statistics.* New York: W.W. Norton.

Humphreys, Laud. 1975. *Tearoom Trade: Impersonal Sex in Public Places.* Chicago: Aldine.

Improbable Research. 2012. "The 2011 Ig Noble Prize Winners." Retrieved January 24, 2012. from http://improbable.com/ig/winners/#ig2011.

Jones, James H. 1993. *Bad Blood: The Tuskegee Syphilis Experiment.* New York: The Free Press.

Kleinman, Arthur, Jung-Bao Nie, and Mark Selden. 2010. "Introduction: Medical Atrocities, History, and Ethics," in Nie, Jing-Bao, Nanyan Guo, Mark Selden, and Arthur Kleinman, *Japan's Wartime Medical Atrocities: Comparative Inquiries in Science, History, and Ethics.* New York: Routledge.

Knutson, Ryan. 2010. "Republicans Send Out a 'Census' Form—That's Really a Fundraiser." Propublica.org Web site, February 10. Retrieved January 24, 2012. from http://www.propublica.org/article/republicans-send-out-a-census-form-thats-really-a-fundraiser-210.

Kohn, Melvin L. 1977. *Class and Conformity: A Study in Values.* Chicago: University of Chicago Press.

Kosmin, Barry A., and Ariela Keysar. 2008. "American 'Nones': The Profile of the No Religion Population." A Report Based on the American Religious Identification Survey, Trinity College, Hartford, CT.

Kox, Willem, Wim Meeus, and Harm t'Hart. 1991. "Religious Conversion of Adolescents: Testing the Lofland and Stark Model of Religious Conversion." *Sociological Analysis* 52:227–240.

Latané, Bibb, and John Darley. 1968. "Group Inhibition of Bystander Intervention in Emergencies." *Journal of Personality and Social Psychology* 10:215–221.

Lechner, Frank J. 1991. "The Case Against Secularization: A Rebuttal." *Social Forces* 69:1103–1119.

Lee-Treweek, Geraldine, and Stephanie Linkogle. 2000. "Putting Danger in the Frame," in Lee-Treweek and Linkogle, eds., *Danger in the Field: Risk and Ethics in Social Research.* London: Routledge.

Lillard, Angeline S., and Jennifer Peterson. 2011. "The Immediate Impact of Different Types of Television on Young Children's Executive Function." *Pediatrics*, September 12. Retrieved January 24, 2012, from http://pediatrics.aappublications.org/content/early/2011/09/08/peds.2010–1919.full.pdfhtml.

Lofland, John. 1966. *Doomsday Cult: The Process of Conversion, Proselytization, and Maintenance of Faith.* Englewood Cliffs, NJ: Prentice Hall.

Lofland, John, and Lyn Lofland. 1995. *Analyzing Social Settings: A Guide to Qualitative Observation and Analysis.* Belmont, CA: Wadsworth.

Lofland, John, and Rodney Stark. 1965. "Becoming a World-Saver: A Theory of Conversion to a Deviant Perspective." *American Sociological Review* 30:862–875.

Loftus, Jeni. 2001. "America's Liberalization in Attitudes Toward Homosexuality, 1873–1998." *American Sociological Review* 66(5):762–782.

Luckmann, Thomas. 1967. *The Invisible Religion: The Problem of Religion in Modern Society.* New York: Macmillan.

Magnolia Voice. 2011. "Reckless Driver Causes Major Damage." Retrieved July 19, 2011, from www.magnoliavoice.com/2011/07/19/reckless-driver-causes-major-damage.

Marshall, Catherine, and Gretchen B. Rossman. 1995. *Designing Qualitative Research*. Thousand Oaks, CA: Sage.

McKinney, Jennifer. 1995. "'They Passed Out Guns in the Chapel': The Social Construction of Religious Groups in the National Print Media." Master's thesis, Purdue University, West Lafayette, IN.

McKinney, Jennifer. 2001. "Clergy Connections: The Impact of Networks on Evangelical Renewal Movements." PhD diss., Purdue University, West Lafayette, IN.

McKinney, Jennifer, and Robert Drovdahl. 2007. "Vocation as Discovery: The Contribution of Internship Experiences." *The Journal of Youth Ministry* 5(2):51–71.

McKinney, Jennifer, and Roger Finke. 2002. "Reviving the Mainline: An Overview of Clergy Support for Evangelical Renewal Movements." *Journal for the Scientific Study of Religion* 41(4):773–785.

Myers, David G. 1994. *Exploring Social Psychology*. New York: McGraw-Hill.

National Institutes of Health (NIH). 2008. "Protecting Human Research Participants." Retrieved January 24, 2012, from http://phrp.nihtraining.com.

Nie, Jing-Bao, Nanyan Guo, Mark Selden, and Arthur Kleinman. 2010. *Japan's Wartime Medical Atrocities: Comparative Inquiries in Science, History, and Ethics*. New York: Routledge.

Peters, John F. 1998. *Life Among the Yanomami*. Toronto: University of Toronto Press.

PricewaterhouseCoopers. 2001. "Effect of Census 2000 Undercount on Federal Funding to States and Selected Counties, 2002–2012." U.S. Census Monitoring Board, Presidential Members. Retrieved January 24, 2012, from http://govinfo.library.unt.edu/cmb/cmbp/reports/final_report/fin_sec5_effect.pdf.

Ragin, Charles. 1994. *Constructing Social Research*. Thousand Oaks, CA: Pine Forge Press.

Rathje, W., and C. Murphy. 2001. *Rubbish! The Archeology of Garbage*. Tempe: University of Arizona Press.

Rosenthal, Robert, and Lenore Jacobson. 1968. *Pygmalion in the Classroom*. New York: Holt, Rinehart & Winston.

Scarce, Rik. 2005a. "A Law to Protect Scholars." *The Chronicle of Higher Education* 51(49): B24.

Scarce, Rik. 2005b. *Contempt of Court: A Scholar's Battle for Free Speech from Behind Bars*. Walnut Creek, CA: AltaMira Press.

Simons, Daniel J., and Daniel T. Levin. 1997. "Change Blindness." *Trends in Cognitive Sciences* 1(7):261–267.

Simons, Daniel J., and Daniel T. Levin. 1998. "Failure to Detect Changes to People during a Real-World Interaction." *Psychonomic Bulletin & Review* 5(4):644–649.

Spencer, J. William, and Jennifer McKinney. 1997. "'We Don't Pay for Bus Tickets, But We Can Help You Find Work': The Micropolitics of Trouble in Human Service Encounters." *The Sociological Quarterly* 38(1):185–203.

Stark, Rodney. 2003. *For the Glory of God: How Monotheisms Led to Reformations, Science, Witch-Hunts, and the End of Slavery*. Princeton: Princeton University Press.

Stark, Rodney, and William Sims Bainbridge. 1985. *The Future of Religion*. Berkeley: University of California Press.

Stark, Rodney, and William Sims Bainbridge. 1987. *A Theory of Religion*, vol. 2, Edited by D. Wiebe. New York: Peter Lang.

Stark, Rodney, and Roger Finke. 2000. *Acts of Faith: Explaining the Human Side of Religion*. New York: Oxford University Press.

Stark, Rodney, and Lynne Roberts. 2002. *Contemporary Social Research Methods*. Belmont, CA: Wadsworth/Thomson Learning.

Strauss, Anselm, and Juliet Corbin. 1990. *Basics of Qualitative Research*. London: Sage.

Tanner, Lindsey. 2011. "Study: SpongeBob Not Good for 4-Year-Olds." *The Seattle Times* Retrieved September 27, 2011, from http://seattletimes.nwsource.com/html/nationworld/2012180269_spongebob12.html.

Tierney, Patrick. 2002. *Darkness in El Dorado: How Scientists and Journalists Devastated the Amazon*. New York: W.W. Norton.

Tolnay, Stewart E., and E. M. Beck. 1995. *Festival of Violence: An Analysis of Southern Lynchings 1882–1930*. Chicago: University of Illinois Press.

Tschannen, Olivier. 1991. "The Secularization Paradigm: A Systematization." *Journal for the Scientific Study of Religion* 30:395–415.

Venkatesh, Sudhir. 2008. *Gang Leader for a Day: A Rogue Sociologist Takes to the Streets*. New York: Penguin.

von Hoffman, Nicholas. 1975. "Sociological Snoopers and Journalistic Moralizers: Part I," pp. 177–181 in Laud Humphreys *Tearoom Trade: Impersonal Sex in Public Places*. Chicago: Aldine.

Warner, R. Stephen. 1993. "Work in Progress Toward a New Paradigm for the Sociological Study of Religion in the United States." *American Journal of Sociology* 98:1044–1093.

Warner, R. Stephen. 1994. "The Place of the Congregation in the American Religious Configuration," pp. 54–99 in *American Congregations*, vol. 2, ed. J. P. Wind and J. W. Lewis. Chicago: University of Chicago Press.

Warwick, Donald. 1975. "Tearoom Trade: Means and Ends in Social Research," pp. 191–212 in Laud Humphreys, *Tearoom Trade: Impersonal Sex in Public Places*. Chicago: Aldine.

Webb, E. J., D. T. Campbell, R. D. Schwartz, and L. Sechrest. 1999. *Unobtrusive Measures*. New York: Sage.

Weindling, Paul Julian. 2004. *Nazi Medicine and the Nuremberg Trials: From Medical War Crimes to Informed Consent*. New York: Palgrave Macmillan.

White, Susan, and Casey Langer Tesfaye. 2010. "*Who Teaches High School Physics?*" *Focus On*, a publication of the AIP Statistical Research Center.

Wickström, Gustav, and Tom Bendix. 2000. "The 'Hawthorne Effect': What Did the Original Hawthorne Studies Actually Show?" *Scandinavian Journal of Work Environment and Health* 26(4):363–367.

Yamane, David. 1997. "Secularization on Trial: In Defense of a Neosecularization Paradigm." *Journal for the Scientific Study of Religion* 36:109–122.

Young, Alfred. 2009. " 'Gang Leader for a Day': A Rogue Sociologist Takes to the Streets." *Sociological Forum* 24(1):210–214.

ADDITIONAL WEB SITES

The Association of Religion Data Archives (ARDA). http://www.theARDA.com.

General Social Survey (GSS). http://www3.norc.org/GSSWebsite.

National Survey on Youth and Religion (NSYR). http://youthandreligion.org/.

Random.org. http://www.random.org/.

U.S. Census Bureau. http://www.census.gov/.

INDEX

Abbott, Martin Lee, 50, 272–273
Action research, 49, 51
Aggregate data, 10, 36, 38
  congregations example, 246–250
  correlation example, 129–130, 141–142
  ecological fallacy and, 160–161
  nature of, 236
  outliers, 243–244
  rates, 238–242
  regression example, 200–202
  research design, 234–250
  units of analysis, 236
  using SPSS, 290–297
Aggregates, 38, 234
Alternative hypothesis, 340
American Institute of Physics, 113
American National Election Study (ANES), 115
American Religious Identification Survey (ARIS), 9
American Sociological Association (ASA), 54, 61, 306
Analysis of variance (ANOVA), 270–271, 280–300, 330, 336
  assumptions of, 288–289
  calculation of, 284–285
  components of variance, 282–283
  effect size, 286–287
  eta squared, 286
  example of, 289–290
  F test, 286, 289
  interpretation of results, 331–332
  nature of, 281–282
  one way, 280

post hoc analysis, 288
process of, 283–286
Tukey's HSD, 288
using SPSS for, 290–299
Anchoring questions, 212, 219, 223
  nature of, 218
Animal Liberation Front (ALF), 306
ANOVA. *See* Analysis of Variance
Antecedent variables, 148, 159, 168
  crosstab examples in SPSS, 169–174, 178–183
  illustration of, 169
  nature of, 168–69
  regression examples in SPSS, 187–195
Applied research, 49–50
  types of, 50–51
Areal bias, 119, 120, 230
Areal units, 236
The Association of Religion Data Archives (ARDA), 36–37, 229, 244
Attributes. *See* Variables, attributes
Attrition, 229
  in surveys, 229–230
  in experiments, 255, 256

Balanced categories, 219, 223
  unbalanced categories, 222
Baylor Religion Survey, 37, 125, 231
Beer, 7, 241, 313
Belmont Report, 56, 57, 59
Beneficence, 56, 59, 60
Beta, 158–159
  standardized, 190, 192–193, 195
Biased language, 212, 217–218, 219, 223

Understanding and Applying Research Design, First Edition. Martin Lee Abbott and Jennifer McKinney.
© 2013 John Wiley & Sons, Inc. Published 2013 by John Wiley & Sons, Inc.

Printed and bound by CPI Group (UK) Ltd, Croydon, CR0 4YY

16/04/2025

14658517-0004